T0379494

Handbook of Nutraceuticals

Volume 1

Ingredients, Formulations, and Applications

Handbook of Nutraceuticals

Volume 1

Ingredients, Formulations, and Applications

Edited by

Yashwant Pathak

CRC Press is an imprint of the
Taylor & Francis Group, an informa business

CRC Press
Taylor & Francis Group
6000 Broken Sound Parkway NW, Suite 300
Boca Raton, FL 33487-2742

Printed in the United States of America on acid-free paper
10 9 8 7 6 5 4 3 2 1

International Standard Book Number: 978-1-4200-8221-0 (Hardback)

Library of Congress Cataloging-in-Publication Data

Handbook of nutraceuticals / editor, Yashwant Pathak.
p. ; cm.
Includes bibliographical references and index.
ISBN 978-1-4200-8221-0 (hardcover : alk. paper)
1. Functional foods--Handbooks, manuals, etc. 2. Dietary supplements--Handbooks, manuals, etc. I. Pathak, Yashwant. II. Title.
[DNLM: 1. Dietary Supplements. QU 145.5 H2357 2010]

QP144.F85H359 2010
615'.3--dc22 2009035205

Visit the Taylor & Francis Web site at
http://www.taylorandfrancis.com

and the CRC Press Web site at
http://www.crcpress.com

Dedicated to all the Rushies, Sages, Shamans, Medicine Men and Women, and people of ancient traditions and cultures who contributed to the development of drugs and nutraceuticals worldwide and kept the science of health alive for the past several millennia

Contents

Foreword

Shortly after I received Dr. Pathak's kind invitation to contribute to this timely publication, my dormant nutraceutical ethers quickly returned. Let's start with the definition: "A nutraceutical is a food or parts of a food that have a health benefit including the prevention and treatment of disease." A bona fide nutraceutical has been shown to have some type of clinical benefit. Most of the others are potential ones.

Unfortunately, the nutraceutical movement has not been as vigorous as it deserves to be. Books on nutraceuticals, such as this, are urgently needed to keep the flames flickering. The chapters in this book, authored by international experts, cover a broad, promising range of medical-health topics, ranging from cardiovascular therapeutics to mental health to inexorable aging. The scope of nutraceuticals is as broad as that of pharmaceuticals but still lacks serious consideration from critical segments of the health industry. And searching through the labyrinth of reasons behind this neglect, the principal one is unequivocally attributable to a lack of proper economic incentives.

Vibrant nutraceutical basic and applied laboratory and clinical research are intimately intertwined and dependent on an appropriately structured marketplace, very much unlike our current misinformation-driven one. For example, as a result of massive marketing-driven misinformation, Americans now blindly believe that all antioxidants are good for you. The reality is that some antioxidants are good, some are not, and some may harm you, depending on how much and when you take them. Oxidants aren't all bad and are critical, for example, to fight infections.

The foundation and growth of a truly health-oriented nutraceutical marketplace are based on two major factors: the freedom of a company to make a medical health claim and the availability of sufficiently strong proprietary products to justify corporate basic and clinical research and development investment to make money. It is as simple and unromantic as that!

At the Foundation for Innovation in Medicine conferences, I, in addition to addressing the resolution of the medical-health claims issue, emphatically and repeatedly stressed that the rational choice of ingredients coupled with proprietary nutraceutical formulations were essential for the successful establishment of a research-intensive nutraceutical health sector. It would be, although very complicated to understand, almost equivalent to the protection given by pharmaceutical product patents. We are fundamentally talking about optimum ways to deliver ingredients that are both water soluble and fat soluble and that effectively arrive at intended receptor sites in a way that is acceptable to the gastrointestinal track.

Ironically, and I say this with smiling admiration, the contributors to this book do not realize the magnitude of the importance of their research efforts to bring about the next phase of the "nutraceutical revolution." Hopefully, this book will inspire their colleagues to aggressively pursue much needed ingredient formulation research.

Stephen L. DeFelice, MD
Chairman, The Foundation for Innovation in Medicine

Preface

SOLO VERSUS CONCERT PERFORMANCE

Studying the development of the health sciences and the therapeutics, it appears that mankind learned a lot from other creatures and creations of God, recognizing our interdependence with nature for thousands of years. A balance among the individual, society, nature, and the Creator has always kept the world moving. Human beings, over the ages, used herbal and mineral drugs provided by nature for their ailments and treatments of disease.

The indigenous systems of medicines, which were developed by different communities much before the advent of the allopathic system of medicine, were mostly dependent on using several herbal drugs, always in combination with many ingredients. All of these traditions—including Indian Ayurveda, Chinese and African herbal medicine, and other native approaches—considered this to be a "concert performance." Furthermore, most of these people believed that ailments and disease are caused by many factors, including the mental condition of the person. For instance, Ayurveda developed a specialty known as Manas-ayurveda, which addressed the mental condition of the patient. Similar importance has been given to mental conditions in many other indigenous medicinal systems. The human body was studied at different levels of understanding, and hence all these traditional systems thought that there cannot be just a solo treatment for any diseased condition. One has to address the needs and problems of the physical body, intellectual setting, and mental condition. Finally, they also looked at the spiritual needs of human beings and provided treatment for illnesses using different spiritual practices. A whole science of breathing exercises and meditation was developed; in recent days, it is widely used especially for post-cancer patients, postoperation patients, and mental patients.

The Newtonian–Cartesian worldview of recent centuries, in which the body has been viewed as a machine and health as a purely physical condition, has held that most diseases are caused by infections rather as multidimensional imbalances within the host. This led to attempts to identify, extract, and synthesize the single "active ingredient" from natural herbs or minerals, which is more of a typical "solo" approach. Physicians tried to treat the person with one single drug for one disease. However, they lacked a holistic approach to health; instead, they focused on isolated treatments for symptoms. This led to the development of significant medical sciences that discarded the traditional systems responsible for the treatment of the common man for thousands of years.

By ignoring the web-like framework of health and life and rejecting the fact that natural medicines contain not only an active ingredient but also balancing factors and nature's intelligence, the modern system remains incomplete. To be most effective, medical science needs a more integrative use of the bounty of nature's pharmacy, coupled with healing arts that examine disease as a dynamic and functional process. Ancient wisdom that identifies the foundation of nutraceuticals in the marriage of seed and soil, combined skillfully with the best of modern allopathy, stands the best chance of bringing health and harmony to the people of the world.

> Behold, I have given you every plant yielding seed that is on the surface of all the earth, and every tree which has fruit yielding seed; it shall be for you.
>
> **Genesis 1:29**

In Eastern philosophy, it is mentioned that "Nasti mulum vanaushadhim": no plant created by the God is without medicinal values; however, one has to know how to use these and for which disease.

Nutraceuticals, first defined by Dr. Stephen DeFelice, has combined the words nutritional and pharmaceuticals, which reflects that these products have a potential as treatment and need to be treated similar to pharmaceuticals.

The present book is an attempt to look at nutraceuticals from a pharmacist's perspective. Most of the chapter authors have pharmaceutical backgrounds and have been working in pharmaceutical education and in the pharmaceutical industry.

This book covers various aspects of nutraceuticals. The first two chapters deal with the historical perspective of nutraceuticals, including definitions and challenges. The next three chapters offer insight into the nutraceutical ingredients from plant, animal, and mineral origins. The following chapters cover the characterization of nutraceuticals' physicochemical, analytical, pharmacological, and pharmacokinetic classifications, followed by a chapter that covers the regulatory requirements for nutraceuticals.

Chapters on nutraceutical formulations and applications cover nutraceuticals, applications in cardiovascular area, bone and joint treatments, diabetes management, weight management, skin health, probiotics and prebiotics, tranquilizing medicinal plants, dietary foods, and so on. The field of nutraceuticals is so vast that to cover all the applications was beyond the scope of this book, yet we tried to cover the major applications. There are many more not covered here, but adequate information is available in the literature. I have included two chapters with the leading research for the two important nutraceutical ingredients curcumin and pomegranate juice. The chapter on nutraceutical safety and toxicity gives some insight into this aspect.

I am extremely grateful to Dr. Stephen DeFelice for providing the Foreword to this book, as well as offering a chapter titled "Nutraceuticals: Reflections." I am also thankful to all the chapter authors who have given their contributions in a timely manner to be part of this book.

I would be failing in my duties if I do not mention Stephen Zollo and David Fausel from Taylor & Francis for their kind help and encouragement. Thank you very much to Dr. Vimal Patel and Paul Wilson from Health Synergies for their kind help in writing this Preface.

I have no words to express my sincere thanks to Allison Koch for her kind help in this project, as she assisted with editing; she has been very supportive of my endeavors at the Sullivan University College of Pharmacy.

My family always bears with me in all these ventures, and it always takes their portion of time. I am highly indebted my wife, Seema, and my son, Sarvadaman, for their kind support.

Yashwant Pathak

Contributors

John Adams
School of Pharmacy and Pharmaceutical Sciences
University of Manchester
Manchester, United Kingdom

Sudesh Agrawal
Department of Molecular Cardiology
Lerner Research Institute of the Cleveland Clinic
Cleveland, Ohio

Miriam A. Ansong
Sullivan University College of Pharmacy
Louisville, Kentucky

Meghan Bodenberg
Clinical and Administrative Sciences Department
Sullivan University College of Pharmacy
Louisville, Kentucky

Holly Byrnes
Clinical and Administrative Sciences Department
Sullivan University College of Pharmacy
Louisville, Kentucky

Maria Lourdes Ceballos-Coronel
Department of Pharmaceutical Sciences
Sullivan University College of Pharmacy
Louisville, Kentucky

Amitabha Chakrabarti
Cleveland Leukemia Therapeutics
Cleveland, Ohio

Kimberly K. Daugherty
Department of Clinical and Administrative Sciences
Sullivan University College of Pharmacy
Louisville, Kentucky

Stephen L. DeFelice
The Foundation for Innovation in Medicine
Cranford, New Jersey

Ajoy Koomer
Department of Pharmaceutical Sciences
Sullivan University College of Pharmacy
Louisville, Kentucky

Girish J. Kotwal
InflaMed Inc. and Kotwal BioConsulting LLC
Louisville, Kentucky

Cathy Leet
Integrative Therapeutics Inc.
Louisville, Kentucky

Brian Lockwood
School of Pharmacy and Pharmaceutical Sciences
University of Manchester
Manchester, United Kingdom

Mariel Marder
Instituto de Química y Fisicoquímica Biológicas
Universidad de Buenos Aires
Buenos Aires, Argentina

Pamela Mason
Usk, Monmouthshire, South Wales

Vimal Patel
Health Synergies
Indianapolis, Indiana

Yashwant Pathak
Department of Pharmaceutical Sciences
Sullivan University College of Pharmacy
Louisville, Kentucky

Seema Y. Pathak
Sullivan University College of Pharmacy
Louisville, Kentucky

Gwendolyn W. Pla
Department of Nutritional Sciences
Howard University
Washington, D.C.

Charles Preuss
Department of Molecular Pharmacology and Physiology
University of South Florida College of Medicine
Tampa, Florida

Alan Simon
Prospect, Kentucky

Ram H. Singh
Benares Hindu University
Varanasi, India

Hieu T. Tran
Sullivan University College of Pharmacy
Louisville, Kentucky

Cristina Wasowski
Instituto de Química y Fisicoquímica Biológicas
Universidad de Buenos Aires
Buenos Aires, Argentina

Paul Wilson
Health Synergies
Indianapolis, Indiana

Mike Witt
LouisPharma
Louisville, Kentucky

Raghunandan Yendapally
Department of Pharmaceutical Sciences
Sullivan University College of Pharmacy
Louisville, Kentucky

CHAPTER 1

Nutraceuticals of Antiquity

Vimal Patel, Paul Wilson, and Ram H. Singh

CONTENTS

INTRODUCTION

> Let food be thy medicine and medicine be thy food!
>
> **Hippocrates**

आहारसंभवं वस्तु रोगाश्चाहारसंभवाः
हिताहितविशेषाच्च विशेषः सुखदुःखयोः ४५

Àhārasambhavam vastu rogāścāhārasambhavāh,
Hitāhitaviśesāśca viśesah sukhaduhkhayoh.

Caraka Samhita Sutrasthana **28:45**

The physical body is the product of diet and sensory inputs (i.e., lifestyle). Similarly, all ailments are the product of faulty dietetics and lifestyle. Wholesome and unwholesome diets and lifestyles are fundamentals of health and disease.

The world's wisdom on health is captured by the above prophetic pronouncement from the father of Western medicine, Hippocrates. This statement on food reflects the critical importance of food and lifestyle on one's health. The second quotation, extending the concept, has been the basis of health promotion and disease prevention in the oldest and longest continuously practiced medical system [*Caraka Samhita* 700 BCE, Chapter 1; *Susruta Samhita* 600 BCE, Chapters 27–30] in the world, known as Ayurveda (meaning "science of life").

However, the progress and phenomenal success in the past 150 years or so in the field of medicine, based on a Cartesian/Newtonian biomedical model in terms of communicable diseases, emergency medicine, and technology-driven surgical and other procedures, has practically obliterated the importance of food and lifestyle on one's health. Our fast-paced, stressful lives, processed convenience foods, and overreliance on drugs and the high-tech procedure-driven medical system of "disease care" are simply unable to deal with the chronic disease crisis and ever-increasing healthcare costs [Patel 1998]. The system is crumbling under its own weight. We must look as far back as possible into the world's deepest knowledge of health to find a solution to the crisis of unaffordable and unsustainable healthcare costs and unmanageable chronic diseases.

Chronic diseases, most notably the umbrella categories of overweight and obesity that guarantee so many cascaded chronic diseases, are becoming a norm in modern life. Closer examination of the epidemic of obesity and overweight in recent decades reveals that the change in our diet, the polluted environment, and lives lacking physical activity, love, and intimacy are the roots of this epidemic and high incidence of heart disease, cancer, and other chronic conditions that are consuming more than 75% of U.S. health dollars. Recent investigations in healthy and long-living (100 years and beyond) societies of the world reveal that they have diet, lifestyle, and environment remaining essentially unchanged for millennia and conducive to good health and lifespan [Cox and Guyer 2004].

Thousands of years ago, a comprehensive understanding of life—one that incorporated all that is good in life and all that is bad, all that is useful and all that is harmful, at the levels of the physical body, senses, and soul—emerged in an ancient, knowledge-based (Vedic) culture. Known as Ayurveda, it remains a perennial source of wisdom and perspective, coming to the aid of both the theoretical and applied challenges facing modern allopathic medicine.

WESTERN DISCOVERY OF AN ANCIENT MEDICAL SYSTEM OF INDIA

The invasion of northern India by Alexander in 325 BCE led him to discover a vibrant university education system at both Taxshila (now in Pakistan) and Nalanda (now in Bihar state of India). Particularly, he noted the highly advanced medical system and training at these institutions. Such institutions were also present in other parts of India, for example, Kashi University in Varanasi (now Uttar Pradesh in India), Ujjain University (now in Madhya Pradesh), and Vallabhi University (now in Gujarat). It is widely known from the historical literature of India that Alexander, before his demise in Alexandria, Egypt, established a scholarly exchange program between India and Greece, particularly in the field of natural sciences, including Ayurveda. These scholars of Ayurvedic medicine made invaluable contributions to the field of medicine of ancient Greece and Western countries.

In its long history, India has suffered many setbacks during many invasions. During the British rule of India, the Ayurvedic system was marginalized and slowly got replaced by the British allopathic system, particularly among the population centers of urban areas and educated sectors of the Indian population. However, the medical system has largely remained intact, often as a family tradition of vaidyas (Ayurvedic physicians) and in some kingly states. In post-independence India, the Ayurvedic system of medicine has come back and gained recognition as the most comprehensive/integrative healthcare system in India, specifically in providing affordable management and prevention of ever-increasing chronic diseases. In fact, today nearly 70% of Indian people get their healthcare through the Ayurvedic medical system [Valiathan 2006].

Ayurveda, in its ancient heyday, was not only highly advanced but also had eight clinical specialties that would rival those of modern medicine:

- Internal medicine
- Surgery
- Eye and ears, nose, and throat
- Pediatrics and gynecology
- Toxicology
- Psychiatry
- Sexology and reproductive science
- Rejuvenation and geriatrics

The comprehensive/integrative spirit-mind-body-environment concepts of the Ayurvedic model [Singh 2001, 2005] have been reappearing in many healthcare approaches to the chronic disease crisis the world is experiencing. For instance, in the United States, the emerging integrative model called "functional medicine" is based on physiological system imbalances rather than disease categories, drawing heavily from the concepts of the Ayurvedic healthcare model emphasizing *dosha* (governing psycho-physiological-biogenetic principles of intelligence) imbalances, diet, and personal habits [Bland, Costaerlla, Levin, et al. 1999; Singh 2001, 2005; Liska, Quinn, Lukaczer, et al. 2004; Jones 2005].

THE WORLD HEALTH ORGANIZATION AND AYURVEDIC MEDICINE

The United Nation's World Health Organization (WHO) recognizes the inability of the dominant allopathic medical model to deal with ever-increasing healthcare costs and epidemics of many chronic diseases. In its search for affordable and sustainable healthcare for the nations of the world, the WHO surmises that the comprehensive, spirit-mind-body-environment principles of the Ayurvedic medical model [Patwardhan 2005] need to be adopted and incorporated in the emerging healthcare model(s) of the world. The WHO defines health as follows: "Health is the state of physical, mental, social, and spiritual well-being." This definition is, in fact, adopted from the health definition of Ayurveda:

समदोषः समाग्निश्च समधातुमलक्रियः
प्रसन्नात्मेन्द्रियमनाः स्वस्थ इत्यभिधीयते ।

Sama doshah samāgniś ca sama dhātu malakriya prassannāthemendriya manāh swastha ityabhidhiyate.

Susruta Samhita **15:38**

Health is the state of equilibrium of doshas, *agnis* (transformative physiological system functions), *dhatus* (tissues and organs), and *malas* (metabolic byproducts), along with sensorial, mental, and spiritual well-being.

According to the Ayurvedic model, lifestyle and nutrition in a given environmental context is the glue that holds the equilibrium to bring about optimum health.

NUTRACEUTICALS

This ancient understanding is being reintroduced as "nutraceuticals" by present-day healthcare providers. They recognize the fact that our heavily processed food supply, coming from crops grown with chemical fertilizers, pesticides, herbicides, and often genetically modified seeds, lacks sufficient nutrients necessary for optimum health.

The term "nutraceuticals" was coined by Stephen L. DeFelice, MD, in 1989. The word is a portmanteau of "nutrition" and "pharmaceutical" and refers to extracts of foods claimed to have a medicinal effect on human health [DeFelice 2002].

Nutraceuticals are usually contained in a medicinal format such as a capsule, tablet, powder, or liquid in a prescribed dose. The term further implies that the extract or source food is demonstrated to have a physiological benefit or provide protection against a chronic disease. See the sidebar *The Astounding Apple*.

Since the passage of the U.S. Dietary Supplement Health and Education Act of 1994, there has been an explosion of various kinds of nutraceuticals, in terms of both food supplements and "functional" foods. Nutraceuticals are a multi-billion dollar industry, rivaling the pharmaceutical industry. Many new entrepreneurs and pharmaceutical companies have entered the nutraceuticals field with the intention of improving their bottom line. This enthusiasm is based on epidemiological studies of the prevalence of chronic diseases in different parts of the world and of the beneficial health effects of whole foods, fruits, and vegetables and probable active ingredients of these food stuffs and their predicted influence on biochemical pathways.

Increasing numbers of consumers, concerned about healthcare costs and dissatisfied with pharmaceutical agents in promoting health, are turning to nutraceuticals to improve their health and prevent chronic disease. With few exceptions, the U.S. Food and Drug Administration (FDA) has not approved nutraceuticals for health benefits or disease prevention; nonetheless, the manufacturers of nutraceuticals have been touting them as health-promoting agents. Recent studies have questioned the validity of the use of some nutraceuticals and vitamins in health promotion and disease prevention. However, these supplements may still serve useful purposes, especially in light of our agriculture practices, dwindling food supply, and consumption of highly processed foods that often lack sufficient and appropriate nutrients [Cox and Guyer 2004].

THE ASTOUNDING APPLE

That most American of all-natural foods, the lowly apple, is a phytonutritional bonanza, brimming with health-promoting goodness. A brief sample of the more than 10,000 phytonutrients in an apple includes such exotic chemistry as follows:

- Ethyl-methylbutyrate
- d-Galacturonic-acid 13-54
- *Trans-N*-hex-2-en-1-ol
- Protocatechuic-acid
- Cyanidin-3-galactoside

It is not worth publishing the entire list. It is longer than *War and Peace*. The point is that an apple a day *can* keep the doctor away and might even be suggested by the doctor should he or she show up, in lieu of medicines. The complexity of the apple's chemical cornucopia finds its focal point in fiber, fructose, and flavonoids.

FIBER

Apples are great sources of both soluble and insoluble fiber, the latter being a powerful antidote to cholesterol. According to the February 23, 2004 issue of the *Archives of Internal Medicine*, each 10 g of insoluble fiber consumed daily may cut your risks of heart disease by 14% and your risk of dying of heart disease by 27%. A medium-sized apple provides 5 g of insoluble fiber.

FRUCTOSE

The slow metabolism of this natural sweetener in apples, particularly when paired with the apple's fiber, keeps blood sugars more level. This is technically referred to as the glycemic index of the apple. A rose (and the apple is part of the rose family) by any other name would keep your blood sugar in the zone just as effectively.

FLAVONOIDS

The September 2006 issue of *American Journal of Clinical Nutrition* reports that a study in Finland covering 10,054 Finnish men and women from 1966 onward showed reductions in heart disease, cancer, stroke, type 2 diabetes, and asthma in those who ate the most apples. The study attributes the reductions to the flavonoids in the apples.

AYURVEDIC RASAYANAS

This concept of promoting health and preventing disease through nutrition and special nutritional preparations has been used for millennia in Ayurveda. The concept is known as *rasayana* or rejuvenation therapy, and it is part of one of the eight clinical specialties of Ayurveda listed previously. The Sanskrit roots of rasayana (*rasa+ayana*) essentially refer to acquisition, movement, or circulation of nutrition needed to provide nourishment to the organs, tissues, and tissue perfusion [*Ashtanga Hridaya* 300 AD, Chapter 29; *Sharangdhar Samhita* 1300 AD; Singh 2003, 2007; Singh, Mamgain, Narsimhamurthy, et al. 2006; Singh, Narsimhamurthy, and Singh 2008].* The concept of promotion of health through rasayanas is based on Ayurvedic pathophysiology and its understanding of health and physiological system imbalances that lead to disease development (see Figure 1.1).

The philosophy of the Ayurvedic integrative spirit-mind-body-environment model holds that the optimum health of an individual can be achieved only when all human dimensions are integrated. That means that healthfulness is an outcome of a dynamic interaction among our genetic (physiological), mental (psychological), emotional, spiritual, social, and environmental factors. In other words, the basis of the science of rasayana (*rasayanatantra*) is much more comprehensive than the emerging field of nutraceuticals.

On a fundamental level, rasayanatantra is based on the Ayurvedic cosmology, which proposes that consciousness is the basis of matter, energy, chemico-physical, bio-psychological, and spiritual evolution (see Figure 1.2).† This means that every thought becomes a molecule and that our food, sensory inputs, and belief systems are interconnected and govern our gene expressions and thus our healthfulness. Because our genes have not changed much in thousands of years, our food and way of living in a given environmental context play crucial roles in promotion of health and disease prevention. In fact, the emergence of the high-profile fields of epigenetics (how environment influences gene expression) and nutrigenomics (how nutrition affects gene expression [Nathanietsz 1999; Lipton 2005; Lahiri and Maloney 2006; Lahiri, Maloney, Basha, et al. 2007; Ornish 2007; Bland 2008a,b]) validates the usefulness of the ancient Ayurvedic model and rasayanatantra.

* *Rasa* means the following: taste; elixir; serum; food juice; lymphatic fluid; plasma; essence; purified metal oxide; artistic delight; musical note; circulation; to feel lively; to dance; gravy; sauce; appreciation. *Ayana* means the following: pathway; to circulate; to have a home or abode.

† Rasayanas are formulated to reflect unique chemico-physico-bio-psychological concepts of Ayurveda. That is, all animate and inanimate entities have the same basic constituents (the five quantum mechanical spin types, corresponding to the fivefold classification of earth, water, fire, air, and ether/space). These constituents present themselves as *vata*, *pitta*, and *kapha* (see "Age-Specific Rasayana" below) at the chemico-physical level and as *sattwa* (spiritually pure), *rajas* (dynamic), and *tamas* (nonspiritual) at the bio-psychological level. Because Ayurveda categorizes all foodstuffs, medicinal plants, and minerals in the same way, they can be used to bring about balance in vata, pitta, and kapha, as well as sattwa, rajas, and tamas attributes of the various body-mind types recognized by Ayurveda.

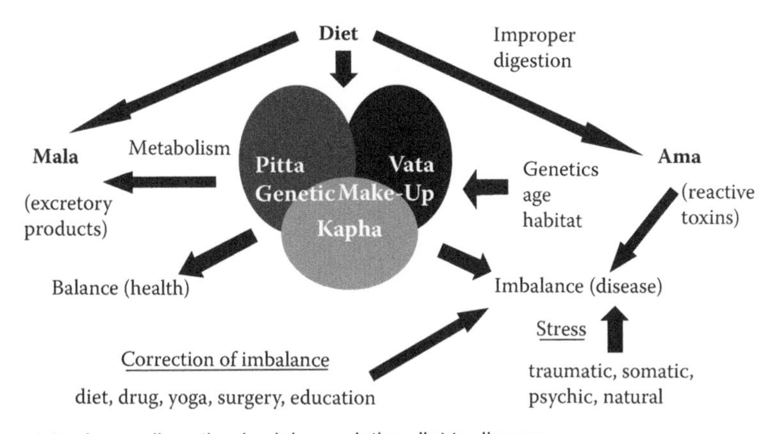

Figure 1.1 Ayurvedic pathophysiology relating diet to disease.

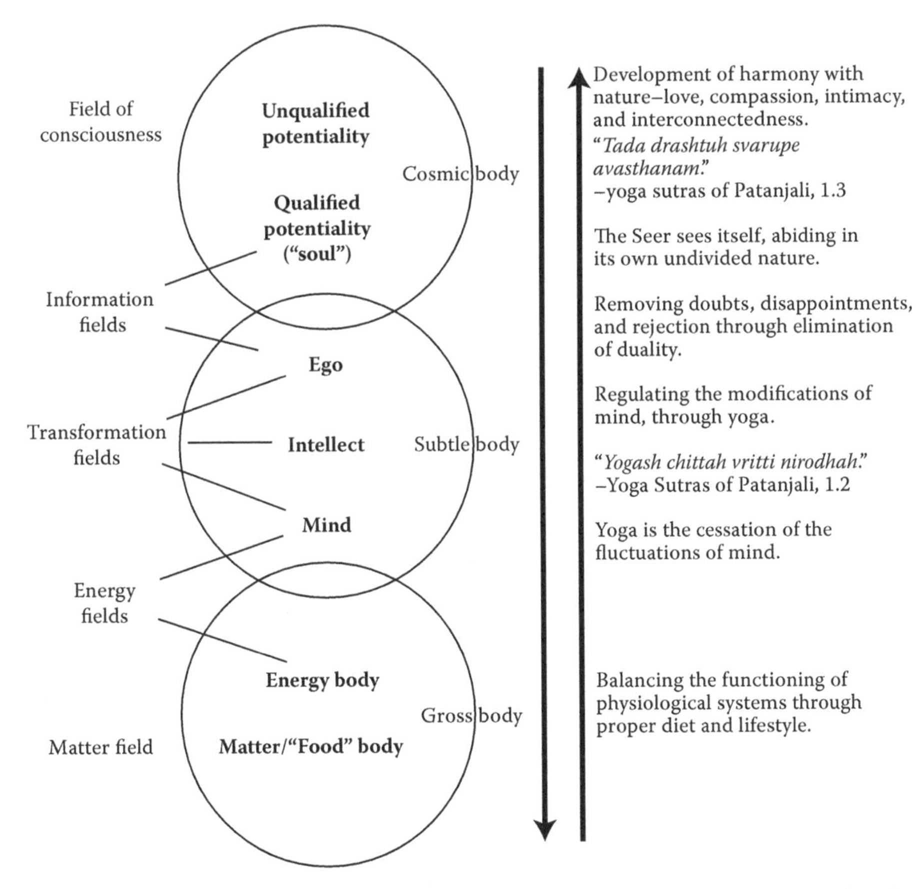

Figure 1.2 Rasayana therapy helps the development of integration of spiritual, mental, and physical fields.

रसायनतन्त्रं नाम वयःस्थापनमायुर्मेधा वलकरं रोगापहरणसमर्थं च

Rasāyanatantram nāma vayahsthāpanamāyurmedhābalakaram rogāpaharanas-amartham ca.

Susruta Samhita **1:7**

Rasayanatantra deals with the methods to maintain youthfulness, to increase longevity, intellectual capacity, and physical strength as well as to enable the person to be free from disease.

रसायनं च तज्ज्ञेयं यज्जराव्याधिनाशनम्

Rasāyanam ca tajjnyeyam yajjaravyādhināśanam.

Susruta Samhita **Part 1, 4:13**

The therapy that helps to retard aging and disease is called rasayana.

Rasayanatantra consists of an amplification and synergy of the natural nutraceutical potential of certain plants and food products and related measures that are supposed to retard aging and to impart longevity, improve immunity and body resistance against disease, improve mental faculties, and add vitality and luster to the body. These rasayanas are individualized and are age, tissue, and organ specific. The rasayana therapies are aimed at bringing a state of equilibrium of doshas, agnis, dhatus, malas, and sensory and physiological system functioning, along with spiritual and mental well-being. In other words, they are designed to bring about balanced functioning of spirit, mind, and body in the context of given environment. Besides promotion of mental and physical health and rejuvenation potential, rasayana therapy affords a preventive role against the entire range of diseases through improved immunity and other physiological system functions. Thus, rasayana is the central consideration in Ayurvedic geriatrics.

MODE OF ACTION

All rasayana measures and remedies produce their effect in the spirit-mind-body system through one or some combination of the following three modes:

- At the level of rasa, by acting directly as a nutrient for the plasma.* Examples include a range of nutrient rasayanas, such as shatavari, sarkara, ghrita, pravala, and mukta.

* Ayurveda posits that the body's tissues develop in a sequence from liquid to solid as food is digested. When the food is exposed to the initial enzymatic transformation stage, i.e., the first agni, it becomes rasa and a corresponding healthy mala. From there, the sequence unfolds linearly, governed by other agnis and producing other corresponding malas: *rakta* (whole blood), *mamsa* (muscle), *medha* (fat), *asthi* (bone), *majja* (bone marrow and nervous tissue), and *shukra* (reproductive tissue). Between each level of transformation and at the end of the process, another substance emerges, *ojas*. Ojas represents biological intelligence and acts as a master coordinator between consciousness and matter.

- At the level of agni, by promoting the enzymatic systems of the body with positive digestive and metabolic functions. Examples include pippali, shunthi, and chitraka.
- At the level of *shrotas*, i.e., microcirculatory and macrocirculatory channels, by inducing a *shrotaprasadana* effect, i.e., improving the competence of the inner transport system, microcirculation, and tissue perfusion. Rasayanas for this purpose include guggulu.

By acting through the above modes, rasayanas establish a positive nutritional status in the body and help in healthier tissue formation, stronger immune status, improved mental power, and long life.

CLASSIFICATION

As comprehensively envisaged in Ayurveda, rasayana is not a mere remedy or a recipe. It is a rejuvenative regimen and an approach to positive health. It encompasses elements of positive living and conduct, healthy dietetics, and rejuvenative herbs and minerals. Rasayanatantra is practiced as a routine way of life or as an intensive indoor regimen, depending on the need and the feasibility for a client. The rasayana therapy can be categorized in the following manner:

- Per method of use
  - *vatatapika* rasayana, or outdoor practice
  - *kutipraveshika* rasayana, or intensive indoor regimen (including the seasonal physiological purification known as *panchakarma*) using a specially designed *trigarbha rasayana kuti* or therapy chamber
- Per scope of application
  - *kamya* rasayana, for promotion of health of the healthy, further subcategorized as
  - shri kamya, to promote luster and beauty
  - prana kamya, to promote longevity
  - medha kamya, to promote mental competence
  - *naimittika* rasayana, to impart biological strength in a person with disease
- Adjunct rasayana, which is non-recipe rejuvenative regimen to be practiced alone or as an adjunct for all forms of rasayana therapy, remedies, and recipes, as follows
  - *achara* rasayana: healthy, rejuvenative lifestyle and conduct
  - *ajashrika* rasayana: daily dietary rasayana approach, consuming sattvic, nourishing elements of diet, such as ghi, milk, milk products, fruits, and vegetables

---

Rasayanas can be operationally defined as nutritives that initiate the maximum outpouring of ojas throughout the entire body-mind. Ojas is an invigorating spiritual essence and is the subtle foundation of immunity, and Ayurveda holds that creating and maintaining ojas is of central importance to health and wholeness.

When digestive processes break down, dhatus and malas are not formed correctly, and another entity, *ama* associated with all manner of ill health, invades the body-mind. In a sense, the polar opposite of ojas, ama, damages the system and lessens the integrated coordination of spirit, mind, and body.

Rasayanas are intended to convert directly to rasa, bypassing ordinary digestive steps. In so doing, they do not produce ama and start the unfolding of dhatus in the best possible manner so that as much ojas as possible can emerge. If rasa is the base of the Ayurvedic nutrition pyramid, ojas could be the pinnacle. Rasayanas ensure a royal path from rasa to ojas.

AGE-SPECIFIC RASAYANA

Aging is the *shwabhawa* or the nature of a living being. The physical body-mind system has been designed to stay for a time-bound tenure of approximately 120 years. During the lifespan, the body undergoes progressive involution and decay, leading ultimately to decadence and death. Ayurveda deliberates on the process of aging and sequential senile changes in different ways in different contexts. Examples include *balyawastha* (childhood), *madhya awastha* (adulthood), and *briddhawastha* (geriatric age), hallmarked by activities of the above-mentioned doshas: *kapha* (structure and fluid balance), *pitta* (heat, digestion, and energy processes), and *vata* (movement and change). Vata is the drying and decaying force and is the master dosha in the aging process.

Two foundational texts of Ayurveda, *Vagbhatta* and *Sarangdhara*, describe a unique scheme of biological aging in a 10-decade frame, speculating the specific sequential loss of certain bio-values relevant to respective decades of life. This information opens the possibility of developing targeted rasayanas to restore the likely losses of the particular decade, ensuring that inevitable change is intelligently and benignly directed at each stage of life to maintain the highest possible level of function and health. Thus, rasayana therapy planned in relation to age creates a possibility of retarding the aging process. Table 1.1 describes the pattern of age-related biological system losses and proposes certain rasayanas for the purpose.

TISSUE- AND ORGAN-SPECIFIC RASAYANA

Although rasayana in general is a holistic restorative and rejuvenative modality, one can visualize some rasayana remedies and recipes for promotion and protection of specific tissues and organs. Such rasayanas can be prescribed in a need-based manner for supportive or even curative purposes for organ protection. Some examples are proposed in Table 1.2.

Table 1.1 The Pattern of Age-Related Biological System Losses

Decades of Life	Natural Bio-Losses	Suggested Rasayana for Restoration
0–10	*Balya*, corpulence	Gambhari, kshira, ghrita
11–20	*Vriddhi*, growth	Bala, amalaki
21–30	*Chhabi*, luster	Amalaki, haridra
31–40	*Medha*, intellect	Brahmi, shankhapushpi
41–50	*Twaka*, skin quality	Bhringaraja, haridra
51–60	*Dristi*, vision	Triphala, jyotishmati
61–70	*Shukra*, virility	Ashwagandha, kapikacchu, shatavari, pippali
71–80	*Vikrama*, physical strength	Amalaki, bala
81–90	*Buddhi*, thinking	Brahmi, shankhapushpi
91–100	*Karmendriya*, locomotion	Bala, sahachara

Table 1.2 Rasayanas

Rasayana Quality	Purpose	Suggested Remedies
Medhya rasayana	Promotion of brain and cognitive functions	Brahmi, shankhapushpi
Hridya rasayana	Cardio-protective	Arjuna, pushkarmula
Mutra janana	Nephro-protective	Punarnava, gokshuru
Twachya rasayana	Skin health	Haridra, somaraji
Chakshushya rasayana	Eye health	Triphala, jyotishmati
Kannthya rasayana	Throat and speech	Vacha, yashtimadhu
Vrishaya rasayana	Virility	Ashwagandha, kapikacchu
Sthanya rasayana	Promotes lactation	Shatavari
Shrotoprasadana	Promotes inner transport	Guggulu
Nasya rasayana	Nose and sinuses	Katphala, apamarga

ACHARA AND AJASHRIKA RASAYANA

Achara rasayana is a unique concept in Ayurveda that implies moral, ethical, and benevolent conduct: truth, nonviolence, personal and public cleanliness, mental and personal hygiene, devotion, compassion, and a yogic life. These behaviors bring about rejuvenation in the body-mind system. One who adopts such conduct gains all benefits of rasayana therapy without physically consuming any material rasayana remedy or recipe, although it can be practiced alone or in a combination with material substance rasayana therapy.

Ajashrika rasayana comprises daily rejuvenating dietetics, with adequate quantities of nourishing sattvic elements of diet, such as ghi, milk, fruits, and vegetables. Ajashrika rasayana is used alone or with material rasayana remedies. Studies conducted in recent years on stress management related to yoga and other nonpharmacological techniques clearly indicate beneficial effects in control and prevention of chronic conditions, such as heart disease, cancer, diabetes, and immune disorders [Innes and Vincent 2007; Innes, Vincent, and Taylor 2007; Masley, Weaver, Peri, et al. 2008].

SAMSHODHANA (PRE-DETOXIFICATION) FOR RASAYANA THERAPY

Besides achara and ajashrika, another important requirement for use of material rasayana therapy is pre-detoxification through appropriate panchakarma procedures. Ayurveda emphasizes that a rasayana remedy yields its full effect only when the body has been therapeutically purified by Ayurvedic cleansing processes (*langhana*, *dipana*, *pachana*, *snehana*, *swedana*, *vamana*, *virechana*, *basti*, *shirovirechana*, and so on). If the shrotas are clean and competent with their physiological integrity at the time of administration of the rasayana remedy, such remedy is used by the system fully, and its bioavailability is ensured. Hence, pre-detoxification processes should be planned accordingly. The most appropriate choice of age for use of rasayana therapy is adulthood to 75 years or so, with the therapy not recommended for old age, when irreversible senile changes might already have occurred.

SUGGESTED RASAYANAS FOR DIFFERENT BODY AND MIND TYPES, SEASONS, AND DIGESTIVE STRENGTH

Please refer to the following for suggestions on rasayanas:

- Vata type: ashwagandha, bala, gambhari, rasona, amrita, shankhapushpi, chyavanaprasha, brahma rasayana
- Pitta type: amalaki, chandana, brahmi, mukta, pravalapishti, amalaka
- Kapha type: pippali, ardraka, shilajatu, bibhitaka, bhallataka
- Dynamic (rajasic) mental type: brahmi, mandukaparni, shankhapushpi, mukta
- Nonspiritual (tamasic) mental type: pippali, amalaki, chitraka, bhallataka
- Lengthening days season: ashwagandha, amalaki, brahmi, chandana, khasa
- Shortening days season: pippali, shilajatu, bhallataka, kasturi, shringa
- Variable digestion: ashwagandha, rasona
- Sharp digestion: apamarga, shankha, pravala, kumari, brahmi
- Weakened digestion: pippali, shunthi, ghrita, chitraka, lavana

A physician should select a suitable rasayana in consideration of different individual and environmental factors, taking into account the principles of *samanya* and *vishesa* (homology versus heterology).

SINGLE, GROUP, AND COMPOUND RASAYANAS

A range of single, group, and compound rasayanas has been described in Ayurvedic classics in different contexts. Some are listed below:

- Popular single rasayanas: amalaki (*Phyllanthus emblica*), haritaki (*Terminalia chebula*), pippali (*Piper longum*), ashwagandha (*Withania somnifera*), brahmi (*Bacopa monniera*), shankhapushpi (*Convolvulus pluricaulis*), guduchi (*Tinospora cordifolia*), madhuyasti (*Glycyrrhiza glabra*), mandukparni (*Centella asiatica*), bala (*Sida coridfolia*), shatavari (*Asparagus recemosus*), bhallataka (*Semecarpus anacardium*), punarnava (*Boerhaavia diffusa*), lauha (iron), swarna (gold), shilajatu (asphaltum)
- Popular compound rasayanas: chyavanaprasha, brahma rasayana, amalaka rasayana, amrita bhallataka, bhallataka kshirapaka, haridra khanda, bala rasayana, amrita rasayana, punarnava rasayana, louhadi rasayana, aindra rasayana, triphala rasayana, shilajatu rasayana, ritu haritaki kalpa, pippali vardhamana kalpa, bhallataka kalpa, panchamrita parpati kalpa

CONCLUSION

The context of rasayana therapy and its ajashrika and achara are largely unexplored parts of ancient wisdom. Searches for drug development based on described properties of Ayurvedic plants have yielded useful drugs that confirm their medicinal usage and the Ayurvedic descriptions [Kapoor 1999]. Experimental and some clinical studies on some rasayanas also validate their expected use [Patel 1998; Nagarathna, Nagendra, and Telles 1999; Douillard 2000; Sing 2003; Ornish 2007;

Sharma, Chandola, Singh, et al. 2007; Sharma, Puri, Agrawal, et al. 2009; Singh, Narsimhamurthy, and Singh 2008]. In addition, yoga techniques for balancing stress responses also confirms the expected outcome [Nagarathna et al. 1999; Innes and Vincent 2007; Innes, Vincent, and Taylor 2007]. The above observations strengthen trust in the power of the time-tested, comprehensive integrative healthcare system of Ayurveda, based on lifestyle and dietary habits that the disease-care model does not adequately address. The prevalent rasayana procedures and recipes may be of great value in promotive, preventive, and therapeutic aspects of geriatric healthcare.

In a larger sense, our health and behavior are expressions of our genes, designed to work in harmony with the natural environment. Rasayanas have their origin in plants and minerals and have played import roles in health promotion and maintenance in Ayurveda. As an ancient Indian Vedic text [Gabhirananda 1983] surmises:

> *Esam bhutanam prithavi rasa, prithavya apo rasso-pam osadhayo rasa, osadhinam purusho rasa*
>
> *Chandogya Upanisad* **1.1.2**

> The essence of all beings is earth. The essence of earth is water. The essence of water is plants. The essence of plants is human beings.

This suggested to us that plants are at the evolutionary center of our being, and, therefore, preservation of our earth, water, air, and plants amounts to preservation of the human race and its health.

REFERENCES

Ashtanga Hridaya. 300 AD. Uttar Tantra. Chapter 39 on rasayana. Edited by Sharma, P. V. Varanasi, India: Choukhamba Prakashana.

Bland, J. 2008a. The future of nutritional pharmacology. *Alt. Ther. Health Med.* 14:12–14.

Bland, J. 2008b. Functional somatic syndromes, stress pathologies and epigenetics. *Alt. Ther. Health Med.* 14:14–16.

Bland, J. S., L. Costaerlla, B. Levin, D. Liska, D. Lukaczer, B. Schiltz, M. A. Schmidt, and R. H. Lerman. 1999. *Clinical Nutrition: A Functional Approach*, 1st ed. Gig Harbor, WA: The Institute for Functional Medicine.

Caraka Samhita. 700 BCE. Chikitsa sthana. Chapter 1. Parts 1–4 on rasayana. Edited and translated to English by Sharma, P. V. Varanasi, India: Choukhamba Prakashan.

Cox, W. and D. Guyer. 2004. *Getting Well*. Bloomington, IN: AuthorHouse, pp. 203–211.

DeFelice, S. L. *FIM Rationale and Proposed Guidelines for the Nutraceutical Research & Education Act—NREA*, November 10, 2002. Foundation for Innovation in Medicine. http://www.fimdefelice.org/archives/arc.researchact.html.

Douillard, J. 2000. *3-Season Diet: Eat the Way Nature Intended*. New York, NY: Three River Press.

Gabhirananda, S. 1983. *Chandogya Upanishad*. Calcutta, India: Advaita Ashram Publication.

Innes, K. E. and H. K. Vincent. 2007. Chronic stress and insulin-resistance-related indices of cardiovascular disease risk. Part 1, Neurophysiologic responses and pathologic sequelae. *Alt. Ther. Health Med.* 13:46–52.

Innes, K. E., H. K. Vincent, and A. G. Taylor. 2007. Chronic stress and insulin-resistance-related indices of cardiovascular disease risk. Part 2: A potential role for mind-body therapies. *Alt. Ther. Health Med.* 13:44–51.

Jones, D. S. (Ed.). 2005. *Textbook of Functional Medicine*. The Institute for Functional Medicine: Gig Harbor, WA.

Kapoor, M. 1999. *Handbook of Ayurvedic Medicinal Plants*. Boca Raton, FL: CRC Press.

Lahiri, D. K. and B. Maloney. 2006. Genes are not our destiny: the somatic epitype bridges between the genotype and phenotype. *Nat. Rev. Neurosci.* 7:583–590.

Lahiri, D. K., B. Maloney, M. R. Basha, Y. W. Ge, and N. H. Zawia. 2007. How and when environmental agents and dietary factors affect the course of Alzheimer's disease: The "LEARn" model (latent early-life associated regulation) may explain the triggering of AD. *Curr. Alzheimer Res.* 4:219–228.

Lipton, B. H. 2005. *The Biology of Belief: Unleashing the Force of Consciousness Matter and Miracles*. Santa Rosa, CA: Love/Elite Books.

Liska, D., S. Quinn, D. Lukaczer, D. S. Jones, and R. H. Lerman. 2004. *Clinical Nutrition: A Functional Approach*, 2nd ed. Gig Harbor, WA: The Institute for Functional Medicine.

Masley, S. C., W. Weaver, G. D. Peri, and S. E. Phillips. 2008. Efficacy of lifestyle changes in modifying practical markers of wellness and aging. *Alt. Ther. Health Med.* 14:24–29.

Nagarathna, R., H. R. Nagendra, and S. Telles. 1999. Yoga in health and disease. *J. Karnataka Med. Assoc.* 3.1a:103–118.

Nathanietsz, P. W. 1999. *Life in the Womb: The Origin of Health and Disease*. Ithaca, NY: Promethean Press.

Ornish, D. 2007. *The Spectrum*. New York, NY: Ballantine Books, pp. 90–114, 115–140.

Patel, V. 1998. *Understanding the Integration of Alternative Modalities into Emerging Healthcare Model in the United States. Alternative Medicine and Ethics*. Edited by Humber, J. M. and R. F. Almeder. pp. 43–95. Totowa, NJ: Humana Press.

Patwardhan, B. 2005. *Traditional Medicine: Modern Approach for Affordable Global Health*. Geneva, Switzerland: World Health Organization.

Sharangdhar Samhita. 1300 AD. Varanasi, India: Choukhamba Prakashan.

Sharma, H., H. M. Chandola, G. Singh, and G. Basishta. 2007. Utilization of Ayurveda in health care: An approach for prevention, health promotion and treatment of disease. Part 1, Ayurveda, science of life. Part 2, Ayurveda, in primary health care. *J. Alt. Compl. Med.* 9:1011–1119.

Sharma, S., S. Puri, T. Agrawal, and V. Sharma. 2009. Diet based on Ayurvedic constitution: Potential for weight management. *Alt. Ther. Health Med.* 15:44–47.

Singh, R. H. 2001. *The Holistic Principles of Ayurvedic Medicine*. Chapter on rasayana therapy. New Delhi, India: Choukhamba Surbharati.

Singh, R. H. 2003. Psychiatric Disorders in Ayurveda. In *Scientific Basis of Ayurvedic Therapies*. Boca Raton, FL: CRC Press, pp. 439–451.

Singh, R. H. 2005. *Panchkarma Therapy*. Varanasi, India: Choukhamba Prakashan.

Singh, R. H. 2007. Brain aging and Ayurveda: Special reference to rasayana therapy. *Ann. Neurosc.* 14 (Suppl.):15–16.

Singh, R. H., P. Mamgain, K. Narsimhamurthy, and S. Rastogi (Eds.). 2006. *Advances in Ayurvedic Medicine*. Volumes 1–5. Varanasi, India: Choukhamba Vshwabharati Prakashan.

Singh, R. H., K. Narsimhamurthy, and G. Singh. 2008. Neuronutrient impact of Ayurvedic rasayan therapy in brain aging. *Biogerontology* 9:369–374.

Susruta Samhita. 600 BCE. Chikitsa sthana. Chapters 27–30 on rasayana. Edited and translated to English by Singhal, G. D., P. Mamgain, S. N. Tripathi, L. V. Guru, and R. H. Singh. Varanasi, India: Choukhamba Surbharati.

Valiathan, M. S. 2006. *Towards Ayurvedic Biology: A Decadal Vision Document*. Bangalore, India: Academy of Sciences.

CHAPTER 2

Nutraceuticals: Definitions, Formulations, and Challenges

Yashwant Pathak

CONTENTS

INTRODUCTION AND DEFINITIONS

"Nutraceuticals are in their formative years. But make no mistake, the nutraceutical boom is coming and it will be worth billions to the companies who define it," predicted Jim Wagner, Editor of the *Nutritional Outlook*, in June 2002. Since then, the nutraceuticals market has been growing by leaps and bounds, leading to a multibillion dollar industry. The recent numbers show that it is almost reaching the dollar value market as pharmaceuticals in many countries and even worldwide.

Forecasting the size of the nutraceuticals market is very difficult because there is no clear definition of nutraceuticals that is accepted universally by all the stakeholders. Historically, it is observed that every country had its own system of products, which had both nutritional and medicinal values. Looking at the traditional medicinal system of African tribes or Native American Nations, Australian Aboriginals, Chinese, Japanese, Hindus, Maoris, and many other indigenous tribes and cultures worldwide, all of them developed the functional, indigenous system of medicine, in which they extensively used the local herbs and natural minerals for their nutritional and medicinal values. Today in the allopathic medical world, most of the drugs marketed find their origin in the form of a lead compound based on the information acquired from these traditional systems. Chapter 1, "Nutraceuticals in Antiquity," describes the concept of nutraceuticals in Indian traditional systems.

The word "nutraceutical" was first coined by Dr. Stephen DeFelice [2007] who defined it as "a substance that is a food or a part of food and provides medical and health benefits, including prevention and treatment of disease." Such products may range from isolated nutrients, dietary supplements, and specific diets to genetically engineered designer foods, herbal products, and processed food such as cereals, soups, vegetable juices, and beverages. It is important to note that this definition applies to all categories of food and parts of food, including folic acid, antioxidant foods substances, stimulant functional food, and pharma food.

Another definition was suggested by Dr. Lockwood in his book [2007], stating that "Nutraceutical is the term used to describe a medicinal or nutritional component that includes a food, plant, or naturally occurring material which may have been purified or concentrated, and that is used for the improvement of health, by preventing or treating a disease."

Zeisal [1999] suggested another definition, claiming that "Nutraceuticals is a diet supplement that delivers a concentrated form of a presumed bioactive agent from a food, presented in a nonfood matrix and used to enhance health in dosages that exceed those that could be obtained from normal food."

It is interesting to know that many countries have adopted different terminologies for nutraceutical and have defined it in different ways. Most of the countries are trying to establish a regulatory framework for these products. Interestingly, the nutraceuticals revolution is becoming part of the mainstream medical discovery process. Medical practitioners worldwide are now accepting these as part of their mainstream clinical practice as a result of more and more scientific medical research [Adebowale, Liang, and Eddington 2000].

The functional food concept was first introduced in Japan; until 1998, it was the only country that legally defined foods for specified health use. A functional food is natural or formulated food that has enhanced physiological performance or prevents or treats a particular disease [Hardy 2000; Lockwood 2007]. In Canada, functional foods is defined as similar in appearance to conventional food, consumed as a regular part of the diet, whereas the nutraceuticals are defined as a product produced from food but sold as pills, tablets, capsules, and other medicinal forms not generally associated with food. In the United Kingdom, the Department of Environment, Food, and Rural Affairs defines functional food as a food that has a component incorporated into it to give it a specific medical or physiological benefit other than purely nutritional benefit.

In the United States, a new product may be introduced as food or dietary supplement or as a medical food [Litov 1998] under the U.S. Dietary Supplement Health and Education Act. According to this act, the dietary supplement can state that these products can offer nutritional support in nutrient deficient diseases, but the companies are expected to write that this statement has not been evaluated by the FDA. Furthermore, they must state that this product is not intended to diagnose, treat, cure, or prevent any disease. The FDA will be expecting that the nutraceuticals need to be manufactured under the current good manufacturing practices (cGMPs), and gradually the FDA is moving toward stricter and stricter regulations for nutraceuticals.

MODIFICATION IN THE DEFINITION OF NUTRACEUTICALS

To consider these regulations, I put forward my own modification in the present nutraceuticals definition as follows: "nutraceuticals are the products developed from either food or dietary substance or from traditional herbal or mineral substance or their synthetic derivatives or forms thereof, which are delivered in the pharmaceutical dosage forms such as pills, tablets, capsules, liquid orals, lotions, delivery systems, or other dermal preparations, and are manufactured under strict cGMPs. These are developed according to the pharmaceutical principles and evaluated using one or several parameters and in process controls to ensure the reproducibility and therapeutic efficacy of the product." I will be discussing more on this definition in the following paragraphs when I discuss the formulation of nutraceuticals.

NUTRACEUTICALS MARKET SCENARIO

Development of market for nutraceuticals is dependent on several factors, such as growing acceptance of non-allopathic systems of medicines. (These non-allopathic systems of medicines are wrongly mentioned as complementary and alternative medicines [CAMs]; I say they are wrongly mentioned because these nutraceuticals and herbal medicines have been used by humanity for many millennia, whereas the allopathic medicine has come into existence in the past few centuries. So, to be truthful, the allopathic medicine can be an alternative medicine for the ancient traditional medicine systems, in my opinion). Lack of patent protection for the nutraceuticals, need for scientific evidence for the relationship between the diet and health, regulatory uncertainty, and the nutritional supplement are competing with the food itself, and in many cases, the products and the consumer environment. It is observed that, as the baby boomers are reaching their golden ages, the acceptance for the nutraceuticals is growing day by day.

PREVENTION VERSUS CURE

The huge wave that started during 1990s in the United States has drastically changed the market trends for nutraceuticals. Consumers' awareness is ever increasing, and more scientific information is available for making the right choices; finally, the number of well-informed customers is also ever increasing. Consumers have started subscribing to diet regimens that reduce the risk of chronic diseases. People are using the nutraceuticals and functional foods more for the prevention than for the cure of the diseases. People have started using the functional foods and dietary supplements that will have prevention values as well as be part of the recommended dietary allowances (RDAs). It is interesting to observe that people prefer more natural substances than the purely synthetic substances and opt for natural and organic foods, functional beverages, and natural supplements. Interestingly, in the past two decades, the media has played an important role and drawn people's attention to scientific developments in the health and nutrition field. The market for nutraceuticals has tremendously benefited by this media attention.

According to a new technical market research report, the nutraceuticals global market was reported to be worth $117.3 billion in 2007 [Technical Market Research 2007]. It was projected that this will increase to $123.9 billion in 2008 and reach $176.7 billion by 2013. Table 2.1 shows the predicted values for the market according to this report.

The report by Global Industry Marker Analyst Inc. suggested that the rising consumer desire for leading a healthy life and increasing scientific evidence supporting health foods continue to drive the nutraceuticals market. They projected a healthy growth in the market to cross $187 billion by 2010. They predicted that the United States, Europe, and Japan will dominate the global market and will constitute 86% of the total global nutraceutical market. They also suggested that the expanding elderly population, enhanced awareness, high income levels of the consumers, widespread preferences for specialty nutritional and herbal products, increasing trends promoting preventive medicines, and self treatment are responsible for the phenomenal growth prediction of the nutraceuticals market [Global Industry Analyst Inc. 2008].

The other emerging market worldwide [Emerging Nutraceuticals Market] includes India, with $540 million in sales in 2006 and an expected rise of almost 38% per year in the coming years, crossing the billion mark by the end of 2009 [according to New Study 2006]. China is another market that is growing significantly, followed by

Table 2.1 Prediction for the Nutraceuticals Market in $ Billions

Year	Foods	Supplements	Beverages
2007	39.9	39.0	38.4
2008	40.6	40.5	42.8
2013	56.7	48.8	71.3

Source: Technical Market Research Report. 2007. *Nutraceuticals: Global Markets and Processing Technologies.* http://www.chem.info.

Brazil ($881 million in 2006), Turkey ($200 million in 2006), and Australia, New Zealand, and Middle East and African countries ($300 million in 2006).

GALENICALS VERSUS NUTRACEUTICALS

Galenicals are medicines prepared according to Galen's formulae and are now used to denote standard preparations containing one or several organic ingredients, in contrast to pure chemical substances. It is interesting to note that, nearly a century ago, most of the pharmaceuticals were known as galenicals and were prepared using processes such as macerations and percolations; also, several herbal products were used for these purposes. Most of these were nonalcoholic, aqueous, or alcoholic extracts of plant and animal origin substances; in due course, the approach was changed, and most of the pharmaceutical companies went for more purified versions of the drugs, and the galenicals were left on the back burner. Although the galenicals had significant medicinal value, the pharmaceutical profession was moving toward solo performance. Many of the products that are classified as nutraceuticals can very well fit into the definition of galenicals, which means they could have been approved as pharmaceuticals and could have had a place in the pharmacopoeias worldwide.

SOLO VERSUS CONCERT PERFORMANCE

Galenicals and nutraceuticals are the products that offer their therapeutic values by concert performance. There are combinations of several ingredients present in these products that offer the medicinal effect through combination of their actions. Hence, I call their actions "concert performances." It is observed that there are several ingredients present in the herbal product or other nutraceuticals that have either similar therapeutic action or sometimes antagonistic action, and they control the effect of each other, which thereby reduces the side and toxic effects of the products. To clarify, reserpine is an alkaloid present in the plant *Roulfia serpentina*, which is used for its antihypertensive action in heart problems and blood pressure. Interestingly, it also has an ingredient that can be used to increase blood pressure. Therefore, both ingredients work in synchronization, providing a very positive effect. All the major antihypertensive drugs have their lead obtained from this plant, and additional drug development lead to several solo performing drugs, which are either derivatives or modified versions of the reserpine molecule. Similar relationships can be seen in many sols drugs that have been developed based on the lead compound from their herbal analog. When a single drug is used for the treatment in the form of some suitable drug dosage form or delivery system, I call it a "solo performance." Most of the allopathic drugs have solo performances; in recent days, there are increasingly more combinations used with a lot of side effects, creating more complications after the treatments. It is observed that all the statins used for lowering cholesterol found their origin in Chinese red rice yeast; unfortunately, most of the statins have shown enough side effects to question their utility as solo drugs. It might be beneficial to use the concert-performing Chinese red rice yeast in place of solo statins. Then there comes the application of nutraceuticals more for prevention. One can quote several such examples, and that is one of the reasons why the nutraceuticals market is growing significantly: as people become more aware of these facts, they are opting for the natural products for their preventive care rather than looking for a solo-performance cure.

FROM GROCERY STORES TO DRUG STORES

Today, most of the nutraceuticals are available in grocery stores and specialized nutritional stores, including Whole Foods Markets®, Rainbow Markets, and so on. Although the nutraceuticals have such a vast market and a very high dollar value, it is not getting the status as an integral part of therapy that it should have. Hence, there is a need for moving the nutraceuticals from grocery stores to drug stores. To do this, the nutraceuticals market has to opt for clearer definitions and better distinction of their products. They have to make a choice between the nutraceuticals and functional foods and beverages. Once these products are separated, as I discussed previously with the new definition offered above, it will be gradually treated as more on the drug side than on the food side. Today, there is no need for a prescription for the nutraceuticals. There are so many adverse reports against the nutraceuticals (many of times with vested interests) so, obviously, to get the scenario changed from grocery stores to drug stores, the nutraceuticals market's best bet will be to break into the educational curriculum of the pharmacy professional degree and lobby for getting the prescription rights to pharmacists. With adequate knowledge about the drugs from the allopathic system, pharmacists will be the right professionals to deal with nutraceuticals; they can also strongly advise about the drug-nutraceutical synergy and teach how to avoid these interactions. The goal of the nutraceuticals market and manufacturers should be to form an association with pharmacists to get their desired status in the healthcare system. They will significantly and positively contribute to the healthcare system of the United States and worldwide. Their transition from grocery stores to drug stores will be easier with the help of pharmacists and pharmacy educators.

FORMULATION CONSIDERATIONS

To achieve this, nutraceuticals manufacturers must first separate the products and treat nutraceuticals differently from functional foods. Second, the formulation processing of the nutraceuticals needs to follow the norms of the pharmaceutical formulations. Some simple tests can be performed by the pharmaceutical scientists to evaluate the dosage forms and provide enough data for the consistency of the product, reproducibility of the product, and *in vitro* and *in vivo* evaluations of the product.

Table 2.2 provides such evaluations performed by the pharmaceutical industry to ensure the quality of a product using a nonfood matrix as defined previously [The University of Sciences 2005]. These dosage forms can be tablets, capsules, liquid orals, ointments, external products, dermal products, pills, and many other commonly used dosage forms, and it is necessary to explore the newer drug delivery systems such as nanoparticulate drug-delivery systems, microcapsules, and so on. Some of these evaluations can be easily performed and can contribute toward the quality of the products. We have a chapter in this book that covers, in detail, the physicochemical evaluations of the nutraceuticals.

Table 2.2 Evaluation Parameters for the Nutraceutical Products Using Nonfood Matrices

Serial Number	Nutraceutical Nonfood Matrix	Pharmaceutical Evaluation Parameters
1	Tablets can be compressed tablets, film-coated tablets, sugar-coated tablets, enteric-coated tablets and multicoated tablets, controlled-release tablets, tablets for solutions, effervescent tablets, and buccal or sublingual tablets	Color and odor Weight variation Tablet thickness Friability losses Content variation *In vitro* disintegration *In vitro* dissolution Stability at room and accelerated temperature and humidity
2	Capsules	Color and odor Weight variation Content variation *In vitro* disintegration *In vitro* dissolution Stability at room and accelerated temperature and humidity
3	Solutions: aqueous solutions, syrups, alcoholic solutions, juices, drinks	Color and odor Density Viscosity Drug content variability Stability at accelerated temperature and humidity Microbiological testing
4	Emulsions: oil in water or water in oil emulsions	Particle size distribution Color and odor Zeta potential Viscosity Surface tension Content uniformity Stability at accelerated temperatures and humidity Microbiological testing
5	Suspensions	Particle size distribution Color and odor Zeta potential Viscosity Surface tension Content uniformity Stability at accelerated temperatures and humidity Microbiological testing

Table 2.2 (*Continued*)

Serial Number	Nutraceutical Nonfood Matrix	Pharmaceutical Evaluation Parameters
6	Ointments/semisolids	Particle size distribution Color and odor Zeta potential Viscosity Surface tension Content uniformity Stability at accelerated temperatures and humidity Microbiological testing
7	Pills and powders	Color and odor Weight variation Tablet thickness Friability losses Content variation *In vitro* disintegration *In vitro* dissolution Stability at room and accelerated temperature and humidity

Source: The University of Sciences (in Philadelphia). 2005. *Remington: The Science and Practice of Pharmacy,* 21st Edition. Philadelphia, PA: Lippincott Williams and Wilkins.

One problem I envisage is the drug content and drug uniformity in the dosage forms because the nutraceuticals are a cluster of a chemical entity, and it will be comparatively difficult to identify and quantify all the ingredients in the products. In such situations, at least one major ingredient can be identified and quantified to ensure the uniform distribution of the product through the matrix. A second major hurdle the nutraceutical products will face is defining and identifying the impurities and ensuring that these impurities are not harmful to the consumer. There is a need to develop a method to ensure such harmlessness of the products to get into the drug store.

PHARMACOLOGICAL/PHARMACOKINETIC EVALUATIONS

The credibility of the pharmaceutical products has been acquired over a period of time as a result of pharmacological and pharmacokinetic evaluations and providing the reproducibility of the effectiveness of the product. It is now a well-established norm that any new drug entity is rigorously checked for its pharmacological, toxicological, and pharmacokinetic evaluations. Most of these studies are done under Phase I, Phase II, or Phase III. During studies at every level, stringent tests are involved to confirm the efficacy and reproducibility of the product in clinical conditions. This also confirms the dose levels of the drugs, their toxicological effects, and the parameters showing their absorption, distribution, metabolism, and elimination (ADME) in the human body. This involves significant amounts of money (in millions of dollars) to get the evaluations done as required by the FDA before the drug is introduced in the market. During all these studies, the formulations are also tested for their efficacy, drug release patterns, stability over the period of time, and confirmation of the expiry date for the products. Normally, the companies strive to get at least two years of expiry date of their products. There is a separate chapter in this book discussing this in detail.

CHALLENGES FOR NUTRACEUTICALS: ESTABLISHING THE CREDIBILITY OF NUTRACEUTICALS

When it comes to nutraceuticals, whether stringent tests are needed or not is a question for those involved in the business to answer. From a regulatory perspective, we need to provide adequate information with scientific evidence to prove that the product is safe, reproducible, and therapeutically efficient and whether it offers such effects for a definite period of time, say two or three years. There are new regulations coming into effect. Many European countries are ahead of the United States in this regard. All the stakeholders, researchers, formulators, manufacturers, medical practitioners, health professionals, and regulatory affairs people need to establish these regulations as soon as possible. They need to consider that these products have been in use for many years and have been consumed by human beings for a long time; these products, over such time, have not shown many untoward effects. The need is to create a mechanism to prove that the product quality is reproducible, and this mechanism needs to be in place with solid, scientific support experimentally that can be proved using a reliable technique. This will establish the credibility of these nutraceutical products. Many health professionals have used these products in conjunction with allopathic medicines and have seen excellent effects. Such research papers need to publish vigorously, and more scientific exercises need to be performed to establish the credibility of these very useful products for humanity.

cGMP MANUFACTURING

With the new regulations being introduced and many ongoing discussions in academic and industrial circles, it is evident that the nutraceuticals market will very soon be under purview of the FDA. This will lead to more stringent regulations, and it will be in the best interest of the nutraceutical manufacturing companies to start adhering to the cGMP guidelines of the FDA strictly. Because the market share of the nutraceuticals is growing, it is now a concern for the government and the FDA regarding the safety of these products. All pharmaceutical products need to be manufactured under cGMPs, and there is separate chapter in this book addressing the need for this and various codes of federal regulation (CFRs) controlling the regulatory requirements for nutraceuticals. The proper manufacturing of nutraceuticals under required cGMPs will definitely increase the credibility of the products, as well as prove the safety of these products for the consumers. The cGMP manufacturing needs standard operating procedures and many requirements that virtually ensure the high quality of the product.

NICHE AREAS FOR NUTRACEUTICALS

The large customer base nutraceuticals have developed in the past two decades and its significant market value shows that nutraceuticals have their own market segment. It is observed that different customer segments approve the use of the nutraceuticals:

1. People believing more in prevention than a cure: This comprises a large part of the population who is worried about their health. This is a learned group that understands that, because of their genetic system, they can expect some hereditary diseases. They want to avoid these diseases or at least lessen the severity of these diseases, such as diabetes, hypertension, and arthritis. Unfortunately, the allopathic medicines have always concentrated on curing the diseases, and there are no specific drugs available for the prevention of such disease, mostly caused by one's lifestyle and genetic history. Glucosamine-chondroitin products are very popular in people who are suffering from arthritis. Many teas and functional food products are found to be useful to diabetic patients, and they have developed their own customer base in the past two decades. There is a significant market for the products that can build up the immunity of children and patients who might be susceptible to infections or have side effects of infective diseases.
2. People who have chronic diseases and have found no solution in allopathic medicines: This is another group in which nutraceuticals products find their usefulness. Many patients who have suffered from diseases such as cancer, arthritis, diabetes, and many others in which there is little hope for their treatment have been using nutraceutical therapy for their treatment. The physicians who are treating such patients are also experimenting with the available nutraceuticals in the market. This group will increase as the acceptance of the nutraceuticals by mainstream physicians practicing allopathic medicines grows.
3. Pediatric and geriatric patients: This is another group that finds nutraceuticals beneficial. Most parents are looking for safe and effective therapy for their children, and the experience of these parents have been positive when they used nutraceuticals because there are fewer side effects.
4. Economically challenged patients: Nutraceuticals, if made available at cheaper or reasonable prices, will create a large clientele. It is expected that, with the deteriorating economic situation, insurance companies will be looking for some low-cost treatments for their customers. It will be a good idea for the manufacturers of nutraceuticals to create such products and make them available for this group through the insurance companies. This will help increase the credibility and utility of nutraceuticals.

COMPETING WITH OR COMPLEMENTING PHARMACEUTICALS

This dilemma will be always there because the nutraceutical market has to compete with the pharmaceutical market. It appears that the best strategy for nutraceuticals will be to complement the allopathic medicines rather than being the competitors to these products. Nutraceuticals have a long history of their efficient applications in many diseases. There are very few scientific studies to support all the claims; hence, it will be an excellent strategy for the nutraceutical manufacturers to sponsor scientific research in this field, especially the clinical studies, to use all the modern techniques of statistics and analysis, and to make a stronger case for the application of these products for the benefit of humanity.

REFERENCES

Adebowale, A. O., Z. Liang, and N. D. Eddington. 2000. Nutraceuticals, a call for quality control of delivery systems: a case study with chondroitin sulfate and glucosamine. *J. Nutraceut. Funct. Med. Foods* 2:15–30.

DeFelice, S. 2007. *The Foundation for Innovation in Medicine*. http://www.fimdefelice.org.

Emerging Nutraceuticals Market Report, http://www.Nutraingredients-usa.com.

Global Industry Analyst Inc. 2008. Report, *Global Nutraceuticals Market to Cross US $187 Billion by 2010*. http://www.StrategyR.com.

Global Nutraceuticals Market Report. *India's Nutraceuticals Market Should Cross Billion Mark by 2009*. http://www.plethico.com/addons/Global%20Nutraceuticals%20Market.pdf.

Hardy, G. 2000. Nutraceuticals and functional foods: Introduction and meaning. *Nutrition* 16:688–689.

Litov, R. E. 1998. Developing claims for new phytochemical products. In *Phytochemicals: A New Paradigm*. Edited by Bidlack, W. R., S. T. Omaye, M. S. Meskin, and D. Jahner. Lancaster, PA: Technomic Publishing, pp. 173–178.

Lockwood, B. 2007. Nutraceuticals, 2nd Edition. London, UK: Pharmaceutical Press, p. 1.

Technical Market Research Report. 2007. *Nutraceuticals: Global Markets and Processing Technologies*. http://www.chem.info.

The University of Sciences (in Philadelphia). 2005. *Remington: The Science and Practice of Pharmacy*, 21st Edition. Philadelphia, PA: Lippincott Williams and Wilkins.

Wagner, J. (Ed.). 2002. *Nutritional Outlook*, June/July.

Zeisal, S. H. 1999. Regulations of nutraceuticals. *Science* 285:1853–1855.

CHAPTER 3

Potential Nutraceutical Ingredients from Plant Origin

Sudesh Agrawal and Amitabha Chakrabarti

CONTENTS

INTRODUCTION

Plants continue to be a major source of medicine, as they have been throughout human history. Plants are the only source of food and energy in our ecosystem. Food itself is a medicine. Food in our diet either enhances or disturbs the potency of the drug consumed. The human body consists of five elements: ether, water, air, fire, and earth. Every ancient medicine system, such as the Indian medicine system Ayurveda, the Chinese medicine Unani, and the Japanese healing system of Reiki, has described it in one way or another. The key to maintaining good physical and mental health is in keeping these five elements in harmonic balance through proper diet, herbs, and lifestyle; otherwise, early aging and various diseases can manifest. In light of the research in progress on the benefits of various phytochemicals in foods, it appears feasible that the naturally occurring chemical compounds in herbs could be helpful in the prevention or treatment of many chronic diseases, including cancer and cardiovascular disease. Although food has been used for a long time to improve health, around 400 BC, Hippocrates said, "Let your food be your medicine, and your medicine be your food." Now, modern knowledge of health is being used to improve food. In recent years, scientific evidence has revealed that bioactive dietary components benefit health in ways that extend beyond meeting basic nutritional needs. The food and nutrition science has moved from simply identifying and correcting nutritional deficiencies to nutraceuticals, foods that promote optimal health and reduce the risk of certain debilitating diseases.

Around 80% of the world population is using herbs, plants, and other natural products as their first choice of medication for general illnesses. These natural plant products are not only used for primary healthcare in rural areas but in developing countries also, where modern medicines are predominantly used. Whereas the traditional medicines are derived from medicinal plants, minerals, and organic matter, the herbal drugs are prepared from medicinal plants. Natural products as disease remedies have a history of more than 5,000 years (India, China, and Greece). However, it is believed that pure compounds can only be the possible pharmaceutical drugs. Natural products are commonly rejected as drugs by regulatory health agencies because, most of the time, they are presented as crude extract mixtures with doubts about the reproducibility and standard of manufacturing.

Functional foods, nutraceuticals, pharmaco-nutrients, and dietary integrators are all terms used commonly for nutrients or nutrient-enriched foods that can prevent or treat diseases. The so-called "physiologically functional foods," which originated in Japan in the 1980s, were defined as "any food or ingredient that has a positive impact on an individual's health, physical performance, or state of mind, *in addition* to its nutritive value." According to Dr. Stephen DeFelice, nutraceuticals are the "Food, or parts of food, that provide medical or health benefits, including the prevention and treatment of disease" [Kalra 2003]. Health Canada defines nutraceutical as "a product isolated or purified from foods, and generally sold in medicinal forms not usually associated with food and demonstrated to have a physiological benefit or provide protection against chronic disease" [Health Canada 2002].

This chapter is confined to the products associated with medicinal food or the medicinal products originated from the plants used by the global community. We have tried to incorporate the information about the plants, which are regularly used by the people as a remedy for their common illness all over the world. This is also a humble attempt to include the active neutraceuticals and ingredients present in the plants.

SPICES AND SEASONINGS OF NUTRACEUTICAL VALUES

Since time immemorial, Indianas, Chinese, Japanese, Egyptians, and many other ancient cultures adopted several spcies as part of a regular diet. This food seasoning showed advantages in slowing the aging process, helping to prevent cancer (tuermeric in colon cancer) and in helping with cardiovascular disease, Alzheimer's disease (AD), diabetes, immune disorders, and obesity. Spices and flavoring plants rich in phytochemicals are receiving much attention as a possible source of cancer chemopreventive compounds. Some of the compelling reasons as to why we should eat more spices include the following:

- Spices contain enormous quantities of extremely valuable disease-preventing phytochemicals. They contain multiple micronutrients and trace minerals required for physiological balance and activity of the human body and growth.
- Antioxidants, such as saffron, onion, garlic, and turmeric, all have multiple antioxidants in them. Individual spices, such as ginger, contain more than 25 antioxidants.
- Cancer rates in spice-consuming nations are up to 40 times lower than those of the United States and other western countries; turmeric, black pepper, cumin, caraway, cloves, ginger, anise, basil, chillies, fennel, mustard, rosemary, and garlic all contain potent anticancer compounds.
- Cardiovascular disease can be assisted with garlic, rosemary, cinnamon, coriander, fenugreek, ginger, oregano, mustard, and thyme, which protect against high cholesterol levels, "sticky platelets," atherosclerosis, and high blood pressure.
- Diabetes care can include cinnamon and fenugreek, which can lower the blood glucose and cholesterol levels of diabetic patients by 25%.
- Aging spices slow the aging process by protecting DNA against oxidative damage and decay; garlic shows a promising role in skin care.

Food seasonings not only contains up-to-date scientific research into the healing properties of spices but also gives a fascinating account of the historical reasons why spices have become incorporated into the cuisines of different nations. It also explains why those people who eat large quantities of spices benefit from their health-promoting properties. In this part of the chapter, we are describing general spices mainly used in India for many years.

Turmeric (*Curcuma longa*)

Curcuma longa is a perennial herb and is a member of the Zingiberaceae (ginger) family. The plant grows to a height of three to five feet and is cultivated extensively in Asia, India, China, and other countries with a tropical climate. The parts used are the rhizomes, which are ovate, oblong, pyriform, or cylindrical and often short-branched. They are yellow to yellowish-brown in color. Turmeric has been used for centuries in Ayurvedic medicine as a treatment for inflammatory disorders, including arthritis. On the basis of this traditional usage, dietary supplements containing turmeric rhizome and turmeric extracts are also being used in the western world for arthritis treatment and prevention.

Curcuma is a rhizome used as a common food ingredient in Indian curries and as food ingredients in many South Asian countries. Curcumin is an important ingrediant in that Curcuma has been reported to have antioxidant properties.

The chemical constituents are as follows: moisture, 13.1%; protein, 6.3%; fat, 5.1%; mineral matter, 3.5%; and carbohydrates, 69.4%. The essential oil (5.8%), obtainable by steam distillation of the rhizomes, has the following constituents: phellandrene, 1%; sabinene, 0.6%; cineol, 1%; borneol, 0.5%; zingiberene, 25%; and sesquiterpenes, 53% [Kapoor 1990]. Curcumin (3–4%) is responsible for the yellow color. In addition, the monodemethoxy and bisdemethoxy derivatives of curcumin have been isolated from the rhizome [Vopel, Gaisbaure, and Winkler 1990].

Interest has greatly increased recently in the pharmaco-therapeutic potential of curcumin. In addition to its reported role in inhibiting tumorigenesis, metastasis, platelet aggregation, inflammatory cytokine production, oxidative processes, and myocardial infarction, curcumin has been shown to correct cystic fibrosis defects, lower cholesterol, suppress diabetes, enhance wound healing, modulate multiple sclerosis and AD, and block human immunodeficiency virus (HIV) replication. Furthermore, reports also support curcumin's role in protecting against cataract formation, alcohol-induced liver injury, adriamycin-induced nephrotoxicity, drug-induced lung injury, and inflammatory bowel disease (IBD), and it has no apparent toxicity in extremely large oral doses (8 g/day) in humans [Okada et al. 2001; Egan et al. 2004]. Curcumin (1,7-bis[4-hydroxy-3-methoxyphenyl]-1,6-heptadiene-3,5-dione) is a naturally occurring plant product (major pigment in the Indian culinary spice turmeric) that acts as a natural nonsteroidal anti-inflammatory molecule. Curcumin has antioxidant, anti-inflammatory, and anticarcinogenic properties, scavenging reactive oxygen and nitrogen free radicals [Barik et al. 2007]. Previous work showed curcumin's antioxidant role at various levels. Primarily, it has been shown to effectively scavenge free radicals and also inhibit the formation of proinflammatory cytokines. Curcumin has been shown to inhibit oxidation of oxidized LDL [Aggarwal, et al 2004]. Other data show that curcumin may be responsible for the lower incidence of colorectal cancer in Asian countries [Sharma, Bani, and Singh 1989].

Mustard (*Brassica juncea*)

There are several plant species in the genera *Brassica* and *Sinapis,* which belong to family Brassicaceae, whose small mustard seeds are used as a spice in India, China, and many parts of the world. It is a perennial herb, usually grown as an annual or biennial, up to 1 m or more tall; branches are long. Mild white mustard (*Sinapis hirta*) grows wild in North Africa, the Middle East, and Mediterranean Europe and has spread farther by long cultivation; brown or Indian mustard (*B. juncea*), originally from the foothills of the Himalaya, is grown commercially in the UK, Canada, and the US; black mustard *(B. nigra)* is grown commercially in Argentina, Chile, the US, and some European countries. Mustard is a nutritious food containing 28% to 36% protein. Its higher protein content is of particular interest when applied to processed meats. The vegetable oil of mustard is nutritionally similar to other oils and makes up 28% to 36% of the seed. Erucic acid is a significant component of mustard oil. Mustard oils are the characteristic flavor components of whole seed, ground mustard, and mustard flour (powder). The essential oil tocopherols present in mustard inhibits growth of certain yeasts, molds, and bacteria, enabling mustard to function as a natural preservative to help protect the oil from rancidity, thus contributing to a long shelf life. When mixed with water (or chewed), a chemical reaction occurs between an enzyme and a glucoside from the seeds, resulting in the production of the oil allyl isothiocyanate. Mustard is widely known for its sharp flavor. This characteristic flavor is an essential component of many dressings and sauces worldwide. Unlike other "hot" flavors, the flavor profile of mustard does not linger. Rather, it presents itself quickly, dissipates, and leaves little or no after-taste. Mustard greens are extremely high in vitamin A, vitamin E, vitamin C, vitamin K and beta-carotene. They also contain vitamin B_6, folic acid, magnesium, calcium, iron, niacin and are an excellent source of phytochemicals thought to prevent cancer [Duke and Waine 1981]. An Indian variety of mustard (*Brassica nigra*), have been reported to be hypoglycaemic and helps in Diabetes mellitus [Srinivasan 2005].

Chili (*Capsicum annum*)

The chili pepper, or chili (*C. annum*), is the fruit of the plants from the genus *Capsicum*, which are members of the nightshade family, Solanaceae. The plant is an annual herb; leaves are alternate, simple, smooth margined; flowers are small, solitary, axillary, white, or greenish, 5-parted; fruit is a shiny, tapered berry of various colors.

The substance that gives chili peppers their intensity when ingested or applied topically is capsaicin (8-methyl-*N*-vanillyl-6-nonenamide) and several related chemicals, collectively called *capsaicinoids*. Capsaicin is the primary ingredient in pepper spray. When consumed, capsaicinoids bind with pain receptors in the mouth and throat that are normally responsible for sensing heat. Once activated by the capsaicinoids, these receptors send a message to the brain that the person has consumed something hot. The brain responds to the burning sensation by raising the heart rate, increasing perspiration, and releasing endorphins.

Red chilis contain high amounts of vitamin C and carotene. In addition, peppers are a good source of vitamin B_6. They are very high in potassium and high in magnesium and iron. Their high vitamin C content can also substantially increase the uptake of non-heme iron from other ingredients in a meal, such as beans and grains. The fruit of the *C. annum* is hot, pungent and is antihemorrhoidal when taken in small amounts. Furthermore, it is considered antirheumatic, antiseptic, diaphoretic, digestive, irritant, rubefacient, sialagogue, and tonic [Pradeep and Geervani 1994; Chiej 1984]. It is taken internally in the treatment of the cold stage of fevers, debility in convalescence or old age, varicose veins, asthma, and digestive problems [Bown 1995]. Externally, it is used in the treatment of sprains, unbroken chilblains, neuralgia, pleurisy, etc. It is an effective sea-sickness preventative [Chiej 1984].

Cumin (*Cuminum cyminum*)

Cumin is a small annual herb native to the Mediterranean region. Primary cultivation of cumin is in Europe, Asia, the Middle East, North Africa, and India. The seeds of *Cuminum cyminum* (family-Apiaceae), commonly known as cumin, are used in food as a vegetable seasoning in India and in South Asian countries and as folk (herbal) medicine all over the world for the treatment and prevention of a number of diseases and conditions that include asthma, diarrhea, and dyslipidaemia.

C. cyminum is widely used in Ayurvedic medicine for the treatment of dyspepsia, diarrhea, and jaundice. Studies had been done to investigate the role of

C. cyminum supplementation on the plasma and tissue lipids in alloxan diabetic rats. Oral administration of *C. cyminum* for six weeks to diabetic rats resulted in significant reduction in blood glucose and an increase in total haemoglobin and glycosylated haemoglobin [Dhandapani et al. 2002]. It also prevented a decrease in body weight. *C. cyminum* treatment also resulted in a significant reduction in plasma and tissue cholesterol, phospholipids, free fatty acids, and triglycerides. In another study, the active component isolated from *C. cyminum* seeds against aldose reductase and b-glucosidase was identified as cuminaldehyde. The inhibitory responses varied with different concentrations. It has been reported that *C. cyminum* seed-derived materials have antimicrobial activity, a food spice, a fungicide, and a tyrosinase inhibitor [Boelens 1991]. It might be expected then that the active component isolated from *C. cyminum* seeds has a range of pharmacological actions for antidiabetic therapeutics. The pharmacological actions of the crude extracts of the seeds (and some of its active constituents, e.g. volatile oil and thymoquinone) that have been reported include protection against nephrotoxicity and hepatotoxicity, induced by either disease or chemicals. The seeds/oil have anti-inflammatory, analgesic, antipyretic, antimicrobial, and antineoplastic activity. The oil decreases blood pressure and increases respiration. Treatment of rats with the seed extract for up to 12 weeks has been reported to induce changes in the haemogram that include an increase in both the packed cell volume (PCV) and haemoglobin (Hb), and a decrease in plasma concentrations of cholesterol, triglycerides, and glucose.

The seeds contain both fixed and essential oils, proteins, alkaloids, and saponin. Much of the biological activity of the seeds has been shown to be attributable to thymoquinone, the major component of the essential oil, but which is also present in the fixed oil.

Fenugreek (*Trigonella foenum-graecum*)

The plant belongs to the family Fabaceae. It is commonly known as methi in India. It is an annual herb. The rhombic is frequently used in the preparation of Indian curries, paste, pickles, etc. The green and dried leaves of plant are also used as vegetable in several South Asian countries including India. Fenugreek may affect blood sugar levels by decreasing the activity of an enzyme that is involved in releasing stored sugar from the liver into the blood. Fenugreek seed contains only minute quantities of an essential oil. In the essential oil, 40 different compounds were found; furthermore, *n*-alkanes, sesquiterpenes, alkanoles, and lactones were reported. The dominant aroma component in fenugreek seeds is a hemiterpenoid b-lactone, sotolone (3-hydroxy-4,5-dimethyl-2(5H)-furanone). Also, fenugreek contains an amino acid called 4-hydroxyisoleucine, which appears to increase the body's production of insulin when blood sugar levels are high [Saxena and Vikram 2004]. For many individuals, higher insulin production decreases the amounts of sugar that stay in the blood in some studies of animals and humans with both diabetes and high cholesterol levels. Fenugreek lowered cholesterol levels as well as blood sugar levels. However, no blood-sugar lowering effect was seen in nondiabetic animals. Similarly, individuals with normal cholesterol levels showed no significant reductions in cholesterol while taking fenugreek. Double blind placebo controlled study in mild to moderate type 2 diabetes mellitus patients show adjunct use of fenugreek seeds improves glycemic control and decreases insulin resistance in mild type-2 diabetic patients. There is also a favorable effect on hyper-triglyceridemia [Gupta, Gupta, and Lal 2001].

Black Cumin (*Nigella sativa*)

In addition to black cumin, it is also called fennel flower, blackseed, black caraway, and Kalonji. It is an annual flowering plant, native to southwest Asia and belongs to the family Ranunculaceae. The fruit is a large and inflated capsule composed of 3–7 united follicles, each containing numerous seeds. *Nigella sativa* has a pungent aromatic taste with the seed used as a spice. Seeds have 0.5 to 1.4% of an essential oil and a saponin like glucoside, melanthin. Nigellone is also isolated from essential oils. The oil and seed constituents of *N. sativa*, in particular thymoquinine (TQ), have shown potential medicinal properties in traditional medicine. Oil also poses antioxidant effects via enhancing the oxidant scavenger system, which as a consequence lead to antitoxic effects. The oil and TQ have shown also potent anti-inflammatory effects on several inflammation-based models including experimental encephalomyelitis, colitis, peritonitis, oedama, and arthritis through suppression of the inflammatory mediators, prostaglandins and leukotriens. The oil and certain active ingredients showed beneficial immunomodulatory properties, augmenting the T cell- and natural killer cell-mediated immune response. Researchers at the Kimmel Cancer at Jefferson in Philadelphia have found that thymoquinone, an extract of nigella sativa seed oil, blocked pancreatic cancer cell growth and killed the cells by enhancing the process of programmed cell death. Although the studies are in the early stages, the findings suggest that thymoquinone could eventually have some use as a preventative strategy in patients who have gone through surgery and chemotherapy or in individuals who are at a high risk of developing cancer [Yi et al. 2008]. The pharmacological actions of the crude extracts of the *N. sativa* seeds that have been reported include protection against nephrotoxicity and hepatotoxicity induced by either disease or chemicals. The oil decreases blood pressure and increases respiration. It would appear that the beneficial effects of the use of the seeds and thymoquinone may be related to their cytoprotective and antioxidant actions, and also effect on some mediators of inflammation.

Coriander (*Coriandrum sativum*)

This soft annual herb belongs to the family Apiaceae. The shape of the leaves is variable, broadly lobed at the base of the plant, slender and feathery on the top. Its flowers are borne in small umbels and are white. The fruit is a small globular dry schizocarp. All parts of the plant are edible, but the fresh leaves and the dried seeds are the most commonly used in cooking.

From the water-soluble portion of the methanol extract of coriander, which has been used as a spice and medicine since antiquity, 33 compounds, including two new monoterpenoids, four new monoterpenoid glycosides, two new monoterpenoid glucoside sulfates and two new aromatic compound glycosides were obtained [Ishikawa, Kondo, and Kitajima 2003]. By gas chromatography (GC) and GC-MS analysis, 64 compounds were isolated and revealed great qualitative and quantitative differences between the analyzed parts of coriander. In all organs, the main compound was (E)-2-dodecenal, followed by (E)-2-tridecenal, gamma-cadinene, (Z)-myroxide, neryl acetate, and eugenol [Msaada, et. al 2007]. Some of the nematicidal activity was found in coriander essential oil. Coriander has been documented as a traditional treatment for cholesterol and diabetes patients. Recently mechanism of some active compounds of coriander has been elucidated [Dhanapakiam et al. 2008].

Fennel (*Foeniculum vulgare*)

This hardy, highly aromatic, and flavorful perennial herb belongs to the family Apiaceae. Leaves are feathery, finely dissected and flowers are yellow on short pedicels. Fresh or dry leaves and dry grooved seeds are sweet and edible. Fennel is well known for its essential oil. Essential oils found in fennel have antimicrobial and alternative to larvicidal activity for mosquito [Schelz, Molnar, and Hohmann 2006; Pitasawat, et. al 2007]. Phenolic and total flavonoid content of wild fennel is higher than the cultivated ones and this aromatic plant has antioxidant activity. Different biologically active compounds like umbelliferone, forpsoralen, and foreugenol, etc. have been isolated from this plant [Dhalwal et al. 2007]. Ethanol extracts from fennel was found to be apoptotic on human leukaemia cell lines [Bogucka-Kocka, Smolarz, and Kocki 2008]. The aqueous extract of *Foeniculum vulgare* possesses significant oculohypotensive activity, which was found to be comparable to that of timolol. Additional investigations into the mechanism of action, possible toxicity, and human clinical trials are warranted before the *Foeniculum vulgare* finds place in the arsenal of antiglaucoma drugs prescribed by physicians. The activation of nuclear transcription factor κB (NF-κB) has now been linked with a variety of inflammatory diseases, including cancer, atherosclerosis, myocardial infarction, diabetes, allergy, asthma, arthritis, Crohn's disease, multiple sclerosis, AD, osteoporosis, psoriasis, septic shock, and AIDS. Anethol, a phytochemicals found in this plant can suppress NF-κB [Aggarwal and Shishodia 2004].

Asafoetida (*Ferula assa-foetida*)

It is usually abbreviated to Asafoetida, and it originates in Persia and Afghanistan. It belongs to the Apiaceae family of plants, and is related to Parsley. This plant grows wild in Kashmir (India), Iran, and Afghanistan. It has an unpleasant smell, is herbaceous and perennial, and grows up to 2 m high. The part used is an oleogum resin, obtained by incision from the root, and called asa-foetida. It is extremely sulphur-rich and is very beneficial as a de-toxicating agent. Even more effective than garlic, it is used in very small quantities only a pinch being required. It is also credited with being an antiflatulent. Glucuronic acid, galactose, arabinose and rhamnose have been isolated from the gum [Kapoor 1990]. Taste and smell are attributable to sulfur-containing compounds. Disulfides as well as symmetric tri- and tetrasulfides have been isolated [Rajanikanth, Ravindranath, and Shankarananrayana 1984]. Umbelliferone, the farnesiferoles A, B, and C, ferulic acid, and the cumarin derivatives foetidin and kamolonol are also present [Caglioti et al. 1958; Caglioti et al. 1959; Hofer, Widhalm, and Greger 1984]. Multiple studies elicited and identified sesquiterpenes from the roots of Asafoetida. They are phenylpropanoid derivatives, coumarin derivatives or chromone derivatives. These sesquiterpene derivatives inhibits nitric oxide (NO) production and inducible NO synthase (iNOS) gene expression [Motai and Kitanaka 2005]. Inhibition of NO production has multiple implications in human disease development especially in respiratory and cardiovascular diseases.

In Nepal asa-foetida is considered to be sedative, carminative, antispasmodic, diuretic, anthelmintic, and emmenagogue, as well as an expectorant. It is an aphrodisiac, and increases the sexual appetite [Eigner and Scholz 1990]. Daily dose is around 0.2–0.5g.

Asa-foetida has not been studied much. It produces slight inhibition of the growth of *Staphylococcus aureus* and *Shigella sonnei*, and some of the sulfur compounds show pesticidal activity. Higher doses taken orally cause diarrhoea, meteorism, headaches, dizziness, and enhanced libido [Kapoor 1990].

Garlic (*Allium sativum*)

Garlic is a small herb that belongs to the family Liliaceae. It is believed to thin the blood, reduce cholesterol, decrease blood pressure, inhibit atherosclerosis, and improve circulation. Epidemiologic studies show an inverse correlation between garlic consumption and progression of cardiovascular disease. Multiple *in vitro* studies have confirmed the ability of garlic to reduce cardiovascular risk factors, such as total cholesterol, raised low-density lipoprotein (LDL) and LDL oxidation, reduced platelet aggregation, and hypertension. Several studies have indicated that garlic and its constituents inhibit key enzymes involved in cholesterol and fatty acid synthesis [Nies et al. 1984; Sovová and Sova 2004]. Randomized clinical trials have been conducted for antihypertensive, anti-atherosclerotic, and antiplatelet actions, and intermittent claudication. A recent summary of the data supporting garlic's potential in modifying cardiovascular risk, although generally supporting its usefulness, also emphasized the lack of knowledge about active compounds and mechanisms of action. Although garlic is believed to be beneficial for conditions for which approved drugs are available, such as hyperlipidemia and hypertension [Tattelman 2005; Knox and Gaster 2007], few studies compare it with pharmacologic treatments. The active substance is allicin, formed by the action of alliinase on alliin when garlic is crushed. Alliinase is inactivated by acid pH, heat, and extraction in organic solvents [Agarwal 1996]. Thus, garlic's effects are dependent on whether it is cooked or in aqueous, oil, or organic extracts. In addition to allicin, other active compounds in garlic include methyl allyl trisulfide, diallyltrisulfide, diallyl disulfide, and ajoene. Its characteristic odor is associated with the active compounds, which limits blinding in clinical trials. Few trials indicate whether placebo could be differentiated from active treatment. Garlic is generally safe and well tolerated; however, serious adverse events, including central nervous system (CNS) bleeding and skin burns from topical application, have been reported [Bent 2008]. Effects such as flatulence, dyspepsia, allergic dermatitis, and asthma have been described in literature. Increases in both the prothrombin time and international normalized ratio in subjects previously stable on warfarin have been attributed to garlic; however, there is little to substantiate the mechanism of the interaction.

Ginger (*Zingiber officinale*)

The common cooking ginger is an herbaceous perennial plant that belongs to the family Zingiberaceae. Ginger grows from an aromatic tuberlike rhizome, which is warty and branched. Dried ginger powder, 500–1,000 g, or fresh ginger, 2–4 m, is used for nausea. Ginger's alleged vitalizing effect on the heart and blood is attributed to decreased platelet aggregation and inhibition of thromboxane synthesis observed in *in vitro* studies [Chrubasik, Pitller, and Roufogalis 2005]. Clinical studies, however, using raw, cooked, or dried ginger do not show an effect on bleeding time, platelet aggregation, or thromboxane production. There is supportive evidence from one randomized controlled trial and an open-label study that ginger reduces the severity and duration of chemotherapy-induced nausea/emesis. Compounds isolated from ginger, including shogaol and gingerol, have been studied for positive inotropic and pressor effects; however, no clinical trials currently support these effects. Neither adverse effects nor drug interactions have been reported. The anti-inflammatory properties of ginger have been known for centuries. During the past years, many laboratories have provided scientific support for the long-held belief that ginger contains constituents with anti-inflammatory properties. *In vitro* ginger inhibits all the inflammatory markers of osteoarthritis, for example, ginger suppresses prostaglandin synthesis through inhibition of cyclooxygenase-1 [COX1] and COX2 [Grzanna, Lindmark, and Frondoza et al. 2005; Pan et al. 2008]. Evaluation of the effect of a ginger extract on patients suffering from gonarthritis shows the effectiveness of the ginger. Ginger extract was as effective as placebo during the first three months of the study, but, at the end of six months, the ginger extract group showed a significant superiority over the placebo group [Wigler et al. 2003]. The main pharmacological actions of ginger and compounds isolated from them include immunomodulatory, antitumorigenic, anti-inflammatory, antiapoptotic, antihyperglycemic, antilipidemic, and antiemetic actions. Ginger is a strong antioxidant substance and may either mitigate or prevent generation of free radicals [Ali et al. 2008].

Onion (*Allium cepa*)

Allium cepa is the common onion usually used as a vegetable and belonging to the Alliaceae family. Onions are perennials that are cultivated for food worldwide. There are many varieties. Most onion bulbs are white, yellow, or red. The green stems and leaves are hollow and can reach 3 feet (1 m) in height. The plants bear small flowers that are usually white or purple. The fleshy bulb that grows below the ground is used medicinally as well as for food.

Multiple studies have suggested that dietary flavonoids are helpful in the prevention of atherosclerosis and cardiovascular disease. Antioxidant properties of flavonoids are suggested as a mechanism for its effect on cardiovascular diseases [Terao, Kawai, and Murato 2008]. Onion contains several flavonoids, including quercetin (3,3′,4′,5,7-pentahydroxyflavone), a major flavonoid in onion [Murota et al. 2007]. Quercetin metabolites are detected in human atherosclerotic plaques and act as complementary antioxidants, when oxidative stress is loaded in the vascular system [Augusti 1996]. Onion contain many sulfur-containing active principles mainly in the form of cysteine derivatives; these get decomposed into a variety of thiosulfinates and polysulfides by the action of an enzyme alliinase. Decomposed products are volatile and present in the oils of onion and garlic. They possess antidiabetic, antibiotic, hypocholesterolaemic, fibrinolytic, and various other biological actions. In addition to free sulfoxides in onion, there are nonvolatile sulfur-containing peptides and proteins, which possess various activities and thus make these vegetables an important source of therapeutic agents [Augusti 1996].

Clove (*Syzygium aromaticum* or *Eugenia caryophyllata*)

It is the aromatic dried flower buds of a tree in the family Myrtaceae. Cloves are native to Indonesia and India and are used as a spice in cuisine all over the world. A small- to moderate-sized evergreen tree grows up to 15 m of height. Leaves are simple, lanceolate, and fragrant; flower buds are pink, found as clusters in the tip of branches, aromatic, and have pungent taste. Fruits are fleshy dark drupes, enclosing oblong grooved seeds. The essential oil extracted from the dried flower buds of clove is used as a topical application to relieve pain and to promote healing and also finds use in the fragrance and flavoring industries. The main constituents of the essential oil are phenylpropanoids, such as carvacrol, thymol, eugenol, and cinnamaldehyde. The biological activity of plant has been investigated on several microorganisms and parasites, including pathogenic bacteria, *Herpes simplex*, and hepatitis C viruses. In addition to its antimicrobial, antioxidant, antifungal, and antiviral activity, clove essential oil possesses anti-inflammatory, cytotoxic, insect repellent, and anesthetic properties.

One scientific study reported an effect on lung carcinoma cells. Infusion of aqueous clove solution can elicit strong proapoptotic effect during early lesion of lung carcinogenesis and can also affect the *in situ* cell proliferation [Banerjee, Panda, and Das 2006]. Serious studies on the effect of clove on human lung carcinomas are still lacking.

Nutmeg (*Myristica fragrans*)

Nutmegs are a genus of aromatic evergreen trees and belong to the Myristicaceae family, indigenous to tropical Southeast Asia and Australia. The extracts of *Myristica fragrans* can be useful in the treatment of human diarrhea if the etiologic agent is a rotavirus [Gonçalves et al. 2005]. Myristicin, or methoxysafrole, is the principal aromatic constituent of the volatile oil of nutmeg, the dried ripe seed of *M. fragrans*.

Large doses of 60 g or more are dangerous, potentially inducing convulsions, palpitations, nausea, eventual dehydration, and generalized body pain [Demetriades et al. 2005]. In amounts of 10–40 g, it is a mild to medium hallucinogen, producing visual distortions and a mild euphoria.

Bay Leaf (*Laurus nobilis*)

It is the aromatic leaf of several species of the Laurel family (Lauraceae). The leaf of the *Cinnamomum* tree (Indian Tejpatra) is similar in fragrance and taste to cinnamon bark. Bay leaves are a major component of Indian spices used for seasoning the vegetables and curries have a pungent and sharp, bitter taste. *Laurus nobilis* fruit contains essential oils and has high anticancerous activity. Study suggests that the ability of *L. nobilis* essential oils and some identified terpenes to inhibit human tumor cell growth. *L. nobilis* plant also contains the sesquiterpene lactones 5a,9-dimethyl-3-methylene-3,3a,4,5,5a,6,7,8-octahydro-1-oxacyclopenta[c]azulen-2-one and 3β-chlorodehydrocostuslactone isolated by the chromatographic separations on active extracts from fruits and leaves of the plant [Dall'Acqua et al. 2006]. The isolated compounds were found to inhibit NO production. Methanolic extract from plant leaves and bark inhibits the lipid peroxidation. The cytotoxic activity was also evaluated against three different tumor cell lines of human origin and found to be effective in killing the carcinogenic cells.

Cardamom (*Elettaria cardamomum*)

Cardamon is a pungent aromatic herbaceous perennial plant growing to 2–4 m in height and belonging to the family Zingiberaceae. The fruit is a three-sided yellow-green pod 1–2 cm long, containing several black seeds. The black seeds from the pods of the plant are dried and are used in Indian and other Asian cuisines either whole or in a ground form.

In many parts of the world, including Indian and China, cardamom is traditionally used to treat stomachaches, constipation, dysentery, and other digestion problems. The administration of cardamom extract to Chinese hamsters increased fecal moisture contents (148–174%), decreased the activities of β-D-glucuronidase, β-D-glucosidase, mucinase, and urease in feces, and reduced the production of toxic ammonia [Huang et al. 2007]. These findings suggested that the consumption of cardamom extract might exert a favorable effect on improving the gastrointestinal milieu and may help in digestion.

Cinnamon (*Cinnamomum verum*; synonym *Cinnamomum zeylanicum*)

Cinnamon is a small evergreen tree 10–15 m (32.8–49.2 feet) tall, belonging to the family Lauraceae, and is native to Sri Lanka, India, Bangladesh, and Nepal. The bark is widely used as a spice because of its distinct odor. Young branches of tree are dark brown, terete, and glabrous.

Its flavor is attributable to an aromatic essential oil that makes up 0.5–1% of its composition. Chemical components of the essential oil include ethyl cinnamate, eugenol, cinnamaldehyde, β-caryophyllene, linalool, and methyl chavicol.

Several pharmacological studies suggest the antioxidant properties of the bark from plant; seven clinical trials suggest strong evidence for the effect of cinnamon on type 2 diabetes. Two of the randomized clinical trials on type 2 diabetes provided strong scientific evidence that cinnamon demonstrates a therapeutic effect in reducing fasting blood glucose by 10.3–29% [Verspohl 2005]. Study on patients by Khan et al. [2003] demonstrates effects of low levels (1–6 g/day) of cinnamon on the reduction of glucose, triglyceride, LDL cholesterol, and total cholesterol levels in subjects with type 2 diabetes, but in other studies, cinnamon does not appear to improve glycated hemoglobin, fasting blood glucose, or lipid parameters in patients with type 1 or type 2 diabetes [Baker et al. 2008].

Saffron (*Crocus sativus*)

Saffron (*Crocus sativus*) is a spice derived from the flower of the saffron crocus, a species of crocus in the family Iridaceae. The true saffron is a low ornamental plant with grass-like leaves and large lily-shaped flowers. The flower has three stigmas, which are the distal ends of the plant's carpels. Together with its style, the stalk connecting the stigmas to the rest of the plant, these components are often dried and used in cooking as a seasoning and coloring agent. The chemical composition of saffron has attracted the interest of several research groups during the past decades and, among the estimated 150 volatile and several nonvolatile compounds of saffron, some of them have been identified [Winterhalter and Staubinger 2000]. Saffron contains three main pharmacologically active metabolites. (1) Saffron-colored compounds are crocins, which are unusual water-soluble carotenoids. These are mono and diglycosyl esters of a polyene dicarboxylic acid, named crocetin. The digentiobiosyl ester of crocetin, -crocin, is the major component of saffron. (2) Picrocrocin is the main substance responsible of the bitter taste in saffron. (3) Safranal is the volatile oil responsible for the characteristic saffron odor and aroma. Furthermore, saffron contains proteins, sugars, vitamins, flavonoids, amino acids, mineral matter, gums, and other chemical compounds [Rios 1996]. *C. sativus* extract and its major constituent, crocin, significantly inhibited the growth of colorectal cancer cells but did not affect normal cells [Aung et al. 2007]. Recent scientific findings have been encouraging, uniformly showing that saffron and its components can affect carcinogenesis, and currently have been studied extensively as the most promising cancer chemopreventive agents. Extracts from saffron, the dried stigmata from *C. sativus*, are being used more frequently in preclinical and clinical trials for the treatment of cancer and depression. Series of clinical trials on saffron on depressed out-patients show a significant improvement in signs of depression. In a double-blind and randomized trial, patients were randomly assigned to receive capsule of petal of *C. sativus* at 15 mg (Group 1) and fluoxetine at 10 mg two times per day (Group 2) for an eight-week study. At the end of trial, a petal of *C. sativus* was found to be effective in the treatment of mild to moderate depression, similar to fluoxetine [Akhondzadeh et al. 2007]. Double-blind trials and placebo-controlled trials were done to investgate whether saffron could relieve symptoms of premenstrual syndrome (PMS). The results of this study indicate the efficacy of *C. sativus* in the treatment of PMS. However, a tolerable adverse effects profile of saffron may well confirm the application of saffron as an alternative treatment for PMS. Promising and selective anticancer effects have been observed *in vitro* and *in vivo* but not yet in clinical trials. Antidepressant effects were found *in vitro* and in clinical pilot studies. Saffron extracts thus have the potential to make a major contribution to rational phytotherapy [Schmidt, Betti, and Hensel 2007].

NUTRACEUTICALS FROM FRUITS AND VEGETABLES

Mango (*Mangifera indica*)

The mango tree is erect, 30–100 feet (roughly 10–30 m) tall, with a broad, rounded canopy, and belongs to the family Anacardiaceae. It is nearly evergreen. Its fruit is aromatic, varies in size and shape, is usually round, oval, ovoid-oblong, or somewhat kidney shaped, often with a break at the apex.

Mango fruit is rich in antioxidants. Mangiferin is an antioxidant present in the riped fruit, which prevents the lipid peroxidation by decreasing the O_2 concentration, blocks reactive oxygen species production, binds metal ions such as Fe^{3+} and Fe^{2+}, and prevents the generation of hydroxyl radicals [Ghosal 1996]. Mango fruit also contains cycloartenol, 3β-hydroxycycloart-24-en-26-al, 24-methylene-cycloartan-3β,26-diol, C-24 epimers of cycloart-25-en-3β,24-diol, α-amyrin, β-amyrin, dammarenediol II, β-taraxastane-3β, 20-diol, ocotillol, methyl mangiferonate, methyl mangiferolate, methyl isomangiferolate, sitosterol, a mixture of 5-(12-*cis*-heptadecenyl)- and 5-pentadecyl-resorcinols, and vitamins A and C [Rastogi and Mehrotra 1995]. Unriped fruits of mango contain polysaccharides, a triterpene, acetates of cycloartanol, amyrin, lupeol, and homomangiferin-2C-glucopyranosyl-3-methoxy-1,6,7-trihydroxyxanthone. Mango stem bark contains protocatechuic acid, catechin, mangiferin, alanine, glycine, aminobutyric acid, kinic acid, shikimic acid, tetracyclic triterpenoids, cycloart-24-en-3β,26-diol, 3-ketodammar-24(E)-en-20S,26-diol, C-24 epimers of cycloart-25-en-3β,24,27-triol, and cycloartan-3β,24,27-triol. Mango bark has been traditionally used in many countries for the treatment of menorrhagia, diarrhea, syphilis, diabetes, scabies, cutaneous infections, and anaemia, using an aqueous extract obtained by decoction as reported in the NAPRALERT database. The use of mango stem bark extract (MSBE) has been documented on more than 7,000 patients with emphasis on patients with malignant tumors [Tamayo et al. 2001]. *In vitro* tests demonstrated that MSBE had no cytotoxic effects on tumor cells. However, more than 95% of cancer patients treated with MSBE (2,286 patients) evidenced an improvement in terms of their quality of life (appetite, body weight, self-independence for the daily life, etc.); inflammation and/or pain were significantly reduced, and several biochemical markers were improved with time (i.e., haemoglobin and transaminase, being the most significant) [Nunez-Selles et al. 2002]. It was relevant that more than

60% of patients with diabetes mellitus (408 patients) reduced the insulin dose by 20 IU after six months of MSBE oral administration [Padin et al. 2005], 80% of patients with benign prostate hyperplasia (826 patients) improved the urine retention after three months of MSBE administration (oral and rectal), and 95% of patients with different types of dermatitis (1,297 patients) were improved after one-week treatment with topical MSBE. Also significant was that 87% of patients with Lupus erythematosus (675 patients) improved their quality of life after the first month of MSBE treatment (oral and topical administration). Mangiferin is a C-glucosylxanthone, and it has cardiotonic and diuretic properties. Gallic acid and quercetine show a strong antiviral activity. Mangiferin stimulates after 48 h the proliferation of thymocytes and spleenic lymphocytes, with a peak response at 5.0 g/ml and 20.0 μg/ml, respectively [Rastogi and Mehrotra 1993].

Apple (*Malus domestica*)

This cultivated fruit tree belongs to the rose family Rosaceae. The tree is small and deciduous with a broad, often densely twiggy crown. The leaves are simple ovals, alternatively arranged with serrated margins. The flowers are pinkish-white with five petals. The center of the pomaceous fruit contains five carpels arranged in a five-point star, each containing one to three seeds. Apples contain a rich source of both nutrient and non-nutrient components and contain high levels of polyphenols and other phytochemicals. Main phytochemicals include hydroxycinnamic acids, dihydrochalcones, flavonols (quercetin glycosides), catechins and oligomeric procyanidins, as well as triterpenoids in apple peel and anthocyanins in red apples [Gerhauser 2008]. Apple peels have high concentrations of phenolic compounds and may assist in the prevention of chronic diseases. Very recently, He and Liu [2008] isolated 29 phytochemicals from apples skin. They have shown that two compounds (i.e., quercetin and quercetin-3-*O*-β-d-glucopyranoside) showed potent antiproliferative activities against HepG2 and MCF-7 cells, whereas six flavonoids and three phenolic compounds showed potent antioxidant activities. Davis et al. [2006] suggested that flavonoids in apple extract downregulates nuclear factor-κB signaling and thereby shows antioxidant effects. Therefore, greater intake of apple contributes to improved health by reducing the risk of diseases, such as cardiovascular disease and some forms of cancer (i.e., colon, prostate, and lung).

Grapes (*Vitis vinifera*)

This perennial, deciduous woody vine with a flaky bark belongs to the family Vitaceae. Most of them are cultivars of *Vitis vinifera*. The leaves are alternate and palmately lobed. The fruit is a berry. Different phytochemicals, such as polyphenols (stilbenes and anthocyanins), condensed tannins (proanthocyanidins), tetrahydro-β-carbolines, dietary indoleamines, melatonin, and serotonin, have been described in grapes [Iriti and Faoro 2006]. Resveratrol, a stilbene-type aromatic phytoalexin predominantly found in grapes, exhibit several physiological activities, including anticancer and anti-inflammatory activities *in vitro* and in experimental animal models, as well as in humans [Udenigwe et al. 2008]. Resveratrol displayed significant antiproliferative effects *in vitro* on cultured human colon cancer cells [Duessel, Heuertz, and Exekiel 2008]. Anthocyanins, a flavonoid found profusely in grapes, has breast cancer chemopreventive potential attributable in part to their capacity to block carcinogen-DNA adduct formation [Singletar, Jung, and Giusti 2007]. Numerous phytonutrients found in grapes are also beneficial for AD and urinary bladder dysfunction.

Bilwa or Bel (*Aegle marmelos*)

It is a small- to medium-sized aromatic tree, the average height of which is 8.5 m, is deciduous, and belongs to the family Rutaceae.

The bilwa tree is one of the most useful medicinal plants of India and in many parts of South-East Asia. Its medicinal properties have been described in the ancient medical treatise in Sanskrit, *Charaka Samhita.* All parts of this tree, including the stem, bark, root, leaves, and fruit, have medicinal values and have been used as medicine for a long time. A series of phenylethyl cinnamides, which included new compounds named anhydromarmeline, aegelinosides A and B, has been identified in leaves of *Aegle marmelos*, which is an α-glucosidase inhibitor [Phuwapraisirisan et al. 2008]. The leaves and fruits of the tree are used in ethanopharmacological drugs for type 2 diabetes. Recent investigations have reported the exceptional actions of α-glucosidase inhibitors from natural sources. A hydroxyl amide alkaloid from *A. marmelos* leaves is reported to suppres both blood glucose and plasma triglyceride levels [Narender et al. 2007]. Many other uses of bel fruits are reported; for instance, the ripe fruit are regarded as a very good laxative. The unripe or half-ripe fruit is perhaps the most effective remedy for chronic diarrhea and dysentery. An infusion of bel leaves is regarded as an effective remedy for peptic ulcer. The fruit, leaves, and bark of the tree is considered as remedy for multiple illnesses. Extensive research is required to use them on a larger scale.

Awala (*Phyllanthus emblica*)

This large deciduous tree belongs to the family Euphorbiaceae. Fruit are loaded with different bioactive compounds, such as phyllambin, phyllemblic acid, gallic acid, emblicol, ellagic acid [Nizamuddin, Hoffman, and Olle 1982], SOD482 [Fengshu et al. 1992], putranjivain A [El-Mekkawy et al. 1995], emblicanin A and B, punigluconin, and pedunculagin [Ghosal 1996]. The fruit of *Phyllanthus emblica* is used as a powerful rejuvenator in Ayurvedic medicine and is one of the major ingredients of chavanprash (a herbal tonic in Ayurvedic medicine). The chondroprotective potentials of this fruit were reported by Sumantran et al. [2008]. Amla fruit extract works effectively in mitigative, therapeutic, and cosmetic applications through control of collagen metabolism, and it also protect the skin from the damaging effects of free radicals, nonradicals, and transition metal-induced oxidative stress [Fujii et al. 2008].

The fruit was found to contain pyrogallol, an active compound responsible for the anti-inflammatory effect in bronchial epithelial cells. Antioxidant-enriched amla is very useful for oxidative stress-related diseases and may prevent age-related hyperlipidaemia through attenuating oxidative stress in the aging process [Yokozawa et al. 2007]. Superoxide scavenging potential has been proven for this fruit extracts [Saito et al. 2008]. Multiple animal studies showed that the dried fruit powder of *P. emblica* is hypolipidemic and induced partial regression of atherosclerotic lesions in arteries and decreased lipogenesis.

The tannoid principles of the fruits of the plant *P. emblica*, including emblicanin A, emblicanin B, punigluconin, and pedunculagin, have been reported to exhibit antioxidant activity *in vitro* and *in vivo* and supports its use in Ayurveda as hepatoprotectant [Bhattacharya et al. 2000].

Banana (*Musa paradisiaca*)

This herbaceous, tall, upright, fairly sturdy plant belongs to the family Musaceae, cultivated mainly for its fruit. Its upright stem is a pseudo stem with big leaves of several meters in length. Each pseudo stem produces a bunch of banana, which grows in hanging clusters. Bananas are rich in vitamin B_6, vitamin C, and potassium. The potassium content of one medium banana is equivalent to a 12 mmol potassium salt tablet [Hainsworth and Gatenby 2008] and is thus recommended for cardiac health. The antioxidant activity of banana flavonoids has been described previously [Vijayakumar, Presannakumar, and Vijayalakshmi 2008]. One of the varieties of banana *M. sapientum* has antidiabetic activity, which may be attributable to the presence of flavonoids, alkaloids, steroid, and glycoside compounds [Dhanabal et al. 2005].

Broccoli (*Brassica oleracea*)

This plant belongs to the family Brassicaceaee. Fleashy green flowerheads are edible and surrounded by leaves. It is high in vitamin C and has several other vitamins and minerals. Selenium is an essential trace element found in this plant. Selenium-enriched broccoli may be a useful dietary ingredient for preventing cancer. Sulforaphane, present in broccoli, mediates growth arrest and apoptosis in human prostate cancer cells [Chiao 2002], and it has been extensively studied in an effort to uncover the mechanisms behind this chemoprotection [Juge, Mithen, and Traka 2007]. Several glucosinolates, a class of phytochemicals, are also reported in broccoli. Broccoli leaf is also edible and contains far more beta-carotenes than the florets. Recently, a report showed that broccoli protects mammalian hearts through the redox cycling of the thioredoxin superfamily [Mukherjee, Gangopadhyay, and Das 2008].

Tomato (*Solanum lycopersicum*)

The tomato (*Solanum lycopersicum*) is herbaceous, a usually sprawling plant, and belongs to the Solanaceae family. Tomatoes are rich in vitamin C and contain lycopene. In the area of food and phytonutrient research, nothing has been hotter in the past several years than studies on lycopene in tomatoes. This carotenoid found in tomatoes has been extensively studied for its antioxidant and cancer-preventing properties. Carotenoids are naturally occurring organic pigments that are believed to have therapeutic benefit in treating cardiovascular disease because of their antioxidant properties [McNulty et al. 2008]. The antioxidant function of lycopene—its ability to help protect cells and other structures in the body from oxygen damage—has been linked in human research to the protection of DNA (our genetic material) inside of white blood cells. Prevention of heart disease has been shown to be another antioxidant role played by lycopene. Among 72 studies, 57 of them reported inverse associations between tomato intake or blood lycopene level and the risk of cancer at a defined anatomic site; 35 of these inverse associations were statistically significant [Giovannucci 1999], although the FDA found no credible evidence for an association between tomato consumption and a reduced risk of lung, colorectal, breast, cervical, or endometrial cancer [Kavanaugh, Trumbo, and Ellwood 2007]. The FDA found very limited evidence to support an association between tomato consumption and reduced risks of prostate, ovarian, gastric, and pancreatic cancers.

Bitter Melon (*Momordica charantia*)

This annual herbaceous tendril-bearing vine belongs to the family Cucurbitaceae. The plant bears simple leaf, alternate, palmately five-lobed in shape. Fruits are used as vegetables, bitter in taste, green, oblong or ovate, with distinct warty exterior, and the central cavity of the fruit is filled with large flat seeds and pith. Four cucurbitane glycosides, momordicosides Q, R, S, and T, and stereochemistry-established karaviloside XI, were isolated from the bitter melon. These compounds and their aglycones exhibited a number of biologic effects beneficial to diabetes and obesity [Tan et al. 2008].

Acetone extract of whole fruit powder of bitter melon showed regeneration of beta cells in Islets of Langerhans of pancreas of alloxan diabetic rats [Singh and Gupta 2007]. In fact, bitter melon improves insulin sensitivity and insulin signaling in high-fat-fed rats and may open new therapeutic targets for the treatment of obesity/dyslipidemia-induced insulin resistance [Sridhar et al. 2008]. Laboratory tests suggest that compounds in bitter melon might be effective for treating HIV infection [Rebultan 1995; Jiratchariyakul 2001]. In one study, it was shown that MAP30, an anti-HIV plant protein isolated from bitter melon, is capable of acting against multiple stages of the viral life cycle, on acute infection as well as replication in chronically infected cells. Biologically active recombinant MAP30 provides an abundant source of homogeneous material for clinical investigations, as well as structure-function studies of this novel antiviral and antitumor agent [Lee-Huang et al. 1995]. Active biocompounds of bitter melon showed anti-atherogenic property as well [Jayasooriay et al. 2000].

Bitter Orange (*Citrus aurantium*)

Bitter orange is a tree, 7–8 m tall, and spines are axillary and sharp. There is no adequate evidence for efficacy of the nutraceutical drug. Safety concerns have been raised by mutliple agencies, but few data are available. Bitter orange is a plant that has been claimed to stimulate weight loss and to be an "ephedra substitute." It is also known as Seville orange, sour orange, green orange, neroli oil, and kijitsu. The active ingredients include various alkaloids with selective and agonist activity, including synephrine and octopamine. Synephrine (oxidrine) is a sympathomimetic amine, structurally similar to epinephrine [Dwyer, Allison, and Coates 2005]. The most recent reviews of clinical trials to date found little evidence that *Citrus aurantium* products were effective in weight loss and suggested that more research is needed. Although those reviews reported no adverse events, there has been considerable recent discussion concerning potential harms. FDA approves citrus oils as safe (generally recognized as safe [GRAS]) for natural flavorings in food products. However, the amounts used as supplements may be much higher, and, thus, their GRAS status as flavoring agents may not be relevant to their use in dietary supplements.

Other compounds are sometimes present, such as 6′,7′,dihydroxybergamottin and bergapten, which may inhibit cytochrome P4503A and increase serum levels of many drugs [Gurley et al. 2004]. However, in one recent clinical study, the *C. aurantium* was devoid of the CYP3A4 inhibitor 6′,7′,dihydroxybergamottin. More safety testing to assess hemodynamic effects and drug interactions over the short and long term is needed. In summary, larger, longer, and more rigorous trials of *C. aurantium* and synephrine alkaloids are needed to assess their efficacy and safety for weight loss. An additional four clinical trials have examined the effects of *C. aurantium*-containing products in food alone or in combination with other ingredients on body weight and/or body composition [Kalman et al. 2000; Armstrong, Johnson and Duhme 2001; Jones 2001]. In these short-term studies, body weight and/or fat loss appears to be enhanced by *C. aurantum*. This may be partially attributed to a suppressing effect of appetite and/or a moderate increase in resting energy expenditure, but it should be kept in mind that these trials are of short duration, and sample sizes are frequently inadequate.

NUTRACEUTICAL ANALGESICS

Analgesics are pain relievers in tablets, creams, lotions, gels, and sprays. It has been widely shown that many plant-derived compounds present significant analgesic as well as anti-inflammatory effects. Doctors often recommend these products in addition to other medications to help temporarily ease pain. There are mltiple analgesics available today, among them are capsaicin, a preparation derived from plants, and methylsalicylates, which are derived from willow bark and often combined with menthol. Most of the analgesics and anti-inflammatory food or drugs avilable today are derived from plants or one can find the origin of analgesic drugs from plants. In this part of the chapter, we will discuss a few important nutraceutical analgesics with their origin, use, and possible adverse effects, if any. The list of nutraceutical analgesics are so big; describing all of them in one book chapter is beyond our scope.

Aloe (*Aloe vera, Aloe indica, Aloe perfoliata*)

The common names of the plants are Chinese aloe, Indian aloe, and True aloe.

It is a succulent herb. The stems are short, suckering freely to form dense clumps. Its leaves are subbasal, slightly distichous in seedlings and new shoots, erect, pale green, and conspicuously exserted.

The leaf juices of the aloe plant have important medicinal uses, making aloe one of the most respected medicinal plants found in many gels, creams, and lotions. The leaf juice or gel is anti-inflammatory, antiseptic, and antifungal.

Aloe vera gel consists primarily of water and polysaccharides (pectins, hemicelluloses, glucomannan, acemannan, and mannose derivatives). It also contains amino acids, lipids, sterols (lupeol, campesterol, and β-sitosterol), tannins, and enzymes. Mannose 6-phosphate is a major sugar component. A total of 123 aroma chemicals were identified in the extracts obtained using both gas chromatography and gas chromatography/mass spectrometry. There were 42 alcohols, 23 terpenoids, 21 aldehydes, 9 esters, 8 ketones, 6 acids, 5 phenols, and 9 miscellaneous compounds. The major aroma constituents of this extract by dichloromethane extraction simultaneous purging (DRP) were (Z)-3-hexenol (29.89%), (Z)-3-hexenal (18.86%), (E)-hexenal (7.31%), 4-methyl-3-pentenol (5.66%), and butanol (4.29%). The major aroma constituents of this extract by simultaneous purging and extraction (SPE) were (E)-2-hexenal (45.46%), (Z)-3-hexenal (32.12%), hexanal (9.14%), (Z)-3-hexenol (1.60%), and 3-pentanone (1.41%). Terpenoids were also found as one of the major constituents. The fresh green note of *Aloe* leaves is a result of the presence of these C(6) alcohols and aldehydes as well as terpenoids [Umano et al. 1999]. *Aloe vera* has been traditionally used for burn healing. Cumulative evidence from clinical trials tends to support that aloe might be an effective intervention used in burn wound healing for first- to second-degree burns. However, well-designed trials with sufficient details of the contents of aloe products are lacking to determine the effectiveness of *Aloe vera.*

Poplar Tree (*Populus balsamifera*)

Populus balsamifera is native to North America. The twig has a bitter aspirin taste. Its bark is greenish gray with lighter lenticels when young, later becoming darker and furrowed with long, scaly ridges. Inner bark is often dried, ground into a powder, and then used as a thickener in soups or added to cereals when making bread.

The leaf buds are antiscorbutic, antiseptic, diuretic, expectorant, stimulant, and tonic [Uphof 1959; Grieve 1984]. The leaf buds are covered with a resinous sap that has a strong turpentine odor and a bitter taste. They are boiled to separate the resin, and the resin is then dissolved in alcohol. The resin is a folk remedy, used as a salve and wash for sores, rheumatism, and wounds [Foster and Duke 1990]. It is made into a tea and used as a wash for sprains, inflammation, and muscle pains. The extract of *Populus* tree inhibited arachidonic acid-induced platelet aggregation [Kagawa et al. 1992]. Pyrocatechol and salicyl alcohol were isolated as active constituents, which showed an inhibitory effect on platelet aggregation induced by arachidonic acid, which was 25 times more potent than aspirin. *Populus* is an important ingredient in an herbal medicine, phytodolor, which is used in painful inflammatory or degenerative rheumatic diseases. The mode of action of phytodolor includes anti-inflammatory, strong antioxidant, and analgesic properties. Multiple clinical studies and randomized placebo-controlled double-blind trials, performed in different subtypes of rheumatic diseases, confirm the pharmacological evidence of efficacy, such as by reducing the intake of nonsteroidal anti-inflammatory drugs [Gundermann and Muller 2007]. Smoke from the leaves of *Populus* tree is useful in cutaneous warts. In a clinical trial, smoke from the leaves cures warts completely in 66.7 versus 46.4% compared with conventional cryotherapy [Rahimi, Emad, and Rezaian 2008].

Salai Guggal (*Boswellia serrata*)

Boswellia serrata is a medium-sized tree belonging to family Burseraceae, with ash-colored papery bark. *B. serrata* is a close relative of the aromatic frankincense, and both contain the anti-inflammatory boswellian acid and are astringent and anti-inflammatory. It is known as Salai guggal in India. The oleo-gum-resin from the tree *B. serrata*, termed frankincense, is a traditional Ayurvedic remedy. This tree, abundantly growing in dry hilly tracts of India, has been used for variety of therapeutic purposes [Marinetz, Lohs, and Janzen 1988], such as cancer [Shao et al. 1998], inflammation [Singh and Atal 1986], and arthritis [Sharma, Bani, and Singh 1989], and also including respiratory problems, diarrhea, constipation, flatulence, CNS disorders, rheumatism, liver disease, wound healing, fat reduction, and fevers.

Salai guggal contains essential oil, gum, and resin. Its essential oil is a mixture of monoterpenes, diterpenes, and sesquiterpenes. In addition, phenolic compounds and a diterpene alcohol (serratol) is also found in essential oil. Gum portion of the drug consist of pentose and hexose sugars, with some oxidizing and digestive enzymes. Resin portion is mainly composed of pentacyclic triterpene acid, of which boswellic acid is the active moiety [Kokate, Purohit, and Gokhale 1999]. A new lupane triterpene was isolated from fractionation of methanol extract of *B. serrata* resin together with Boswellic acids [Pardhy and Battacharya 1978]. The fraction on additional purification with ethanol/hexane (1:1) yielded 3a-hydroxy-lup 20(29) ene-24-oic acid, whose structure was confirmed by nuclear magnetic resonance and mass spectroscopy [Culioli et al. 2003].

There are many new studies and emerging data to suggest that *Boswellia* may have a role to play in the management of IBD [Joos et al. 2006; Clarke and Mullin 2008]. In a study involving a rat model of colitis, investigators showed that oral administration of *Boswellia* extract or acetyl-11-keto-β-BA over a two-day period resulted in a dose-dependent decrease in rolling (up to 90%) and adherent (up to 98%) leukocytes. In addition, necropsy showed improvement of inflammatory changes on both a macroscopic and microscopic level [Krieglstein et al. 2001]. Clinical trials using oleo-gum-resin of *B. serrata* in India and Germany reported improvement in at least one end point of IBD patients. The mechanism of action of Boswellic acid and oleo-gum-resin is not clearly understood, but it is indicated that it is attributable to inhibition of 5-lipoxygenase. However, other factors, such as cytokines (interleukins [ILs] and tumor necrosis factor α [TNF-α]) and the complement system, are also candidates for its action.

Camphor (*Cinnamomum camphora*)

The camphor tree is a dense broad-leaf evergreen that is capable of growing 50–150 feet (15.2–45.7 m) tall and spreading twice that wide, with a trunk up to 15 feet. The shiny foliage is made up of alternate 1–4 inch (2.5–10.2 cm) oval leaves dangling from long petioles. Each leaf has three distinct yellowish veins. The inconspicuous tiny cream-colored flowers are borne in the spring on branching 3 inch (7.6 cm) flower stalks. They are followed by large crops of fruit, comprising round pea-sized berries attached to the branchlets by cup-like little green cones. The berries first turn reddish and then ripen to black. Camphor tree can be readily identified by the distinctive odor of a crushed leaf. Camphor is present in every part of the tree but is usually taken from the wood of mature trees by steam distillation. Oil is clear and has a scent similar to eucalyptus. It has a duality of hot and cold actions, cooling at first touch and then stimulating heat and circulation. It is very useful in rheumatic inflammation. Ophthacare clinical trials show that *C. camphora* with other herbal ingredients improves the patients suffering from various ophthalmic disorders, namely conjunctivitis, conjunctival xerosis (dry eye), and acute dacryocystitis [Biswas et al. 2001].

Chamomilla (*Chamomile matricaria recutita*; synonym *Matricaria chamomilla*)

Chamomilla belongs to the Compositae family. It is a perennial herb found in dry fields and around gardens and cultivated grounds. The white ray florets are furnished with a ligule, whereas the disc florets are yellow. The hollow receptacle is swollen and lacks scales.

The flowers of chamomile provide 1–2% volatile oils containing α-bisabolol, α-bisabolol oxides A and B, and matricin (usually converted to chamazulene). Other active constituents include the bioflavonoids apigenin, luteolin, and quercetin [McKay and Blumberg 2006]. These active ingredients contribute to chamomile's anti-inflammatory, antispasmodic, and smooth muscle-relaxing effects, particularly in the gastrointestinal tract [Rodriguez-Fragoso et al. 2008].

Hemp/Cannabis, Marijuana (*Cannabis sativa*)

This plant belongs to the family Urticaceae. *Cannabis sativa* has been cultivated for over 4,500 years for different purposes, as fiber, oil, or narcotics. It is an annual herb, usually erect, and the stems are variable, up to 5 m tall, with resinous pubescence, are angular, and are sometimes hollow. Medicinally, plant is a tonic, intoxicant, stomachic, antispasmodic, analgesic, narcotic, sedative, and anodyne.

Most varieties contain cannabinol and cannabinin. Egyptian variety contains cannabidine, cannabol, and cannabinol, their biological activity being attributable to the alcohols and phenolic compounds. Per 100 g, the seed is reported to contain 8.8 g of H_2O, 21.5 g of protein, 30.4 g of fat, 34.7 g of total carbohydrate, 18.8 g of fiber, and 4.6 g of ash. In Asia, per 100 g, the seed is reported to contain 421 calories,

13.6 g of water, 27.1 g of protein, 25.6 g of fat, 27.6 g of total carbohydrate, 20.3 g of fiber, 6.1 g of ash, 120 mg of Ca, 970 mg of phosphorus, 12.0 mg of iron, 5 mg of beta-carotene equivalent, 0.32 mg of thiamine, 0.17 mg of riboflavin, and 2.1 mg of niacin. A crystalline globulin has been isolated from defatted meal. It contains 3.8% glycocol, 3.6% alanine, 20.9% valine and leucine, 2.4% phenylalanine, 2.1% tyrosine, 0.3% serine, 0.2% cystine, 4.1% proline, 2.0% oxyproline, 4.5% aspartic acid, 18.7% glutamic acid, 14.4% tryptophane and arginine, 1.7% lysine, and 2.4% histidine. Oil from the seeds contains 15% oleic, 70% linoleic, and 15% linolenic and isolinolenic acids. The seed cake contains 10.8% water, 10.2% fat, 30.8% protein, 40.6% N-free extract, and 7.7% ash (20.3% K_2O, 0.8% Na_2O, 23.6% CaO, 5.7% MgO, 1.0% Fe_2O_3, 36.5% P_2O_5, 0.2% SO_3, 11.9% SiO_2, 0.1% Cl, and a trace of Mn_2O_3). Trigonelline occurs in the seed. *Cannabis* also contains choline, eugenol, guaiacol, nicotine, and piperidine (Council for Scientific and Industrial Research, 1948–1976), all listed as toxins by the National Institute of Occupational Safety and Health. Cannabinoids (CBs) are chemical compounds derived from cannabis. Animal models demonstrate that CB receptors on immune cells play a fundamental role in peripheral, spinal, and supra-spinal nociception, and CBs are effective analgesics [Hosking 2008]. Clinical trials of CBs in multiple sclerosis have suggested a benefit in neuropathic pain. CBs are also used as a powerful pain management drug for cancer patients. Experience from multiple case studies and clinical trials in Phases I–III demonstrate marked improvement in subjective sleep parameters in patients with a wide variety of pain conditions, including multiple sclerosis, peripheral neuropathic pain, intractable cancer pain, and rheumatoid arthritis, with minimum adverse events. In another study, CBs are no more effective than codeine in controlling pain and have depressant effects on the CNS. Use of CBs into clinical practice for pain management is therefore undesirable. Cannabis is traditionally used in the treatment of multiple sclerosis. Anecdotal evidence suggests that it may be beneficial in controlling symptoms such as spasticity, pain, tremor, and bladder dysfunction in multiple sclerosis patients. Recent research in animal models of multiple sclerosis has showed that CBs are able to control disease-induced symptoms such as spasticity and tremor and help patients suffering from the severity of the disease. CBs from *C. sativa* present an interesting therapeutic potential as antiemetics, appetite stimulant, analgesics, and in the treatment of spinal cord injuries, Tourette's syndrome, epilepsy, and glaucoma.

OTHER NUTRACEUTICALS FROM PLANTS: PLANTS WITH NO BOUNDARIES

Tulsi (*Ocimum sanctum*)

Tulsi has been used in India for thousands of years and has been described as queen of all herbs. This most sacred aromatic plant of India belongs to the family Lamiaceae. It is an erect, many-branched subshrub, with simple opposite green or purple ovate leaves that are strongly scented with hairy stems. Flowers are purplish in elongate racemes in close whorls [Warrier 1995].

In traditional Indian medicine, this plant is used to treat several diseases such as bronchitis, bronchial asthma, malaria, diarrhea, dysentery, skin diseases, arthritis, painful eye diseases, chronic fever, and insect bites. This plant has also been suggested to possess antifertility, anticancer, antidiabetic, antifungal, antimicrobial, hepatoprotective, cardioprotective, antiemetic, antispasmodic, analgesic, adaptogenic, and diaphoretic actions [Prakash and Gupta 2005]. Different chemical compounds, such as eugenol, luteolin, ursolic acid, and oleanolic acid, was isolated from the leaf of green and black varieties of tulsi [Anandjiwala, Kalola, and Rajani 2006]. Kaul et al. [2005] showed that high-performance liquid chromatography purified polyphenolic fraction IV of tulsi may have a profound antiatherogenic effect. The two water-soluble flavonoids orientin and vicenin isolated from the leaves of the Indian plant *Ocimum sanctum* was found to provide radioprotective effect in mice [Uma et al. 1999]. Studies have also shown tulsi to be effective for diabetes [Kapoor 2008], and the beneficial effect on blood glucose levels is attributable to the antioxidant properties of this plant [Sethi et al. 2004].

Danshen (*Salvia miltiorrhiza*)

Danshen is a perennial herb that grows on sunny hillsides and stream edges mainly in China. Its violet-blue flowers bloom in the summer, and the leaves are oval, with finely serrated edges. It belongs to the family Lamiaceae. Remedies containing danshen are used traditionally to treat a diversity of ailments, particularly cardiac (heart) and vascular (blood vessel) disorders such as atherosclerosis or blood clotting abnormalities. The ability of danshen to "thin" the blood and reduce blood clotting is well documented, although the herb's purported ability to "invigorate" the blood or improve circulation has not been demonstrated in high-quality human trials. According to multiple randomized clinical trails of danshen on patients of acute myocardial infarction (AMI), the evidence to support use of danshen preparations is too weak to make any judgment about its effects. Evidence from randomized clinical trials is insufficient and of low quality [Wu, Ni, and Wu 2008].

Danshen is used in traditional Chinese medicine (TCM) to promote blood flow and treat cardiovascular diseases. It is sold in over-the-counter herbal preparations, prescribed by TCM doctors, and administered in Chinese hospitals for angina pectoris [Lei and Chiou 1986a], AMI [Liu et al. 1992], and ischemic and thrombotic disorders. *In vitro* and animal studies suggest that it may be vasoactive, scavenge free radicals [Ji, Tan, and Zhu 2000], and inhibit platelet aggregation [Han et al. 2008]. This Chinese herbal treatment for AMI is widely used in China in addition to usual western forms of therapy in the treatment of AMI. However, there is no strong evidence to support its use, and few rigorous studies have been conducted. Well-designed and randomized trials are needed to provide adequate evidence of its role in the treatment of AMI. The active compounds in danshen are tanshinones and phenolic compounds [Hu et al. 2005; Cao et al. 2008]. Danshen has been studied in China for AMI and ischemic heart disease [Hu et al. 2005]. Most studies are neither placebo controlled nor blinded and often use danshen combined with other herbs. No differences in cardiac contractility, compliance, inotropy, blood viscosity, or fibrinogen were found. Danshen has been studied for artery vasodilation, the mechanism underlying its use in angina [Zhou et al. 2005]. At low dose, it causes generalized vasodilation and decreases blood pressure [Lei and Chiou 1986b]. At higher doses, however, it causes vasoconstriction in noncoronary arteries [Lei and Chiou 1986; Zhou et al. 2005]. Animal studies confirm that danshen decreases warfarin clearance and increases bioavailability [Chan et al. 1995].

Horse Chestnut Seed (*Aesculus hippocastanum*)

Aesculus hippocastanum is a large deciduous tree, commonly known as horse chestnut or conker tree, and belongs to the family Sapindaceae. Seeds of horse chestnut contain saponins, known collectively as "aescin," which have a gentle soapy feel and are potent anti-inflammatory compounds. Saponins, such as aescin, also reduce capillary fragility and therefore help to prevent leakage of fluids into surrounding tissues, which can cause swelling. An extract of horse chestnut has been shown recently to have one of the highest "active-oxygen" scavenging abilities of 65 different plant extracts tested. Such extracts are more powerful antioxidants than vitamin E and also exhibit potent cell-protective effects that are linked to the well-known antiaging properties of antioxidants [Wilkinson and Brown 1999]. The active compound aescin is a mixture of triterpene glycosides [Kapusta et al. 2007]. Aescin decreases lower-extremity edema by decreasing capillary permeability via inhibition of endothelial lysosomal enzymes and preservation of capillary wall glycocalyx [Wilkinson and Brown 1999] and vesoconstriction via prostaglandin F2 [Fujimura et al. 2007]. Extracts standardized for aescin content are available, and the Commission E recommends 50 mg of aescin twice daily [Loew et al. 2000]. A review of eight placebo-controlled trials for venous insufficiency reported that lower-extremity circumference and volume decreased, and leg pain and pruritus improved [Pittler and Ernst 1998, 2004, 2006]. The most frequent adverse effects were gastrointestinal symptoms, dizziness, headache, and pruritus. Trials that compared horse chestnut with hydroxyethylrutosides, a semisynthetic mixture of flavonoid compounds [Pittler and Ernst 2004], found little difference between treatments. In a partially blind study of horse chestnut seed versus compression stockings or placebo, the volume of lower-extremity edema decreased 45 ml for each active treatment compared with a 10 ml increase for placebo. Side effects include pruritus, nausea, headache, and dizziness [Pittler and Ernst 1998]. There is one reported case of hepatitis from a commercial preparation of horse chestnut extract, Venoplant [Takegoshi et al. 1986]. Venocuran, an herbal preparation that contained horse chestnut, plus phenopyrazone, extracts of white squill, convallaria, oleander, and adonis, was removed from the market because of a systemic lupus-like syndrome.

Feverfew (*Tanacetum parthenium*; synonym, *Chrysanthemum parthenium*)

Feverfew is a small herb that grows from a few inches to 2 feet and belongs to the family Asteraceae. It is leaf lobed, the margins are entire or dentate, and the flowers capitulate. It is a contracted raceme composed of numerous individual sessile flowers, called the florets, and shares the same receptacle. Fruits are specialized achene, sometimes called cypsela, with one seed per fruit.

Feverfew is primarily used for migraine prophylaxis [Evans and Taylor 2006]. It inhibits platelet release of serotonin [Silberstein 2001] and may have vasoactive effects [Silberstein 2001]. The active compound in feverfew, parthenolide, is a sesquiterpene lactone. Other compounds have been investigated for biologic activity, most notably flavonoids for a potential anti-inflammatory effect [O'Hara et al. 1998]. Feverfew is one of a few herbs for which data on the content of the active compounds in commercial preparations are available [Wu et al. 2007]. Typical daily doses for migraine prophylaxis are 50–100 mg of whole or powdered dried leaves, corresponding to 500 g of parthenolide [Maizels, Blumenfeld, and Burchette 2006; Wu et al. 2007]. *In vitro* studies demonstrate that feverfew and parthenolide inhibit platelet aggregation and platelet and leukocyte release of serotonin [Till et al. 1989]. *In vitro* vasoactive effects vary with the formulation. Chloroform extracts of fresh leaves inhibit smooth muscle contractility, but extracts of dried leaves elicit a contractile response. Because chloroform extracts of dried feverfew do not contain measurable amounts of parthenolide, other vasoactive compounds may be present [Heptinstall et al. 1985]. Feverfew's inhibition of platelet serotonin release *in vitro* raises concerns about interaction with anti-serotonin migraine prophylactic drugs [Haaz et al. 2006] and potentiation of bleeding with antiplatelet agents. However, serious adverse events or interactions have not been reported.

Ephedra (*Ephedra sinica*)

Ephedra is an evergreen shrub that grows up to 0.5 m, and it belongs to the family Ephedraceae. Ma huang is a natural source of ephedrine and has potent sympathomimetic activity. Herbal remedies and soft drinks used for energy or weight loss often contain ma huang. In past few years, this plant has been extensively studied for its medicinal uses, there are 35 clinical trials, and 47 observational studies had been done on Ephedra showing its effect on weight loss [Lenz and Hamilton 2004; Pittler and Ernst 2004; Haaz et al. 2006; Norris et al. 2005]. Ephedra contains the chemicals ephedrine and pseudoephedrine, which are bronchodilators used for nasal allergies and asthama [Bielory 2004]. It has been used and studied to treat asthma and chronic obstructive pulmonary disease in both children and adults [Lanski et al. 2003; Avois 2006]. Other treatments, such as beta-agonist inhalers, are more commonly recommended as a result of safety concerns with ephedra or ephedrine. One hundred five different studies on human and animals show its adverse effects and toxicity. Between 1997 and 1999, a total of 140 reports of adverse events were related to ma huang; 13 caused permanent impairment, and 10 resulted in death. Many reports concern healthy young people without known cardiac disease. The majority had new-onset hypertension [Ernst 2003; Richard and Jurgens 2005], and other findings included cerebrovascular accidents, arrhythmias and myocardial infarction [Rogers, Shin, and Wang 1997; Bohn, Khodaee, and Schwenk 2003]. Several reports link the adverse response of ma huang to concurrent use of caffeine or exercise [Avois et al. 2006; Chitturi and Farrell 2008].

Dong Quai (*Angelica sinensis*)

Angelica sinensis (commonly known as dong quai) is a fragrant, perennial herb found in mainland China, Japan, and Korea. It belongs to family the Umbelliferae and grows to 1 × 0.7 m. The dried root is valued for its therapeutic properties. Its flavor is a distinct blend of bitter, sweet, and pungent, and its overall effect is warming in nature.

Dong quai has been called the "female ginseng" and is excellent as an all-purpose women's herb. Dong quai is also considered a TCM remedy for menstrual symptoms and menopause [Haines et al. 2008]. Clinical studies found no significant difference between dong quai and placebo in the treatment of vasomotor symptoms in Hong Kong Chinese women. The frequency of mild, moderate, and severe hot flushes decreased in both treatment and placebo groups, but dang quai was statistically superior to placebo only in the treatment of mild hot flashes. There were no serious adverse effects found during the course of the study [Haines et al. 2008].

Dong quai root contains 0.4–0.7% volatile oil, the key components of which are *n*-butylidenephthalide, ligustilide, *n*-butylphthalide, ferulic acid, nicotinic acid, and succinic acid [Duke 1992]. Significant amounts of vitamin A and carotenoids (0.675%), vitamin B_{12} (0.25–0.40 mcg/100 g), vitamin E, ascorbic acid, folinic acid, biotin, various phytosterols (e.g., β-sitosterol), calcium, magnesium, and other essential macrominerals are also found in dong quai root. This plant is also used for antithrombotic, antiasthmatic, and analgesic effects [Chang et al. 2005; Dong et al. 2006; Gao et al. 2006]. In humans, dong quai has been evaluated for estrogenic effects [Gao et al. 2007]. Antithrombotic effects are attributed to coumarin derivatives and ferulic acid contained in the oil of the root [Chang et al. 2005]. Ferulic acid may cause platelet dysfunction by inhibiting production of thromboxane A2. In a controlled trial of 96 subjects with new cerebral thrombosis or embolism, there was no difference in improvement rate with dong quai [Liao et al. 1989]. The root, dang quai, is valuable in anemia and menstrual pain or as a general tonic after childbirth [Shen et al. 2005]. It clears liver stagnation (of both energy and toxins) and can relieve constipation, especially in the elderly.

Kava (*Piper methysticum*)

Kava is a shrub about 6 feet high, somewhat resembling the bamboo in growth, and belongs to the family Piperaceae. The root is the part recommended for use in medicine. The main root seems to grow horizontally beneath the surface of the ground, sending up stalks at intervals of from 2 to 4 inches. Each stalk is from 0.5 to 3 inches in diameter at the base and is hollow. Externally, the main root is brown and covered with a thin bark. From the sides and lower part are secondary roots, about 0.5–0.75 inches in diameter. These appear to be arranged about the bases of the stalks; in some cases, they are quite long and send out rootlets at a distance of

6 inches from the main root. Aboriginal peoples of the South Pacific, as an anxiolytic, use kava, a member of the black pepper family, and it has been promoted to treat anxiety, depression and muscle tension [Saeed, Bloch, and Antonacci 2007; van der Watt, Laugharne, and Janca 2008]. Kava pyrones, the active compounds in kava [Schulze, Raasch, and Siegers 2003], may inhibit cyclooxygenase and thromboxane synthase. A small observational study of an aboriginal community found that high-density lipoproteins (HDLs) were higher in kava users [Clough, Rowley, and O'Dea 2004]. Adverse effects include rash, elevated hepatic enzymes, pulmonary hypertension, and hepatitis [Schulze, Raasch, and Siegers 2003; Stickel and Schuppan 2007]. Kava may also interact with benzodiazepines, inducing coma [Izzo and Ernst 2001; Hu et al. 2005].

Licorice (*Glycyrrhiza glabra*)

Licorice, an extract of the root of *Glycyrrhiza glabra*, is used as a sweetening and flavoring agent. This plant is used as an herbal remedy for gastritis and upper respiratory tract infections [Fiore et al. 2008]. The active constituent of licorice is glycyrrhizic acid [Baltina 2003]. A metabolite, glycyrrhetinic acid, inhibits renal 11-hydroxysteroid dehydrogenase and causes a state of mineralocorticoid excess by impeding the inactivation of cortisol [Baltina 2003]. Case reports link licorice to hypertension, hypertensive encephalopathy, pulmonary edema, edema, hypokalemia, arrhythmias, congestive heart failure, muscle weakness, and acute renal failure [Schambelan 1994; Thyagarajan et al. 2002; Coon and Ernst 2004]. Dilated cardiomyopathy resulting from excessive use of licorice and glycyrrhizin for gastritis has been reported [Shintani et al. 1992]. Fifty to 100 g of confectionary licorice, or 50–300 mg of glycyrrhetinic acid, over weeks may cause adverse effects [Cosmetic Ingredient Review Expert Panel 2007]. A study of 30 healthy, normal volunteers reported that 100 g/day of licorice (270 mg of glycyrrhizic acid) over four weeks increased systolic blood pressure 6.5 mm Hg and decreased plasma potassium 0.24 mmol/L from baseline [Schambelan 1994; Ferrari et al. 2001]. Susceptibility to licorice varies greatly; subjects with underlying hypertension and women may be more sensitive [Mattarello et al. 2006; Sigurjonsdottir et al. 2006]. Dietary consumption of licorice acts as an antioxidant and is reported to help in reducing cardiovascular inflammation and athrosclerosis; licorice-root extract by hypercholesterolemic patients may act as a moderate hypocholesterolemic nutrient and a potent antioxidant agent and hence fight cardiovascular disease [Fuhrman et al. 2002].

Ginseng (*Panax ginseng*)

Ginseng refers to the root of *Panax* species and belongs to the family Araliaceae. The most commonly examined species are *Panax ginseng* (Asian ginseng), *Panax quinquefolius* (American ginseng), and *Panax japonicus* (Japanese ginseng). The terms "red" and "white" refer to different methods of ginseng preparation, not different species. Ginseng is believed to promote vigor, potency, well-being, and longevity. In China, it is used for angina pectoris, myocardial infarction, and congestive heart failure. It has been evaluated for many other indications, most notably for use as an antihyperglycemic. It is administered as a whole dried root, extract, tea, or capsule. The active compounds are heterogeneous triterpene saponin glycosides, collectively termed ginsenosides. The exact ginsenosides vary by *Panax* species, root age, and preparation method [Valli and Giardina 2002]. The actions of specific ginsenosides vary and, in some instances, are inconsistent.

In multiple double-blind clinical trials on the effect of ginseng on erectile dysfunction and on female sexual function and libido [Choi, Seong, and Rha 1995; Ito et al. 2006], ginseng shows marginal improvements in patients' conditions. In a total of 90 patients with 30 patients in each group, changes in symptoms such as frequency of intercourse, premature ejaculation, and morning erections after treatment were not changed in all three groups.

However, in the group who received ginseng, changes in erectile parameters such as penile rigidity and girth, libido, and patient satisfactions were significantly higher than that of other groups. The overall therapeutic efficacies on erectile dysfunction were 60% for the ginseng group and 30% for the placebo- and trazodone-treated groups, statistically confirming the effect of ginseng. Ginseng is useful in viral myocarditis, which is a heart disease when the muscles in the walls of heart become infected with a virus. Ginseng preparation showed significant effects on reducing myocardial enzymes and improving cardiac function with no serious adverse effect reported in a clinical trial on viral myocarditis [Liu, Yang, and Du 2004]. Ginseng has been studied in some depth as an antifatigue agent; preclinical evidence shows some immune-stimulating activity [Block and Mead 2003].

CONCLUSION

It is clear from clinical trials and studies that nutraceuticals are more preventive than curative. If people use them wisely and in their food regularly, medicinal compounds present in the food will keep them healthy and disease free. How to make use of these food and spices full of medicinal values in our daily life is always a challenge for people and dieticians. In some ancient cultures and traditions, as in Indian and Chinese traditions, these food materials with medicinal values are integrated into the lifestyle in many ways. For example, spices such as turmeric, chili, and all the mentioned spices in this chapter are used as regular seasoning in curries and soups. Regular daily diets in many cultures contain fruits and milk in everyday use. Extensive epidemiological studies and investigations are required to confirm the occurrence of the major diseases and dietary habits in different parts of the world.

Not only are these plants and plant products used in regular diets, but these plants are also tactfully connected to the people by involving them in people's faith (i.e., inculcating plants and trees in religious practices). Involving plants and plant products in religious faiths not only shows the importance of plants in human lives but also protects and conserves the ecosystem by respecting and preserving them.

Ancient art and science of traditional medicine or Ayurveda (ayus, meaning "life" and veda, meaning "science") was developed about 5,000 years ago in ancient

India. The *Sushruta Samhita* and the *Charaka Samhita* were influential works on traditional medicine of India during that period. Both are now identified worldwide as important early sources of medical understanding and practice, independent of ancient Greek civilization. Ayurvedic medicine provided clues that lead to the discovery of thousands of phytochemicals from medicinal plants. The active compounds have physiological benefits or provide protection against a chronic disease. Food also plays a major role in the concepts of illness and curing. Therapeutic actions are maximally effective only if appropriate dietary measures are taken to support the restoration of physiological balance. Furthermore, food and spices themselves constitute an integral part of traditional medical prescriptions. These traditional prescriptions contain small amounts of micronutrients and active biocompounds that are consumed daily in much higher quantities.

"Plants are a source of many biologically active products and nowadays they are of great interest to the pharmaceutical industry. The study of how people of different culture use plants in particular ways has led to the discovery of important new medicines" [Borges et al. 2005]. Therefore, ancient literature on plant use, medicinal plants used by different tribes around the world, could provide a more vital clue to the discovery of many more phytochemicals from the nutraceutical herbs. Massive monetary investment and intense research is required. Because of irresponsible human acts of mass destruction of forests worldwide, we are losing flora at an alarming rate. Unless we act immediately to preserve the medicinal plants, plants with nutraceutical values, future generations will loose tremendous health and wellness benefits from nutraceutical herbs that we are enjoying now. We can expect that, in the near future, most active biocompounds will be taken along with vitamin pills everyday to prevent and cure disease.

FOR MORE INFORMATION AND RESEARCH

Chan, T. Y. K. 1998. Drug interactions as a cause of over anticoagulation and bleedings in Chinese patients receiving warfarin. *Int. J. Clin. Pharmacol. Ther.* 36:403–405.

Keji, C. 1981. Certain progress in the treatment of coronary heart disease with traditional medicinal plants in China. *Am. J. Chin. Med.* 9:193–196.

Natural Products Alert Database, NAPRALERT, http://napralert.org.

Onitsuka, M., M. Fujiu, N. Shinma, and H. B. Maruyama. 1983. New platelet aggregation inhibitors from tanshen. *Chem. Pharm. Bull. (Tokyo)* 31:1670–1675.

Pittler, M. H. and E. Ernst. 2002. Horse chestnut seed extract for chronic venous insufficiency. *Cochrane Database Syst. Rev.* 2002:CD003230.

Qiu, R.-X., Z.-Q. Luo, H.-C. Luo. 1997. Effect of xinmaitong capsule on damage of lipid peroxidation in coronary heart disease patients with myocardial ischemia (translated from Chinese). *Chung Kuo Chung Hsi I Chieh Ho Tsa Chih* 17:342–344.

Shanghai Cooperative Group for the Study of Tanshinone IIA. 1984. Therapeutic effect of sodium tanshinone IIA sulfonate in patients with coronary heart disease. *J. Trad. Chin. Med.* 4:20–24.

Wang, Z., J. M. Roberts, P. G. Grant, R. W. Colman, and A. D. Schreiber. 1982. The effect of a medicinal Chinese herb on platelet function. *Thromb. Haemost.* 48:301–306.

Wu Y.-J., C.-Y. Hong, S.-J. Lin, P. Wu, and M.-S. Shiao. 1998. Increase of vitamin E content in LDL and reduction of atherosclerosis in cholesterol-fed rabbits by a water-soluble antioxidant-rich fraction of *Salvia miltiorrhiza. Arterioscler. Thromb. Vasc. Biol.* 18:481–486.

Yu C. M., J. C. N. Chan, and J. E. Sanderson. 1997. Chinese herbs and warfarin potentiation by "danshen." *J. Intern. Med.* 241:337–339.

REFERENCES

Agarwal, K. C. 1996. Therapeutic actions of garlic constituents. *Med. Res. Rev.* 16:111–124.

Aggarwal, B. B., A. Kumar, M. S. Aggarwal, and S. Shishodia. 2004. Curcumin derived from turmeric (*Curcuma longa*): a spice for all seasons. In *Phytochemicals in Cancer Chemoprevention.* Edited by Bagchi D. and H. G. Preuss. Boca Raton, FL: CRC Press, pp. 349–387.

Aggarwal, B. B. and S. Shishodia. 2004. Suppression of the nuclear factor-kappaB activation pathway by spice-derived phytochemicals: reasoning for seasoning. *Ann. NY Acad. Sci.* 1030:434–441.

Agha-Hosseini, M., L. Kashani, A. Aleyaseen, A. Ghoreishi, H. Rahmanpour, A. R. Zarrinara, and S. Akhondzadeh. 2008. *Crocus sativus* L. (saffron) in the treatment of premenstrual syndrome: A double-blind, randomised and placebo-controlled trial. *Br. J. Obstet. Gyn.* 115:515–519.

Akhondzadeh Basti, A., E. Moshiri, A. A. Noorbala, A. H. Jamshidi, S. H. Abbasi, and S. Akhondzadeh. 2007. Comparison of petal of *Crocus sativus* L. and fluoxetine in the treatment of depressed outpatients: A pilot double-blind randomized trial. *Prog. Neuropsychopharmacol. Biol. Psychiatry* 31:439–442.

Ali, B. H., G. Blunden, M. O. Tanira, and A. Nemmar. 2008. Some phytochemical, pharmacological and toxicological properties of ginger (*Zingiber officinale* Roscoe): A review of recent research. *Food Chem. Toxicol.* 46:409–420.

Anandjiwala, S., J. Kalola, and M. Rajani. 2006. Quantification of eugenol, luteolin, ursolic acid, and oleanolic acid in black (Krishna Tulasi) and green (Sri Tulasi) varieties of Ocimum sanctum Linn. using high-performance thin-layer chromatography. *J. AOAC Int.* 89:1467–1474.

Armstrong, W. J., P. Johnson, and S. Duhme. 2001. The effect of commercial thermogenic weight loss supplement in body composition and energy expenditure in obese adults. *J. Exer. Physiol.* 4:28–35.

Augusti, K. T. 1996. Therapeutic values of onion (*Allium cepa* L.) and garlic (*Allium sativum* L.), *Indian J. Exp. Biol.* 34:634–640.

Aung, H. H., C. Z. Wang, M. Ni, A. Fishbein, S. R. Mehendale, J. T. Xie, C. Y. Shoyama, and C. S. Yuan. 2007. Crocin from *Crocus sativus* possesses significant anti-proliferation effects on human colorectal cancer cells. *Exp. Oncol.* 29:175–180.

Avois, L., N. Robinson, C. Saudan, N. Baume, P. Mangin, and M. Saugy. 2006. Central nervous system stimulants and sport practice. *Br. J. Sports Med.* 40 (Suppl. 1):i16–i20.

Baker, W. L., G. Gutierrez-Williams, C. M. White, J. Kluger, and C. I. Coleman. 2008. Effect of cinnamon on glucose control and lipid parameters. *Diabetes Care* 31:41–43.

Baltina, L.A. 2003. Chemical modification of glycyrrhizic acid as a route to new bioactive compounds for medicine. *Curr. Med. Chem.* 10:155–171.

Banerjee, S., C. K. Panda, and S. Das. 2006. Clove (*Syzygium aromaticum* L.), a potential chemopreventive agent for lung cancer. *Carcinogenesis* 27:1645–1654.

Barik, A., B. Mishra, A. Kunwar, R. M. Kadam, L. Shen, S. Dutta, S. Padhye, A. K. Satpati, H. Y. Zhang, and K. Indira Priyadarsini. 2007. Comparative study of copper(II)-curcumin complexes as superoxide dismutase mimics and free radical scavengers. *Eur. J. Med. Chem.* 42:431–439.

Bent, S. 2008. Herbal medicine in the United States: review of efficacy, safety, and regulation: grand rounds at University of California, San Francisco Medical Center. *J. Gen. Intern. Med.* 23:854–859.

Bhattacharya, A., M. Kumar, S. Ghosal, and S. K. Bhattacharya. 2000. Effect of bioactive tannoid principles of *Emblica officinalis* on iron-induced hepatic toxicity in rats. *Phytomedicine* 7:173–175.

Bielory, L. 2004. Complementary and alternative interventions in asthma, allergy, and immunology. *Ann. Allergy Asthma Immunol.* 93 (2 Suppl. 1):S45–S54.

Biswas, N. R., S. K. Gupta, G. K. Das, N. Kumar, P. K. Mongre, D. Haldar, and S. Beri. 2001. Evaluation of Ophthacare eye drops: a herbal formulation in the management of various ophthalmic disorders. *Phytother. Res.* 15:618–620.

Block, K. I. and M. N. Mead. 2003. Immune system effects of echinacea, ginseng, and astragalus: a review. *Integr. Cancer Ther.* 2:247–267.

Boelens, M. H. 1991. Spices and condiments II. In *Volatile Compounds in Foods and Beverages.* Edited by Maarse, H. New York, NY: Marcel Dekker, pp. 449–482.

Bogucka-Kocka, A., H. D. Smolarz, and J. Kocki. 2008. Apoptotic activities of ethanol extracts from some Apiaceae on human leukaemia cell lines. *Fitoterapia* 79:487–497.

Bohn, A. M., M. Khodaee, and T. L. Schwenk. 2003. Ephedrine and other stimulants as ergogenic aids. *Curr. Sports Med. Rep.* 2:220–225.

Borges, M. H., D. L. Alves, D. S. Raslan, D. Piló-Veloso, V. M. Rodrigues, M. I. Homsi-Brandeburgo, and M. E. de Lima. 2005. Neutralizing properties of *Musa paradisiaca* L. (Musaceae) juice on phospholipase A2, myotoxic, hemorrhagic and lethal activities of crotalidae venoms. *J. Ethnopharmacol.* 98:21–29.

Bown, D. 1995. *Encyclopaedia of Herbs and their Uses.* London, UK: Dorling Kindersley.

Caglioti, L., H. Naef, D. Arigoni, and O. Jeper. 1958. Über die Inhaltsstoffe der *Asa foetida.* I. Farnesiferol A. *Helvetica Chimica Acta* 41:2278–2287.

Caglioti, L., H. Naef, D. Arigoni, and O. Jeper. 1959. Über die Inhaltsstoffe der *Asa foetida.* II. Farnesiferol B und C. *Helvetica Chimica Acta* 42:2557–2570.

Cao, J., Y. J. Wei, L. W. Qi, P. Li, Z. M. Qian, H. W. Luo, J. Chen, and J. Zhao. 2008. Determination of fifteen bioactive components in Radix et Rhizoma Salviae miltiorrhizae by high-performance liquid chromatography with ultraviolet and mass spectrometric detection. *Biomed. Chromatogr.* 22:164–172.

Chan, K. A. C. T. Lo, J. H. K. Yeung, and K. S. Woo. 1995. The effects of danshen (*Salvia miltiorrhiza*) on warfarin pharmacodynamics and pharmacokinetics of warfarin enantiomers in rats. *J. Pharm. Pharmacol.* 47:402–406.

Chang, G. T., S. K. Kang, J. H. Kim, K. H. Chung, Y. C. Chang, and C. H. Kim. 2005. Inhibitory effect of the Korean herbal medicine, Dae-Jo-Whan, on platelet-activating factor-induced platelet aggregation. *J. Ethnopharmacol.* 102:430–439.

Chiao, J. W., F. L. Chung, R. Kancherla, T. Ahmed, A. Mittelman, and C. C. Conaway. 2002. Sulforaphane and its metabolite mediate growth arrest and apoptosis in human prostate cancer cells. *Int. J. Oncol.* 20:631–636.

Chiej, R. 1984. *Encyclopaedia of Medicinal Plants.* Edinburgh, UK: MacDonald.

Chitturi, S. and G. C. Farrell. 2008. Hepatotoxic slimming aids and other herbal hepatotoxins. *J. Gastroenterol. Hepatol.* 23:366–373.

Choi, H. K., D. H. Seong, and K. H. Rha. 1995. Clinical efficacy of Korean red ginseng for erectile dysfunction. *Int. J. Impot. Res.* 7:181.

Chrubasik, S., M. H. Pittler, and B. D. Roufogalis. 2005. *Zingiberis rhizoma*: a comprehensive review on the ginger effect and efficacy profiles. *Phytomedicine* 12:684–701.

Clarke, J. O. and G. E. Mullin. 2008. A review of complementary and alternative approaches to immunomodulation. *Nutr. Clin. Pract.* 23:49–62.

Clough, A. R., K. Rowley, and K. O'Dea. 2004. Kava use, dyslipidaemia and biomarkers of dietary quality in Aboriginal people in Arnhem Land in the Northern Territory (NT). *Australia Eur. J. Clin. Nutr.* 58:1090–1093.

Coon, J. T. and E. Ernst. 2004. Complementary and alternative therapies in the treatment of chronic hepatitis C: a systematic review. *J. Hepatol.* 40:491–500.

Cosmetic Ingredient Review Expert Panel. 2007. Final report on the safety assessment of glycyrrhetinic acid, potassium glycyrrhetinate, disodium succinoyl glycyrrhetinate, glyceryl glycyrrhetinate, glycyrrhetinyl stearate, stearyl glycyrrhetinate, glycyrrhizic acid, ammonium glycyrrhizate, dipotassium glycyrrhizate, disodium glycyrrhizate, trisodium glycyrrhizate, methyl glycyrrhizate, and potassium glycyrrhizinate. *Int. J. Toxicol.* 26 (Suppl. 2):79–112.

Culioli G., C. Mathe, P. Archier, and C. Vieillescazes. 2003. A lupane triterpene from frankincense (Boswellia sp., Burseraceae). *Phytochemistry* 62:537–541.

Dall'Acqua, S., G. Viola, M. Giorgetti, M. C. Loi, and G. Innocenti. 2006. Two new sesquiterpene lactones from the leaves of *Laurus nobilis*. *Chem. Pharm. Bull. (Tokyo)* 54:1187–1189.

Davis, P. A., J. A. Polagruto, G. Valacchi, A. Phung, K. Soucek, C. L. Keen, and M. E. Gershwin. 2006. Effect of apple extracts on NF-kappaB activation in human umbilical vein endothelial cells. *Exp. Biol. Med.* 231:594–598.

Demetriades, A. K., P. D. Wallman, A. McGuiness, and M. C. Gavalas. 2005. Low cost, high risk: accidental nutmeg intoxication. *Emerg. Med. J.* 22:223–225.

Dhalwal, K., V. M. Shinde, K. R. Mahadik, and A. G. Namdeo. 2007. Rapid densitometric method for simultaneous analysis of umbelliferone, psoralen, and eugenol in herbal raw materials using HPTLC. *J. Sep. Sci.* 30:2053–2058.

Dhanabal, S. P., M. Sureshkumar, M. Ramanathan, and B. Suresh. 2005. Hypoglycemic effect of ethanolic extract of *Musa sapientum* on alloxan induced diabetes mellitus in rats and its relation with antioxidant potential. *J. Herb. Pharmacother.* 5:7–19.

Dhanapakiam, P., J. M. Joseph, V. K. Ramaswamy, M. Moorthi, and A. S. Kumar. 2008. The cholesterol lowering property of coriander seeds (*Coriandrum sativum*): Mechanism of action. *J. Environ. Biol.* 29:53–56.

Dhandapani, S., V. R. Subramanian, S. Rajagopal, and N. Namasivayam. 2002. Hypolipidemic effect of *Cuminum cyminum* L. on alloxan-induced diabetic rats. *Pharmacol. Res.* 46:251–255.

Dong, T. T., K. J. Zhao, Q. T. Gao, Z. N. Ji, T. T. Zhu, J. Li, R. Duan, A. W. Cheung, and K. W. Tsim. 2006. Chemical and biological assessment of a chinese herbal decoction containing Radix Astragali and Radix Angelicae Sinensis: Determination of drug ratio in having optimized properties. *J. Agric. Food Chem.* 54:2767–2774.

Duessel, S., R. M. Heuertz, and U. R. Ezekiel. 2008. Growth inhibition of human colon cancer cells by plant compounds. *Clin. Lab. Sci.* 21:151–157.

Duke, J. A. 1992. *Handbook of Phytochemical Constituents of GRAS Herbs and Other Economic Plants*. Boca Raton, FL: CRC Press.

Duke, J.A. and K. K. Wain. 1981. *Medicinal Plants of the World.* Computer index with more than 85,000 entries. Three volumes.

Dwyer, J. T., D. B. Allison, and P. M. Coates. 2005. Dietary supplements in weight reduction. *J. Am. Dietetic Assoc.* 105 (Suppl. 1):80–86.

Egan, M. E., M. Pearson, S. A. Weiner, V. Rajendran, D. Rubin, J. Glockner-Pagel, S. Canny, K. Du, G. L. Lukacs, and M. J. Caplan. 2004. Curcumin, a major constituent of turmeric, corrects cystic fibrosis defects. *Science* 304:600–602.

Eigner, D. and D. Scholz. 1990. The magic book of Gyani Dolma. *Pharm. Unserer Zeit.* 19:141–152.

El-Mekkawy, S., M. R. Meselhy, I. T. Kusumoto, S. Kadota, M. Hattori, and T. Namba. 1995. Inhibitory effects of Egyptian folk medicines on human immunodeficiency virus (HIV) reverse transcriptase. *Chem. Pharmacol. Bull. (Tokyo)* 43:641–648.

Ernst, E. 2003. Cardiovascular adverse effects of herbal medicines: A systematic review of the recent literature. *Can. J. Cardiol.* 19:818–827.

Evans, R. W. and F. R. Taylor. 2006. "Natural" or alternative medications for migraine prevention. *Headache* 46:1012–1018.

Fengshu, L., H. Kaiwei, L. Shaojia, Y. Chenwu, and Z. Ping. 1992. Antisenescent effect of *Phyllanthus emblica* fruits. I. Analysis of super oxide dismutase activity in fruits. *Chem. Abstr.* 116:127–273.

Ferrari, P., A. Sansonnens, B. Dick, and F. J. Frey. 2001. In vivo 11beta-HSD-2 activity: Variability, salt-sensitivity, and effect of licorice. *Hypertension* 38:1330–1336.

Fiore, C., M. Eisenhut, R. Krausse, E. Ragazzi, D. Pellati, D. Armanini, and J. Bielenberg. 2008. Antiviral effects of Glycyrrhiza species. *Phytother. Res.* 22:141–148.

Foster, S. and J. A. Duke. 1990. *A Field Guide to Medicinal Plants. Eastern and Central N. America.* Boston, MA: Houghton Mifflin.

Fuhrman, B., N. Volkova, M. Kaplan, D. Presser, J. Attias, T. Hayek, and M. Aviram. 2002. Antiatherosclerotic effects of licorice extract supplementation on hypercholesterolemic patients: Increased resistance of LDL to atherogenic modifications, reduced plasma lipid levels, and decreased systolic blood pressure. *Nutrition* 18:268–273.

Fujii, T., M. Wakaizumi, T. Ikami, and M. Saito. 2008. Amla (*Emblica officinalis* Gaertn.) extract promotes procollagen production and inhibits matrix metalloproteinase-1 in human skin fibroblasts. *J. Ethnopharmacol.* 119:53–57.

Fujimura, T., K. Tsukahara, S. Moriwaki, M. Hotta, T. Kitahara, and Y. Takema. 2007. A horse chestnut extract, which induces contraction forces in fibroblasts, is a potent anti-aging ingredient. *Int. J. Cosmet. Sci.* 29:140.

Gao, Q. T., J. K. Cheung, J. Li, G. K. Chu, R. Duan, A. W. Cheung, K. J. Zhao, T. T. Dong, and K. W. Tsim. 2006. A Chinese herbal decoction, Danggui Buxue Tang, prepared from Radix Astragali and Radix Angelicae Sinensis stimulates the immune responses. *Planta Med.* 72:1227–1231.

Gao, Q., J. Li, J. K. Cheung, J. Duan, A. Ding, A. W. Cheung, K. Zhao, W. Z. Li, T. T. Dong, and K. W. Tsim. 2007. Verification of the formulation and efficacy of Danggui Buxue Tang (a decoction of Radix Astragali and Radix Angelicae Sinensis): An exemplifying systematic approach to revealing the complexity of Chinese herbal medicine formulae. *Chin. Med.* 2:12.

Gerhauser, C. 2008. Cancer chemopreventive potential of apples, apple juice, and apple components. *Planta Med.* 74:1608–1624.

Ghosal, S. 1996. A plausible chemical mechanism of bioactivities of mangiferin. *Indian J. Chem.* 35B:561–566.

Giovannucci, E. 1999. Tomatoes, tomato-based products, lycopene, and cancer: Review of the epidemiologic literature, *J Natl Cancer Inst.* 91:317–331.

Gonçalves, J. L., R. C. Lopes, D. B. Oliveira, S. S. Costa, M. M. Miranda, M. T. Romanos, N. S. Santos, and M. D. Wigg. 2005. In vitro anti-rotavirus activity of some medicinal plants used in Brazil against diarrhea. *J. Ethnopharmacol.* 99:403–407.

Grieve, M. 1984. *A Modern Herbal*. New York, NY: Penguin.

Grzanna, R., L. Lindmark, and C. G. Frondoza. 2005. Ginger: An herbal medicinal product with broad anti-inflammatory actions. *J. Med. Food* 8:125–132.

Gundermann, K. J. and J. Müller. 2007. Phytodolor: Effects and efficacy of a herbal medicine. *Wien Med. Wochenschr.* 157:343–347.

Gupta, A., R. Gupta, and B. Lal. 2001. Effect of *Trigonella foenum-graecum* (fenugreek) seeds on glycaemic control and insulin resistance in type 2 diabetes mellitus: A double blind placebo controlled study. *J. Assoc. Physicians India* 49:1057–1061.

Gurley, B. J., S. F. Gardner, M. A. Hubbard, K. Williams, B. Gentry, J. Carrier, D. Edwards, and I. Khan. 2004. Assessment of botanical supplementation on human cytochrome P450 phenotype Citrus aurantium, Echinacea, milk thistle, saw palmetto (abstract PI 124). *Clin. Pharmacol. Ther.* 78:35.

Haaz, S., K. R. Fontaine, G. Cutter, N. Limdi, S. Perumean-Chaney, and D. B. Allison. 2006. Citrus aurantium and synephrine alkaloids in the treatment of overweight and obesity: An update. *Obes. Rev.* 7:79–88.

Haines, C. J., P. M. Lam, T. K. Chung, K. F. Cheng, and P. C. Leung. 2008. A randomized, double-blind, placebo-controlled study of the effect of a Chinese herbal medicine preparation (Dang Gui Buxue Tang) on menopausal symptoms in Hong Kong Chinese women. *Climacteric* 11:244–251.

Hainsworth, A. J. and P. A. Gatenby. 2008. Oral potassium supplementation in surgical patients. *Int. J. Surg.* 6:287–288.

Han, J. Y., J. Y. Fan, Y. Horie, S. Miura, D. H. Cui, H. Ishii, T. Hibi, H. Tsuneki, and I. Kimura. 2008. Ameliorating effects of compounds derived from *Salvia miltiorrhiza* root extract on microcirculatory disturbance and target organ injury by ischemia and reperfusion. *Pharmacol. Ther.* 117:280–295.

He, X. and R. H. Liu. 2008. Phytochemicals of apple peels: Isolation, structure elucidation, and their antiproliferative and antioxidant activities. *J. Agric. Food Chem.* 56:9905–9910.

Health Canada. 2008. Policy Paper, *Nutraceuticals/Functional Foods and Health Claims on Foods*. http://www.hc-sc.gc.ca/fn-an/label-etiquet/claims-reclam/nutra-funct_foods-nutra-fonct_aliment-eng.php.

Heptinstall, S., A. White, L. Williamson, and J. R. Mitchell. 1985. Extracts of feverfew inhibit granule secretion in blood platelets and polymorphonuclear leucocytes. *Lancet* 1:1071–1074.

Hofer, O., M. Widhalm, and H. Greger. 1984. Circular dichroism of ses-quiterpene-umbelliferone ethers and structure elucidation of a new derivative isolated from the gum resin "asa foetida". *Monatsh. Chem.* 115:1207–1218.

Hosking, R. D., and J. P. Zajicek. 2008. Therapeutic potential of cannabis in pain medicine. *Br. J. Anaesth.* 101:59–68.

Hu, P., Q. L. Liang, G. A. Luo, Z. Z. Zhao, and Z. H. Jiang. 2005. Multi-component HPLC fingerprinting of Radix Salviae Miltiorrhizae and its LC-MS-MS identification. *Chem. Pharm. Bull. (Tokyo)* 53:677–683.

Hu, Z., X. Yang, P. C. Ho, S. Y. Chan, P. W. Heng, E. Chan, W. Duan, H. L. Koh, and S. Zhou, 2005. Herb-drug interactions: A literature review. *Drugs* 65:1239–1282.

Huang, Y. L., G. C. Yen, F. Sheu, J. Y. Lin, and C. F. Chau. 2007. Dose effects of the food spice cardamom on aspects of hamster gut physiology. *Mol. Nutr. Food Res.* 51:602–608.

Iriti, M. and F. Faoro. 2006. Grape phytochemicals: A bouquet of old and new nutraceuticals for human health. *Med. Hypotheses* 7:833–838.

Ishikawa, T., K. Kondo, and J. Kitajima. 2003. Water-soluble constituents of coriander. *Chem. Pharm. Bull. (Tokyo)* 51:32–39.

Ito, T. Y., M. L. Polan, B. Whipple, and A. S. Trant. 2006. The enhancement of female sexual function with ArginMax, a nutritional supplement, among women differing in menopausal status. *J. Sex Marital Ther.* 32:369–378.

Izzo, A. A. and E. Ernst. 2001. Interactions between herbal medicines and prescribed drugs: A systematic review. *Drugs* 61:2163–2175.

Jayasooriya, A. P., M. Sakono, C. Yukizaki, M. Kawano, K. Yamamoto, and N. Fukuda. 2000. Effects of *Momordica charantia* powder on serum glucose levels and various lipid parameters in rats fed with cholesterol-free and cholesterol-enriched diets. *J. Ethnopharmacol.* 72:331–336.

Ji, X. Y., B. K. Tan, Y. Z. Zhu. 2000. *Salvia miltiorrhiza* and ischemic disease. *Acta Pharmacol. Sin.* 21:1089–1094.

Jiratchariyakul, W., C. Wiwat, M. Vongsakul, A. Somanabandhu, W. Leelamanit, I. Fujii, N. Suwannaroj, and Y. Ebizuka. 2001. HIV inhibitor from Thai bitter gourd. *Planta Med.* 67:350–353.

Jones, D. 2001. Regulation of appetite, body weight, and athletic function with materials derived from citrus varieties. U. S. Patent 6,224,873, filed May 2001.

Joos, S., T. Rosemann, J. Szecsenyi, E. G. Hahn, S. N. Willich, and B. Brinkhaus B. 2006. Use of complementary and alternative medicine in Germany: A survey of patients with inflammatory bowel disease. *BMC Complement. Altern. Med.* 6:19.

Juge, N., R. F. Mithen, and M. Traka. 2007. Molecular basis for chemoprevention by sulforaphane: A comprehensive review. *Cell. Mol. Life Sci.* 64:1105–1127.

Kagawa, K., K. Tokura, K. Uchida, H. Kakushi, and T. Shike. 1992. Platelet aggregation inhibitors from *Populus sieboldii* Miquel. *Chem. Pharm. Bull. (Tokyo)* 40:2191–2192.

Kalman, D. S., C. M. Colker, Q. Shi, and M. A. Swain. 2000. Effects of a weight-loss aid in healthy overweight adults: Double-blind, placebo-controlled clinical trial. *Curr. Ther. Res.* 61:199–205.

Kalra, E. K. 2003. Nutraceutical: Definition and introduction. *AAPS Pharm. Sci.* 5:25.

Kapoor, L. D. 1990. *Handbook of Ayurvedic Medicinal Plants.* Boca Raton, FL: CRC Press, p. 185.

Kapoor, S. 2008. *Ocimum sanctum*: A therapeutic role in diabetes and the metabolic syndrome. *Horm. Metab. Res.* 40:296.

Kapusta, I., B. Janda, B. Szajwaj, A. Stochmal, S. Piacente, C. Pizza, F. Franceschi, C. Franz, and W. Oleszek. 2007. Flavonoids in horse chestnut (*Aesculus hippocastanum*) seeds and powdered waste water byproducts. *J. Agric. Food Chem.* 55:8485–8490.

Kaul, D., A. R. Shukla, K. Sikand, and V. Dhawan. 2005. Effect of herbal polyphenols on atherogenic transcriptome. *Mol. Cell. Biochem.* 278:177–184.

Kavanaugh, C. J., P. R. Trumbo, and K. C. Ellwood. 2007. The U.S. Food and Drug Administration's evidence-based review for qualified health claims: Tomatoes, lycopene, and cancer. *J. Natl. Cancer Inst.* 99:1074–1085.

Khan, A., M. Safdar, M. M. Ali Khan, K. N. Khattak, and R. A. Anderson. 2003. Cinnamon improves glucose and lipids of people with type 2 diabetes. *Diabetes Care* 26:3215–3218.

Knox, J. and B. Gaster. 2007. Dietary supplements for the prevention and treatment of coronary artery disease. *J. Altern. Complement. Med.* 13:83–95.

Kokate, C. K., A. P. Purohit, and S. B. Gokhale. 1999. Resin and resin combinations. In *Pharmacognosy*. Pune, India: Nirali Prakashan; pp. 13, 455.

Krieglstein, C. F., C. Anthoni, E. J. Rijcken, M. Laukötter, H. U. Spiegel, S. E. Boden, S. Schweizer, H. Safayhi, N. Senninger, and G. Schürmann. 2001. Acetyl-11-keto-beta-boswellic acid, a constituent of a herbal medicine from *Boswellia serrata* resin, attenuates experimental ileitis. *Int. J. Colorectal Dis.* 16:88–95.

Lanski, S. L., M. Greenwald, A. Perkins, and H. K. Simon. 2003. Herbal therapy use in a pediatric emergency department population: Expect the unexpected. *Pediatrics* 111: 981–985.

Lee-Huang, S., P. L. Huang, H. C. Chen, P. L. Huang, A. Bourinbaiar, H. I. Huang, and H. F. Kung. 1995. Anti-HIV and anti-tumor activities of recombinant MAP30 from bitter melon. *Gene* 161:151–156.

Lei, X. L. and G. C. Y. Chiou. 1986a. Cardiovascular pharmacology of *Panax notoginseng* (Burk) F.H. Chen and *Salvia miltiorrhiza*. *Am. J. Chin. Med.* 14:145–152.

Lei, X. L. and G. C. Y. Chiou. 1986b. Studies on cardiovascular actions of *Salvia miltiorrhiza*. *Am. J. Chin. Med.* 14:26–32.

Lenz, T. L. and W. R. Hamilton. 2003. Supplemental products used for weight loss. *J. Am. Pharm. Assoc.* 44:59–67; quiz 67–68.

Liao, J. Z., J. J. Chen, Z. M. Wu, W. Q. Guo, L. Y. Zhao, L. M. Qin, S. R. Wang, and Y. R. Zhao. 1989. Clinical and experimental studies of coronary heart disease treated with yi-qi huoxue injection. *J. Trad. Chin. Med.* 9:193–198.

Liu, G. T., T.-M. Zhang, B.-E. Wang, and Y.-W. Wang. 1992. Protective action of seven natural phenolic compounds against peroxidative damage to biomembranes. *Biochem. Pharmacol.* 43:147–152.

Liu, J. P., M. Yang, and X. M. Du. 2004. Herbal medicines for viral myocarditis. *Cochrane Database Syst. Rev.* 2004:CD003711.

Loew, D., A. Schrödter, W. Schwankl, and R. W. März. 2000. Measurement of the bioavailability of aescin-containing extracts. *Methods Find. Exp. Clin. Pharmacol.* 22:537.

Loizzo, M. R., R. Tundis, F. Menichini, A. M. Saab, G. A. Statti, and F. Menichini. 2007. Cytotoxic activity of essential oils from labiatae and lauraceae families against *in vitro* human tumor models. *Anticancer Res.* 27:3293–3299.

Maizels, M., A. Blumenfeld, and R. Burchette. 2006. A combination of riboflavin, magnesium, and feverfew for migraine prophylaxis: A randomized trial. *Headache* 46:531.

Marinetz, D., K. Lohs, and J. Janzen. 1988. Weihrauch and myrrhe: Kulturgeschichtliche and wirtschaftl. *Bedeutung Botanik Chemir Wiss Vertges Stuttgast.* 153.

Mattarello, M. J., S. Benedini, C. Fiore, V. Camozzi, P. Sartorato, G. Luisetto, and D. Armanini. 2006. Effect of licorice on PTH levels in healthy women. *Steroids* 71:403–408.

McKay, D. L. and J. B. Blumberg. 2006. A review of the bioactivity and potential health benefits of chamomile tea (*Matricaria recutita* L.). *Phytother. Res.* 20:519–530.

McNulty, H., K. Pentieva, L. Hoey, and M. Ward. 2008. Homocysteine, B-vitamins and CVD. *Proc. Nutr. Soc.* 67:232–237.

Motai, T. and S. Kitanaka. 2005. Sesquiterpene chromones from Ferula fukanensis and their nitric oxide production inhibitory effects. *J. Nat. Prod.* 68:1732–1735.

Msaada, K., K. Hosni, M. B. Taarit, T. Chahed, and B. Marzouk. 2007. Variations in the essential oil composition from different parts of *Coriandrum sativum* L. cultivated in Tunisia. *Ital. J. Biochem.* 56:47–52.

Mukherjee, S., H. Gangopadhyay, and D. K. Das. 2008. Broccoli: A unique vegetable that protects mammalian hearts through the redox cycling of the thioredoxin superfamily. *J. Agric. Food Chem.* 56:609–617.

Murota, K., A. Hotta, H. Ido, Y. Kawai, J.-H. Moon, K. Sekido, H. Hayashi, T. Inakuma, and J. Terao. 2007. Antioxidant capacity of albumin-bound quercetin metabolites after onion consumption in humans. *J. Med. Invest.* 54:370–374.

Narender, T., S. Shweta, P. Tiwari, K. P. Reddy, T. Khaliq, P. Prathipati, A. Puri, A. K. Srivastava, R. Chander, S. C. Agarwal, and K. Raj. 2007. *Bioorg. Med. Chem. Lett.* 17:1808–1811.

Nies, L. K., A. A. Cymbala, S. L. Kasten, D. G. Lamprecht, and K. L. Olson. 2006. Complementary and alternative therapies for the management of dyslipidemia. *Ann. Pharmacother.* 40:1984–1992.

Nizamuddin, M., J. Hoffman, and L. Olle. 1982. Fractionation and characterization of carbohydrates from *Emblica officinalis* G fruit. *Chem. Abstr.* 96:196557m.

Norris, S. L., X. Zhang, A. Avenell, E. Gregg, C. H. Schmid, and J. Lau. 2005. Pharmacotherapy for weight loss in adults with type 2 diabetes mellitus. *Cochrane Database Syst. Rev.* 2005:CD004096.

Nunez-Selles, A. J., E. Paez-Betancourt, D. Amaro-González, J. Acosta-Esquijarosa, J. Aguero-Aguero, and R. Capote-Hernández. 2002. Oficina Cubana de la Propiedad Industrial. Cuba Patent 1814.

O'Hara, M., D. Kiefer, K. Farrell, and K. Kemper. 1998. A review of 12 commonly used medicinal herbs. *Arch. Fam. Med.* 7:523–536.

Okada, K., C. Wangpoengtrakul, T. Tanaka, S. Toyokuni, K. Uchida, and T. Osawa. 2001. Curcumin and especially tetrahydrocurcumin ameliorate oxidative stress-induced renal injury in mice. *J. Nutr.* 131:2090–2095.

Padin, C., G. Fernández-Zeppenfeldt, F. Yegres, and N. Richard-Yegres. 2005. *Scytalidium dimidiatum:* An opportunistic fungus for both man and Mangifera indica trees in Venezuela. *Rev. Iberoam. Micol.* 22:172–173.

Pan, M. H., M. C. Hsieh, P. C. Hsu, S. Y. Ho, C. S. Lai, H. Wu, S. Sang, and C. T. Ho. 2008. 6-Shogaol suppressed lipopolysaccharide-induced up-expression of iNOS and COX-2 in murine macrophages. *Mol. Nutr. Food Res.* 52:1467–1477.

Pardhy. R. S. and S. C. Bhattacharya. 1978. Tetracyclic triterpenic acids from the resin of *Boswellia serrata* Roxb. *Indian J. Chem.* 16B:174–175.

Phuwapraisirisan, P., T. Puksasook, J. Jong-Aramruang, and U. Kokpol. 2008. Phenylethyl cinnamides: a new series of alpha-glucosidase inhibitors from the leaves of *Aegle marmelos*. *Bioorg. Med. Chem. Lett.* 18:4956–4958.

Pitasawat, B., D. Champakaew, W. Choochote, A. Jitpakdi, U. Chaithong, D. Kanjanapothi, E. Rattanachanpichai, P. Tippawangkosol, D. Riyong, B. Tuetun, and D. Chaiyasit. 2007. Aromatic plant-derived essential oil: An alternative larvicide for mosquito control. *Fitoterapia* 78:205–210.

Pittler, M. H. and E. Ernst. 1998. Horse-chestnut seed extract for chronic venous insufficiency. A criteria-based systematic review. *Arch. Dermatol.* 134:1356–1360.

Pittler, M. H. and E. Ernst. 2004. Horse chestnut seed extract for chronic venous insufficiency. *Cochrane Database Syst. Rev.* 2004:CD003230.

Pittler, M. H. and E. Ernst. 2006. Horse chestnut seed extract for chronic venous insufficiency. *ACP J. Club* 145:20.

Pradeep, K. U. and P. Geervani. 1994. Influence of spices on protein utilisation of winged bean (*Psophocarpus tetragonolobus*) and horsegram (*Dolichos biflorus*). *Plant Foods Hum. Nutr.* 46:187–193.

Prakash, P. and N. Gupta. 2005. Therapeutic uses of *Ocimum sanctum* Linn (Tulsi) with a note on eugenol and its pharmacological actions: a short review. *Indian J. Physiol. Pharmacol.* 49:125–131.

Rahimi, A. R., M. Emad, and G. R. Rezaian. 2008. Smoke from leaves of *Populus euphratica* Olivier vs. conventional cryotherapy for the treatment of cutaneous warts: A pilot, randomized, single blind, prospective study. *Int. J. Dermatol.* 47:393–397.

Rajanikanth, R., B. Ravindranath, and M. L. Shankarananrayana. 1984. Volatile polysulphides of *Asa foetida*. *Phytochemistry* 23:899.

Rastogi, R. P. and B. N. Mehrotra. 1993. *Compendium of Indian Medicinal Plants*, Central Drug Research Institute. New Delhi, India: Lucknow and Publications and Information Directorate.

Rastogi, R. P. and B. N. Mehrotra. 1995. *Compendium of Indian Medicinal Plants*, Central Drug Research Institute. New Delhi, India: Lucknow and Publications and Information Directorate.

Rebultan, S. P. 1995. Bitter melon therapy: An experimental treatment of HIV infection. *AIDS Asia* 2:6–7.

Richard, C. L. and T. M. Jurgens. 2005. Effects of natural health products on blood pressure. *Ann. Pharmacother.* 39:712–720.

Rios, J. L., M. C. Recio, R. M. Giner, and S. Mañez. 1996. An update review of saffron and its active compounds. *Phytother. Res.* 10:189–193.

Rodriguez-Fragoso, L., J. Reyes-Esparza, S. W. Burchiel, D. Herrera-Ruiz, and E. Torres. 2008. Risks and benefits of commonly used herbal medicines in Mexico. *Toxicol. Appl. Pharmacol.* 227:125–135.

Rogers, P. L., H. S. Shin, and B. Wang. 1997. Biotransformation for L-ephedrine production. *Adv. Biochem. Eng. Biotechnol.* 56:33–59.

Saeed, S. A., R. M. Bloch, and D. J. Antonacci. 2007. Herbal and dietary supplements for treatment of anxiety disorders. *Am. Fam. Physician* 76:549

Saito, K., M. Kohno, F. Yoshizaki, and Y. Niwano. 2008. Extensive screening for edible herbal extracts with potent scavenging activity against superoxide anions. *Plant Foods Hum. Nutr.* 63:65–70.

Saxena, A. and N. K. Vikram. 2004. Role of selected Indian plants in management of type 2 diabetes: A review. *J. Complement. Altern. Med.* 10:369–378.

Schambelan, M. 1994. Licorice ingestion and blood pressure regulating hormones. *Steroids* 59:127.

Schelz, Z., J. Molnar, and J. Hohmann. 2006. Antimicrobial and antiplasmid activities of essential oils. *Fitoterapia* 77:279–285.

Schmidt, M., G. Betti, and A. Hensel. 2007. Saffron in phytotherapy: Pharmacology and clinical uses. *Wien Med. Wochenschr.* 157:315–319.

Schulze, J., W. Raasch, and C. P. Siegers, 2003. Toxicity of kava pyrones, drug safety and precautions: A case study. *Phytomedicine* 10 (Suppl. 4):68–73.

Sethi, J., S. Sood, S. Seth, and A. Talwar, 2004. Evaluation of hypoglycemic and antioxidant effect of *Ocimum sanctum*. *Indian J. Clin. Biochem.* 19:152–155.

Shao, Y., C. T. Ho, C. K. Chin, V. Badmaev, W. Ma, and M. T. Huang. 1998. Inhibitory activity of Boswellic acids from *B. serrata* against human leukemia HL-60 cells in culture. *Planta Med.* 64:328–331.

Sharma, M. L., S. Bani, and G. B. Singh. 1989. Anti-arthritic activity of boswellic acids in bovine serum albumin (BSA)-induced arthritis. *Int. J. Immunopharmacol.* 11: 647–652.

Shen, A. Y., T. S. Wang, M. H. Huang, C. H. Liao, S. J. Chen, and C. C. Lin. 2005. Antioxidant and antiplatelet effects of dang-gui-shao-yao-san on human blood cells. *Am. J. Chin. Med.* 33:747.

Shintani, S., H. Murase, H. Tsukagoshi, and T. Shiigai. 1992. Glycyrrhizin (licorice)-induced hypokalemic myopathy. Report of 2 cases and review of the literature. *Eur. Neurol.* 32:44–51.

Sigurjonsdottir, H. A., M. Axelson, G. Johannsson, K. Manhem, E. Nyström, and S. Wallerstedt. 2006. The liquorice effect on the RAAS differs between the genders. *Blood Press.* 15:169–172.

Silberstein, S. D. 2001. Migraine: Preventive treatment. *Curr. Med. Res. Opin.* 17 (Suppl 1): s1–s3.

Singh, G. B. and C. K. Atal. 1986. Pharmacology of an extract of Salai guggal ex-*Boswellia serrata*, a new non-steroidal anti-inflammatory agent. *Agents Actions* 18:407–412.

Singh, N. and M. Gupta. 2007. Regeneration of beta cells in islets of Langerhans of pancreas of alloxan diabetic rats by acetone extract of *Momordica charantia* (Linn.) (bitter gourd) fruits. *Indian J. Exp. Biol.* 45:1055–1062.

Singletary, K. W., K. J. Jung, and M. Giusti. 2007. Anthocyanin-rich grape extract blocks breast cell DNA damage. *J. Med. Food.* 10:244–251.

Sovová, M. and P. Sova. 2004. Pharmaceutical importance of *Allium sativum* L. 5. Hypolipemic effects in vitro and in vivo. *Ceska. Slov. Farm.* 53:117–123.

Sridhar, M. G., R. Vinayagamoorthi, V. Arul Suyambunathan, Z. Bobby, and N. Selvaraj. 2008. Bitter gourd (*Momordica charantia*) improves insulin sensitivity by increasing skeletal muscle insulin-stimulated IRS-1 tyrosine phosphorylation in high-fat-fed rats. *Br. J. Nutr.* 99:806–812.

Srinivasan, K. 2005. Plant foods in the management of diabetes mellitus: Spices as beneficial antidiabetic food adjuncts. *Int. J. Food Sci. Nutr.* 56:399–414.

Stickel, F. and D. Schuppan. 2007. Herbal medicine in the treatment of liver diseases. *Dig. Liver Dis.* 39:293–304.

Sumantran, V. N., A. Kulkarni, R. Chandwaskar, A. Harsulkar, B. Patwardhan, A. Chopra, and U. V. Wagh. 2008. Chondroprotective potential of fruit extracts of *Phyllanthus emblica* in osteoarthritis. *Evid Based Complement. Alternat. Med.* 5:329–335.

Takegoshi, K., T. Tohyama, K. Okuda, K. Suzuki, and G. Ohta. 1986. A case of Venoplant-induced hepatic injury. *Gastroenterol. Jpn.* 21:62–65.

Tamayo, D., E. Mari, S. González, M. Guevara, G. Garrido, and R. Delgado. 2001. Vimang as natural antioxidant supplementation in patients with malignanttumors. *Minerva Medica* 92:95–97.

Tan, M. J., J. M. Ye, N. Turner, C. Hohnen-Behrens, C. Q. Ke, C. P. Tang, T. Chen, H. C. Weiss, E. R. Gesing, A. Rowland, D. E. James, and Y. Ye. 2008. Antidiabetic activities of triterpenoids isolated from bitter melon associated with activation of the AMPK pathway. *Chem. Biol.* 15:263–273. (Erratum: *Chem. Biol.* 2008 15:520)

Tattelman, E. 2005. Health effects of garlic. *Am. Fam. Physician.* 72:103–106.

Terao, J., Y. Kawai, and K. Murato. 2008. Vegetable flavonoids and cardiovascular disease. *Asia Pac. J. Clin. Nutr.* 17 (Suppl. 1):291–293.

Thyagarajan, S. P., S. Jayaram, V. Gopalakrishnan, R. Hari, P. Jeyakumar, and M. S. Sripathi. 2002. Herbal medicines for liver diseases in India. *J. Gastroenterol. Hepatol.* 17 (Suppl. 3):S370.

Till, U., I. Bergmann, K. Breddin, S. Heptinstall, W. Lösche, A. Mazurov, and G. Pescarmona. 1989. Sulfhydryl/disulfide-status of blood platelets: A target for pharmacological intervention? *Prog. Clin. Biol. Res.* 301:341–345.

Udenigwe, C. C., V. R. Ramprasath, R. E. Aluko, and P. J. Jones. 2008. Potential of resveratrol in anticancer and anti-inflammatory therapy. *Nutr. Rev.* 66:445–454.

Uma Devi, P., A. Ganasoundari, B. S. Rao, and K. K. Srinivasan. 1999. In vivo radioprotection by ocimum flavonoids: Survival of mice. *Radiat. Res.* 151:74–78.

Umano, K., K. Nakahara, A. Shoji, and T. Shibamoto. 1999. Aroma chemicals isolated and identified from leaves of *Aloe arborescens* Mill. Var. Natalensis Berger. *J. Agric. Food Chem.* 47:3702–3705.

Uphof, J. C. T. 1959. *Dictionary of Economic Plants.* Weinheim, Germany: H.R. Engelman (J. Cramer).

Valli, G. and E.-G. Giardina. 2002. Benefits, adverse effects and drug interactions of herbal therapies with cardiovascular effects. *J. Am. Coll. Cardiol.* 39:7.

van der Watt, G., J. Laugharne, and A. Janca. 2008. Complementary and alternative medicine in the treatment of anxiety and depression. *Curr. Opin. Psychiatry* 21:37–42.

Verspohl, E. J. 2005 Treatment of diabetes mellitus. New developments and hopes. *Med. Monatsschr. Pharm.* 28:193–202.

Vijayakumar, S., G. Presannakumar, and N. R. Vijayalakshmi. 2008. Antioxidant activity of banana flavonoids. *Fitoterapia* 79:279–282.

Vopel, G., M. Gaisbauer, and W. Winkler. 1990. *Phytotherapie in der Praxis.* Köln, Germany: Deutscher Ärzteverlag, p. 74.

Warrier, P. K. 1995. *Indian Medicinal Plants.* Chennai, India: Orient Longman, p. 168.

Wigler. I., I. Grotto, D. Caspi, and M. Yaron. 2003. The effects of Zintona EC (a ginger extract) on symptomatic gonarthritis. *Osteoarthritis Cartilage* 11:783–789.

Wilkinson, J. A. and A. M. Brown. 1999. Horse chestnut—*Aesculus hippocastanum:* Potential applications in cosmetic skin-care products. *Int. J. Cosmet. Sci.* 21:437–447.

Winterhalter, P. and M. Straubinger. 2000. Saffron-renewed interest in an ancient spice. *Food Rev. Int.* 16:39–59.

Wu, C., F. Chen, X. Wang, Y. Wu, M. Dong, G. He, R. D. Galyean, L. He, and G. Huang. 2007. Identification of antioxidant phenolic compounds in feverfew (*Tanacetum parthenium*) by HPLC-ESI-MS/MS and NMR. *Phytochem. Anal.* 18:401–410.

Wu, T., J. Ni, and J. Wu. 2008. Danshen (Chinese medicinal herb) preparations for acute myocardial infarction. *Cochrane Database Syst. Rev.* 16:CD004465.

Yi, T., S. G. Cho, Z. Yi, X. Pang, M. Rodriguez, Y. Wang, G. Sethi, B. B. Aggarwalm, and M. Liu. 2008. Thymoquinone inhibits tumor angiogenesis and tumor growth through suppressing AKT and extracellular signal-regulated kinase signaling pathways. *Mol. Cancer Ther.* 7:1789–1796.

Yokozawa, T., H. Y. Kim, H. J. Kim, T. Okubo, D. C. Chu, and L. R. Juneja. 2007. Amla (*Emblica officinalis* Gaertn.) prevents dyslipidaemia and oxidative stress in the ageing process. *Br. J. Nutr.* 97:1187–1195.

Zhou, L., Z. Zuo, and M. S. Chow. 2005. Danshen: An overview of its chemistry, pharmacology, pharmacokinetics, and clinical use. *J. Clin. Pharmacol.* 45:1345–1359.

CHAPTER 4

Nutraceuticals with Animal Origin

Raghunandan Yendapally

CONTENTS

INTRODUCTION

This chapter outlines the nutraceuticals derived from animal origin. Nutraceuticals derived from animals include, but are not limited to, fish oils, conjugated linoleic acids (CLAs), choline, chitin and chitosan, glucosamine, chondroitin, and L-carnitine. Nutraceuticals that are obtained from animals have a gamut of applications. Typically, they have beneficial effects relating to cardiovascular diseases, inflammation, tumors, obesity, joint pains, diabetes, convulsions, and hypercholesterolemia. They are commonly formulated as soft-gel capsules, tablets, and powders.

OMEGA-3 FATTY ACIDS FROM FISH

Typical Properties and Description

Omega-3 fatty acids are the essential fatty acids, meaning the fatty acids that cannot be biosynthesized in the human body. Consequently, they must be obtained by supplementation in diet. Fish oils and cold-water fish such as tuna, salmon, catfish, sardines, and mackerel are great sources of omega-3 fatty acids. Omega-3 fatty acids from the fish primarily contain eicosapentaenoic acid (EPA) and docosahexaenoic acid (DHA).

Structural Formula

See Figures 4.1 and 4.2 for the structural formula.

Functional Category

Fatty acids are long-chain aliphatic carboxylic acid groups. These aliphatic systems are either saturated or unsaturated. Typically, fatty acids contain an even number of carbon atoms. Omega-3 fatty acids are unsaturated fatty acids, and the first carbon-carbon double bond (–C=C) is present on the third carbon from the terminal methyl group ($-CH_3$). The chain length in fish oils consist of either 20 carbons, as in EPA, or 22 carbons, as in DHA.

Applications in Nutraceuticals

Omega-3 fatty acids have a wide variety of applications in nutraceuticals. Omega-3 fatty acids are recommended by several scientific organizations, such as the American Heart Association and the European Society for Cardiology, to prevent or reduce the risk of cardiovascular diseases [Harris 2007]. In a study conducted by Gruppo Italiano per lo Studio della Sopravvivenza nell'Infarto-Prevenzione investigators, the supplementation of omega-3 fatty acids after myocardial infarction has led to the reduction in rate of death, nonfatal myocardial infarction and stroke [Gruppo Italiano per lo Studio della Sopravvivenza nell'Infarto-Prevenzione Investigators 1999]. Omacor, a drug composed of 85% of omega-3 fatty acid ethyl esters, when administered, has significantly lowered triglycerides, very-low-density lipoprotein (VLDL) cholesterol and increased high-density lipoprotein (HDL) cholesterol in people suffering from hypertriglyceridemia [Harris et al. 1997]. High doses of omega-3 fatty acids at 3 g/day have significantly reduced the blood pressure in older (> 45 years of age) and hypertensive patients [Geleijnse et al. 2002].

Maxepa, a drug composed of EPA (171 mg/capsule) and DHA (114 mg/capsule), when administered, has reduced the requirement of nonsteroidal anti-inflammatory

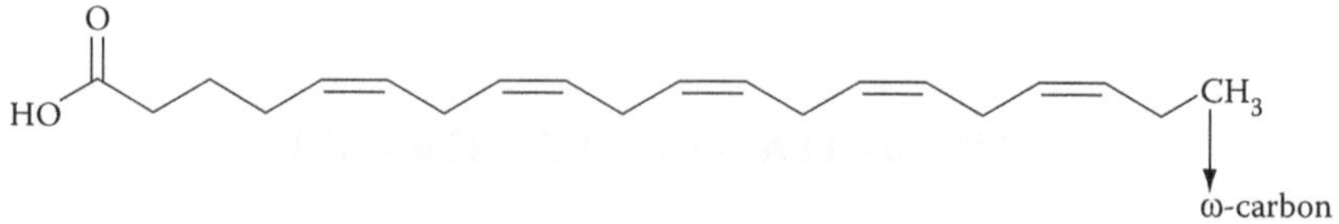

Figure 4.1 The structure of EPA.

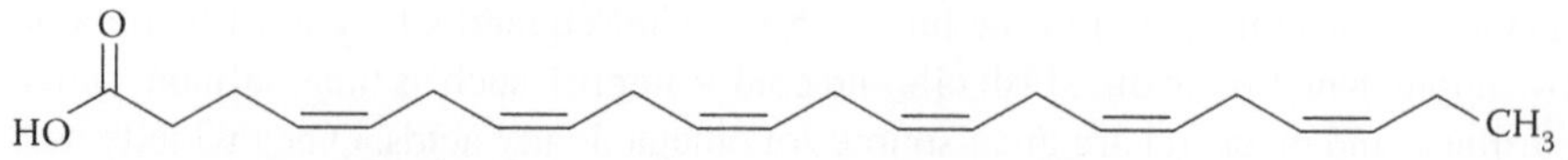

Figure 4.2 The structure of DHA.

drugs (NSAIDs) for the treatment of rheumatoid arthritis (RA), indicating the potential application of omega-3 fatty acids in the treatment of RA. However, the maximum benefit was observed after 11 months, implying the need for long-term treatment [Lau, Morley, and Belch 1993].

Stability and Storage Conditions

Omega-3 fatty acids contain double bonds, which makes them unstable and prone to oxidation. The unique chemical structure causes rapid deterioration during handling and storage. A chemical quality study was conducted using the fish oils extracted from shad, horse mackerel, garfish, and golden mullet at different temperatures and time periods. All these oils had acceptable characteristics for 90 days when stored at 4°C. The acceptable characteristics for these oils (except shad oil) were increased from 90 to 150 days when the temperature was decreased from 4 to –18°C. However, the acceptable characteristics for the shad oil were found to be only 120 days when stored at –18°C [Borana, Karaçam, and Boran 2006].

Interactions

In a case study, it was observed that the international normalized ratio (INR) significantly increased when fish oil dose was doubled from 1 to 2 g/day during anticoagulation therapy with warfarin. This indicates that omega-3 fatty acids when administered along with warfarin may increase the risk of bleeding [Buckley, Goff, and Knapp 2004]. However, large clinical studies have to be conducted to critically examine the anticoagulant effect of omega-3 fatty acids and warfarin. Patients who are under warfarin therapy may be supplemented with safe doses of omega-3 fatty acids only under medical supervision.

Method of Manufacture

Omega-3 fatty acids obtained from fish have a strong odor and are unstable in the atmosphere. To overcome these problems, fish oils are usually manufactured by micro-encapsulation. Many of the fatty acids are marketed in the form of soft-shell capsules and oils.

Safety

Industrial contaminants and pesticide residues lead to contamination of dietary fish oil supplements. It has been reported that some of the fish oils that are used as dietary supplements contain high levels of organochlorine residues [Jacobs et al. 1998]. Therefore, during the usual manufacturing process, the chemical residual levels are checked before marketing these supplements. Cod liver contains high amounts of vitamins A and D apart from omega-3 fatty acids. Therefore, vitamins A and D should be consumed only in recommended doses during the concurrent supplementation of cod liver oil to avoid vitamin toxicities.

Handling Precautions

Because fish oils are prone to oxidation, special handling precautions are taken to minimize their exposure to atmosphere, and they are always stored under low temperatures. In fact, some of the fish products are incorporated with antioxidants.

Regulatory Status

In 2004, the FDA approved soft-gel capsules of omacor for the treatment of elevated blood triglyceride levels. Typically, 1 g of omacor capsule contains approximately 465 mg of EPA ethyl ester and 375 mg of DHA ethyl ester [FDA 2004]. Recently, as per the FDA's request in response to dispensing and prescribing errors attributable to close similarity in names between omacor and amicar (an antifibrinolytic), the reliant pharmaceuticals changed the name of omacar to lovaza (Reliant Pharmaceuticals Inc.).

Related Substances

Omega-3 fatty acids are also obtained from plants. For example, alpha-linoleic acid (ALA), an omega-3 fatty acid, is predominantly present in vegetable oils such as soybean and canola oil. Other sources of ALA include nuts, seeds, legumes, vegetables, fruits, and grains. Omega-3 fatty acids are related to other polyunsaturated systems such as omega-6 fatty acids, in which the first carbon-carbon double bond (–C=C) is present on the sixth carbon from the terminal methyl group ($-CH_3$). Linoleic acid and arachidonic acid are the typical examples of omega-6 fatty acids.

CONJUGATED LINOLEIC ACIDS

Typical Properties and Description

CLAs were first identified by Pariza and Hargraves [1985] while investigating the extracts from ground beef. Other sources of CLAs include lamb and dairy products obtained from ruminants [Lin et al. 1995].

Structural Formula

See Figures 4.3 and 4.4 for the structural formula.

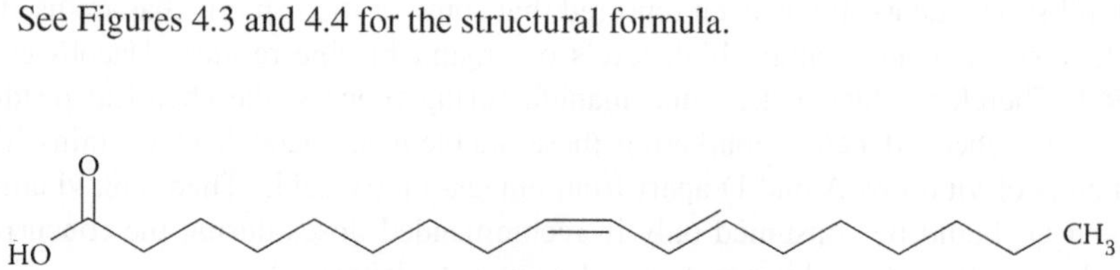

O
HO
CH_3

Figure 4.3 The structure of conjugated (9Z,11E)-linoleic acid.

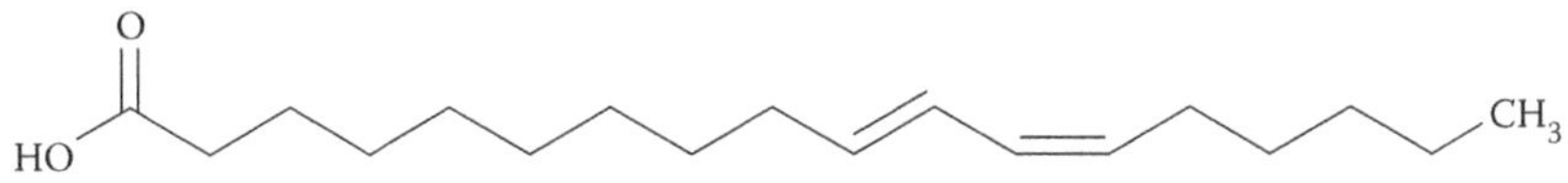

Figure 4.4 The structure of conjugated (10E,12Z)-linoleic acid.

Functional Category

CLAs are a group of naturally occurring stereoisomers and positional isomers of ocatadecadienoic acid, a polyunsaturated fatty acid. The term conjugated refers to the presence of alternating single and double bonds. These molecules exhibit *cis* or *trans* stereoisomerism based on the orientation of the functional groups attached to the double bonds. They also exhibit positional isomerism based on the location of the double bonds. In most of the CLAs, the double bonds are located at C8 and C10, C9 and C11, C10 and C12, or C11 and C13 [Bhattacharya et al. 2006]. However, the 9-*cis*, 11-*trans* CLA is the major isomer present in the food materials.

Applications in Nutraceuticals

CLAs have a wide array of beneficial effects on health [Bhattacharya et al. 2006]. The anticarcinogenic properties of CLAs were first discovered from fried ground beef extracts [Pariza and Hargraves 1985]. Although the individual CLA isomers were not tested for anticarcinogenic properties in humans, based on the data that were obtained from animals, there is a speculation that various isomers of CLA may regulate tumor growth by different mechanisms [Bhattacharya et al. 2006].

In a recent meta-analysis in humans, it was demonstrated that CLAs, when administered at 3.2 g/day, resulted in reduction of fat mass [Whigham, Watras, and Schoeller 2007]. CLAs have been shown to reduce the body weight by decreasing the body fat and increasing the lean mass in mice [Park et al. 1997]. Furthermore, it was also identified that 10-*trans*, 12-*cis* isomer is responsible for decreasing body fats [Navarro et al. 2006; Silveira et al. 2007] by inhibiting the functions of lipoprotein lipase and stearoyl-coenzyme A (CoA) desaturase and thereby resulting in reduction of deposition of lipids in adipocytes [Pariza, Park, and Cook 2001].

In a study on healthy humans, a dietary supplement of the CLA isomers 9-*cis*, 11-*trans* CLA and 10-*trans*, 12-*cis* CLA has been shown to reduce mitogen-induced T-cell lymphocyte activation [Albers et al. 2003]. However, these two isomers had no effect on lymphocyte subpopulations, cytokine production, or serum concentration of C-reactive protein [Albers et al. 2003]. In another study, administration of 3 g of soft-gel capsules of CLAs, 1:1 mixture of 9, 11 isomer and 10, 12 isomer, for 12 weeks resulted in decreased levels of the proinflammatory cytokines, TNF-α and IL-1β and increased the levels of anti-inflammatory cytokines, such as IL-10 [Song et al. 2005]. These results indicate that CLA may be useful in increasing the immunity in humans [Song et al. 2005].

CLAs reduce the levels of leptin, an adipose hormone that suppresses bone formation [Yamasaki et al. 2000]. Banu et al. [2006] have studied the effects of dietary supplementation of CLAs in BALB/c mice. It was shown that CLAs have a positive effect on bone mass in cancellous and cortical bones of the proximal tibial metaphysis, cortical bones of the tibia fibular junction.

Dietary supplementation of CLAs in combination with calcium during pregnancy has been shown to decrease pregnancy-induced hypertension [Herrera et al. 2005, 2006]. CLAs supplementation was shown to reduce hypertension and hyperinsulinemia in Zucker diabetic fatty [Nagao et al. 2003] rats, as well as primary hypertension in hypertensive rats [Inoue et al. 2004]. CLAs were also shown to exhibit antiatherogenic properties in rabbits [Kritchevsky et al. 2000]. CLAs, when administered as dietary supplements in rats, reduced the cardiac myocyte hypertrophy by activating peroxisome proliferator-activated receptors α and γ [Alibin, Kopilas, and Anderson 2008].

Stability and Storage Conditions

CLAs are prone to epoxidation when exposed to oxygen because of the double bonds. Many experiments were conducted to study the effect of conjugation in CLAs and auto oxidation. The free form CLAs and triacyl glycerols forms of CLAs are extremely unstable in air at 90°C, and the relative rate of oxidation is similar to DHA. However, CLAs are oxidized at a much faster rate than linoleic acid probably because of the conjugated double bonds [Zhang and Chen 1997].

The rate of oxidations is also different for different isomers. For example, it was found that rate of oxidation is much faster in 10-*trans*, 12-*cis* CLA isomer than in 9-*cis*, 11-*trans* CLA isomer [Minemoto et al. 2003]. Furthermore, the CLA isomers have different oxyradical scavenging capacity at different concentrations. For example, 10-*trans*, 12-*cis* CLA at 2–200 μM displays antioxidant properties, whereas 9-*cis*, 11-*trans* CLA at lower concentration (2 and 20 μM) exhibits antioxidant properties and at higher concentrations (200 μM) it demonstrates pro-oxidant properties [Leung and Liu 2000].

Interactions

A study conducted by the United States Department of Agriculture found that female mice, when fed with 10-*trans*, 12-*cis* isomer, reduced the concentrations of omega-3 fatty acids and omega-6 fatty acids in the liver by greater than 50% and reduced the concentrations of omega-3 fatty acids in the heart by 25%. It has also increased the concentrations of omega-3 fatty acids in the spleen by 700% [Kelley et al. 2006].

Method of Manufacture

CLAs are manufactured in the form of soft-gel capsules and tablets.

Safety

Maternal supplementation of CLAs has decreased milk fat and increased CLA milk concentrations in humans [Masters et al. 2002]. Therefore, it is recommended that CLAs should not be consumed by lactating women [Masters et al. 2002]. It was also observed that supplementation of 10-*trans*, 12-*cis* CLA has induced the oxidative stress and increased very-low-density lipoprotein levels in men with metabolic syndrome [Riserus et al. 2002]. More detailed information on the safety of CLAs have been discussed in a review published previously [Pariza 2004].

Handling Precautions

Because CLAs are prone to oxidation, meticulous handling precautions are taken during the manufacturing and storage process to minimize their exposure to atmosphere.

Regulatory Status

In the United States, CLA has gained GRAS status. However, according to the French food authority, the Agence Française de Sécurité Sanitaire des Aliments, the addition of certain CLA isomers in the form of supplements or food ingredients are not justified.

Related Substances

Chemically, CLAs are related to other polyunsaturated systems. These systems are conjugated containing three conjugated double bonds or unconjugated as in omega-3 and omega-6 fatty acids.

L-CARNITINE

Typical Properties and Description

Carnitine is biosynthesized mainly in the liver and kidneys from the two amino acids lysine and methionine [Rapport and Lockwood 2000]. Carnitine is also known as vitamin B_T. However, it is not categorized as a true vitamin because it can be synthesized in the body. Carnitine and carnitine transporters primarily help in the transfer of fatty acids from the cytosol into mitochondrial matrix. In the mitochondrial matrix, the fatty acid oxidation takes place, which serves as the major source of energy during exercise [Wasserman and Whipp 1975]. Carnitine is present in good quantities in animal sources such as steak, beef, chicken, and eggs [Erfle, Fisher, and Sauer 1970; Rudman, Sewell, and Ansley 1977]. Apart from animal sources, carnitine is also present in plant products such as grains, fruits, and vegetables [Rudman, Sewell, and Ansley 1977].

Functional Category

Chemically, carnitine is β-hydroxy-γ-trimethylaminobutyric acid. Carnitine has one stereocenter and therefore exists as two enantiomers: D and L forms. However, the L form is the naturally occurring enantiomer and it has R-absolute configuration at the 3 position. The L form of carnitine is biologically and physiologically active. The D form of carnitine is physiologically inactive, inhibits the uptake of L-carnitine, and is toxic. Therefore, carnitine must be supplemented only in the L form.

Structural Formula

See Figure 4.5 for the structural formula.

Applications in Nutraceuticals

Adult diet from meat and dairy products provides about 75% of carnitine daily requirements [Rebouche 2004]. Therefore, carnitine supplementation may not be required under usual conditions. Even in people who follow strict vegetarianism, the carnitine levels are normal, indicating the effectiveness of carnitine biosynthesis and carnitine renal reabsorption [Rebouche 2004]. Carnitine supplementation has beneficial effects in people suffering from carnitine deficiency, which arises as a result of disorders of carnitine synthesis and carnitine transport across the mitochondrial membranes [Long, Amat di San Filippo, and Pasquali 2006]. Carnitine is supplemented under different forms, such as free carnitine, acetyl-L-carnitine, and propionyl-L-carnitine.

Acetyl-L-carnitine supplementation decreases the serum ammonia levels and enhances neurophysiological functions during the treatment of minimal hepatic encephalopathy [Malaguarnera et al. 2008]. Carnitine has a beneficial effect in the treatment of valproic acid, a broad-spectrum anticonvulsant, toxicity [Carcione et al. 1991] by reducing hyperammonemia caused by valproic acid poisoning [Chan, Tse, and Lau 2007].

Carnitine deficiency in hemodialysis patients is caused by the loss of carnitine during dialysis and insufficient carnitine synthesis [Bohmer, Bergrem, and Eiklid 1978]. In a study, it was demonstrated that intravenous administration of carnitine in subjects undergoing hemodialysis had a significant improvement in lipid metabolism, red blood cell count, antioxidant abilities, and protein nutrition [Vesela et al. 2001]. L-carnitine supplementation has shown to decrease the left ventricular hypertrophy [Sakurabayashi et al. 2008], an independent predicator of cardiac mortality in patients undergoing hemodialysis [Silberberg et al. 1989].

A review of clinical trials indicated that supplementation of carnitine did not improve the exercise capacity in healthy individuals; however, it is speculated that it may improve the exercise capacity in patients suffering from renal disease and peripheral arterial disease [Brass and Hiatt 1998]. It was also demonstrated that L-carnitine administration has hepatic and cardiovascular antioxidant properties in hypertensive rats [Gómez-Amores et al. 2007].

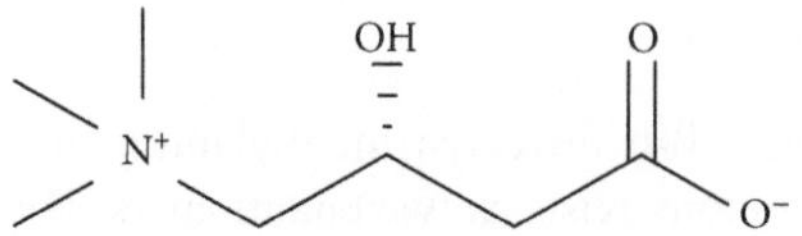

OH O N$^+$ O$^-$

Figure 4.5 The structure of L-carnitine.

L-carnitine, when administered as a continuous infusion, significantly improved the insulin resistance in type II diabetic patients, implying the potential application of L-carnitine in the treatment of diabetes [Mingrone et al. 1999].

Stability and Storage Conditions

Carnitine is stable at room temperature and should be stored away from light to avoid any chemical degradation.

Incompatibilities Known

L-carnitine levels were significantly reduced during the treatment of convulsions with phenobarbital, valproic acid, phenytoin, and carbamazepine [Hug et al. 1991].

Method of Manufacture

L-carnitine is obtained by chemical synthesis coupled with stereochemical biotransformation using bacterial culture and is sold under the trademarked name L-Carnipure [Held and Siebrecht 2003]. L-carnitine and its acetyl and propionyl derivatives are manufactured in the form of tablets, capsules, or powders.

Safety

In general, L-carnitine and its derivatives are considered to be safe supplements. According to the observed safe level (OSL) risk assessment method, L-carnitine is deemed safe at an intake of up to 2000 mg/day for chronic supplementation [Hathcock and Shao 2006]. However, appropriate care has to be taken in patients with low levels of thyroid because it has been reported that L-carnitine inhibits thyroid hormone nuclear uptake [Benvenga, Lakshmanan, and Trimarchi 2000]. Acetyl-L-carnitine may cause gastrointestinal disturbances and should be used under medical supervision in patients suffering from seizure disorders.

Handling Precautions

Carnitine does not need any special handling precautions because of its excellent stability. In fact, carnitine is heat stable up to 120°C, and, even in baking trials, it has minimal degradation. Therefore, carnitine can be added to a variety of food products, including but not limited to bars, cereals, chocolates, bread, and biscuits [Held 2004].

Regulatory Status

In the United States, the two forms of carnitine, L-carnitine crystalline and L-carnitine L-tartrate, hold GRAS status when used at prescribed levels [Held and Siebrecht 2003]. Carnitine was approved by the FDA as an orphan drug for the treatment of pediatric rare and serious diseases.

CHONDROITIN

Typical Properties and Description

Chondroitin is an integral component that is biosynthesized by specialized cells referred to as chondrocytes. Chondroitin provides the cushioning effect to the joints. Despite that chondroitin is synthesized in normal individuals, supplementation of chondroitin is necessary in deficient individuals. Chondroitin deficiency can be caused by external injury, age, or arthritis. Externally, chondroitin is usually obtained from bovine cartilage, bovine trachea, and shark cartilage.

Structural Formula

See Figure 4.6 for the structural formula.

Functional Category

Chondroitin is composed of heteropolysaccharides, known as glycosaminoglycans (GAGs). GAGs are polymers composed of repeated disaccharide units. Chondroitin chain typically contains 60 disaccharides, which is made up of alternating *N*-acetylgalactosamine and D-glucouronic acid. Chondroitin is often supplemented in the sulfate form, and these sulfate groups are located at either the 4 position or the 6 position in the cyclic ring system. The carboxylic acid and the sulfate groups present in chondroitin impart negative charge to the molecule.

Applications in Nutraceuticals

Osteoarthritis is typically characterized by excruciating pain and inflammation in joints caused by of the wearing of cartilage. Pharmacological intervention for osteoarthritis typically consists of analgesics and NSAIDs [Tamblyn et al. 1997]. However, these drugs typically alleviate the symptoms but do not replenish the cartilage, which is the underlying problem. As a result, several studies were conducted by

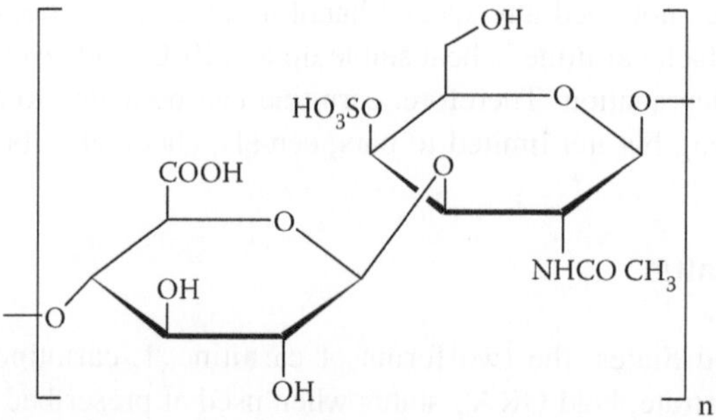

Figure 4.6 The structure of chondroitin 4-sulfate.

supplementing chondroitin externally. However, the interpretations of these results vary greatly. In a series of clinical studies, chondroitin was widely interpreted as a very useful agent for the treatment of osteoarthritis [Bourgeois et al. 1998; Bucsi and Poor 1998; Uebelhart et al. 1998]. The efficacy of glucosamine and chondroitin was reviewed for the treatment of osteoarthritis, and it was concluded that these agents may have some degree of efficacy; however, previous publications might have exaggerated their effects [McAlindon et al. 2000]. In a recent systematic review and meta-analysis of available randomized, controlled studies, it was observed that the benefit attributable to supplementation of chondroitin is minimal or nonexistent, and therefore it has been suggested that the usage of chondroitin should be discouraged [Reichenbach et al. 2007]. However, the methodology of this review was further questioned [Goldberg, Avins, and Bent 2007]. In conclusion, based on the several published studies, chondroitin may be beneficial to a certain extent in osteoarthritis. Additional concrete studies are absolutely necessary to unequivocally claim the beneficial effects of chondroitin for the treatment of osteoarthritis.

Stability and Storage Conditions

Chondroitin and glucosamine tablets, when manufactured and stored at 25°C and modest humidity (15%), were found to retain full potency and quality for a period of two years after the packing. This indicates that these compounds are quite stable [Kennedy et al. 2006].

Interactions

Chondroitin may interact with warfarin and may potentiate the effect of warfarin [Rozenfeld, Crain, and Callahan 2004]. This effect is monitored by increased INR or increased bleeding or bruising [Knudsen and Sokol 2008]. Therefore, patients are required to inform their healthcare provider if they are using these medications in combination.

Method of Manufacture

Chondroitin is manufactured in the form of tablets, capsules, or powders.

Safety

According to the OSL risk assessment method, chondroitin is considered safe at an intake of up to 1200 mg/day [Hathcock and Shao 2007]. It is also suggested that there may be a link between chondroitin and glucosamine supplements and asthma exacerbations [Tallia and Cardone 2002]. Therefore, the patients suffering from asthma should seek medical advice before taking these medications. On March 19, 2008, the FDA identified that heparin, a blood-thinning drug, was being contaminated with an over-sulfated chondroitin sulfate, which is responsible for serious adverse reactions [Food and Drug Administration 2008]. However, chondroitin sulfate (natural origin) should not be mistaken with over-sulfated chondroitin sulfate (synthetic origin).

Regulatory Status

In the United States, chondroitin is available as a dietary supplement and regulated by the Dietary Supplement Health and Education Act of 1994. Recently, the bovine-based OptaFlex™ Chondroitin, manufactured by Cargill, holds a self-affirmed GRAS status [Cargill Health and Food Technologies 2004]. In Europe, chondroitin is approved for the treatment of symptomatic slow-acting drug for osteoarthritis (SYSADOA).

Related Substances

Hyaluronate is the other major heteropolysaccharide containing GAGs. It is an essential component of cartilage and ligaments providing strength and elasticity. In addition, it serves as a lubricant in the synovial fluid joints. Like chondroitin, hyaluronate is composed of D-glucouronic acid and *N*-acetylglucosamine. However, it differs from chondroitin by the presence of about 50,000 disaccharide units per chain.

GLUCOSAMINE

Typical Properties and Description

Glucosamine is found naturally in the body, especially in cartilage, tendons, and ligament tissues. Glucosamine supplements are usually obtained from the shells of shrimps, crabs, and other crustacean exoskeleton. In addition, glucosamine is derived from the fungus [Cargill Health and Food Technologies 2007].

Structural Formula

See Figure 4.7 for the structural formula.

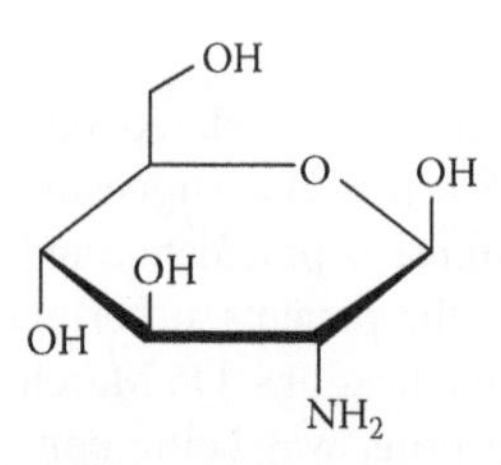

Figure 4.7 The structure of β-D-glucosamine.

Functional Category

Glucosamine is a monosaccharide and hexose derivative containing an amino group at C-2 position. Glucosamine-6-phosphate is biosynthesized from fructose-6-phosphate and amino acid glutamine. Glucosamine is a precursor for the biosynthesis of GAGs and proteoglycans. All the nitrogen-containing sugars obtain nitrogen atom from glucosamine-6-phosphate [Ghosh et al. 1960]. Several scientists believe that glucosamine is the most important substance in the formation of cartilage [Andersen 1998a].

Applications in Nutraceuticals

In a long-term, randomized, placebo-controlled, double-blind clinical trial study, supplementation of glucosamine sulfate has delayed the advancement of knee osteoarthritis in terms of joints structure changes and symptoms [Reginster et al. 2001]. Another study, involving 202 patients, demonstrated that oral administration of glucosamine sulfate, at 1,500 mg once a day, significantly retarded the progression of knee osteoarthritis, possibly by disease modification [Pavelka et al. 2002]. However, in a recent randomized trial for a period of 18 months, it was found that supplementation of glucosamine for the treatment of osteoarthritis combined with exercise did not have any significant additive effect when compared with the exercise itself [Kawasaki et al. 2008].

Combination of chondroitin and glucosamine sulfate demonstrated a beneficial effect in osteoarthritis pathophysiology by decreasing pro-resorptive properties of human osteoarthritis subchondral bone structural changes [Tate et al. 2007]. A recent review suggested that chondroitin and glucosamine sulfate may have symptomatic efficacy in moderate arthritis, probably by interfering with the disease progression [Bruyere and Reginster 2007]. In a study involving 1,583 patients, it was suggested that glucosamine and chondroitin sulfate in combination may be effective in a subgroup of patients with moderate to severe knee pain. However, it did not reduce pain effectively in the overall group of patients with osteoarthritis of the knee [Clegg et al. 2006].

Interactions

Similar to chondroitin, glucosamine may interact with warfarin and may potentiate the effect of warfarin [Knudsen and Sokol 2008]. Therefore, the patients are required to inform the healthcare provider if they are using these medications in combination.

Method of Manufacture

Glucosamine is manufactured in the form of tablets, capsules, or powders. Glucosamine is often sold in combination with chondroitin. Glucosamine is supplemented in various forms, such as glucosamine sulfate (stabilized with sodium chloride or potassium chloride), *N*-acetyl glucosamine, and glucosamine hydrochloride [Andersen 1998b]. Special formulations of glucosamine are also used for other purposes. For example, glucosamine-carrying temperature- and pH-sensitive microgels may be useful in targeted drug delivery to tumors [Teng et al. 2008].

Safety

According to the OSL risk assessment method, glucosamine is considered as safe at intake of up to 2,000 mg/day [Hathcock and Shao 2007]. Patients suffering from asthma should seek medical advice before taking these medications. It is indicated that glucosamine may worsen insulin resistance in patients with underlying poorer insulin sensitivity [Pham et al. 2007]. Therefore, diabetic patients should take medical advice before using glucosamine.

Regulatory Status

In the United States, the animal form of glucosamine does not hold GRAS status. However, GRAS has recently recognized the fungus-derived glucosamine as safe. In Europe, glucosamine is used for the treatment of SYSADOA.

Related Substances

Glucosamine is structurally related to other hexose amino sugars such as galactosamine and mannosamine. Glucosamine is also the precursor for the biosynthesis of *N*-acetylmuramic acid, which is an integral component of bacterial cell wall.

CHITIN AND CHITOSAN

Typical Properties and Description

Chitin is obtained from the hard exoskeleton of shrimps, crabs, lobsters, and insects. Unlike glucosamine, chitin is not biosynthesized in the human body and cannot be digested by vertebrates.

Structural Formula

See Figure 4.8 for the structural formula.

Functional Category

Chitin is a homopolysaccharide containing several units of *N*-acetylglucosamine units in β linkage. Chitosan is chemically derived from chitin by *N*-deacetylation reaction. Therefore, chitosan is also referred to as deacetylated chitin.

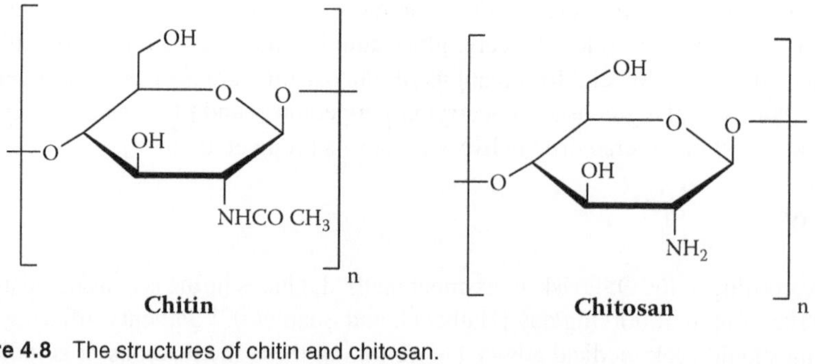

Figure 4.8 The structures of chitin and chitosan.

Applications in Nutraceuticals

Several studies indicate that chitosan has a beneficial effect in weight reduction and cholesterol management [Shields et al. 2003]. Because chitosan has an amino group at C-2, it is ionized into positive charge at gastrointestinal pH, and these cationic groups are believed to interact with the anionic groups of lipids and bile, preventing their absorption and storage by the body [Shields et al. 2003].

In a randomized, double-blind, placebo-controlled dietary intervention study, involving 134 overweight adults, it was found that the chitosan treatment group lost more weight than the placebo and control groups. In this study, chitosan also facilitated the depletion of excess body fat under free-living conditions with minimal loss of fat-free or lean body mass [Kaats, Michalek, and Preuss 2006].

In a multicenter, placebo-controlled, randomized study, eligible patients were treated with HEP-40 low-molecular-weight chitosan at daily doses of 1,200, 1,600, or 2,400 mg or placebo for 12 weeks. LDL cholesterol concentrations were reduced to some extent in patients treated with chitosan, suggesting that chitosan may be beneficial in patients with low to moderate hypercholesterolemia [Jaffer and Sampalis 2007]. However, chitosan was not as effective as statins, the class of drugs commonly used in lowering cholesterol levels [Jaffer and Sampalis 2007].

Chitosan sulfate has strong anticoagulant properties [Bourin and Lindahl 1993]. However, chitosan as such does not have any anticoagulant activity. It was reported that chitosan sulfate prolongs the activated partial thromboplastin time and thrombin time, the two assays involved in measuring the effectiveness of coagulation cascade [Hirano et al. 1985].

Stability and Storage Conditions

Chitin is a highly stable molecule at room temperature. Chitin by itself serves as a protective layer in animals and insects and shields them against pressure variations and changes in external environment. Therefore, chitin does not need any special storage conditions, except it should be stored away from chemicals.

Interactions Known

It was reported recently that chitosan potentiated the warfarin anticoagulant effect probably by binding to the lipids of the intestine and decreasing the absorption of vitamin K.

Method of Manufacture

Chitin and chitosan are manufactured in the form of tablets, capsules, or powders. Chitin and chitosan are also manufactured for other applications attributable to their disintegrant, dissolution, adhesive, and mechanical properties [Schneider et al. 2007].

Safety

Chitin, when given to mice, induces the tissue accumulation of IL-4, eosinophils, and basophils, a characteristic feature of allergic and parasitic immunity [Reese et al. 2007]. This indicates that chitin and chitosan should be used with caution in patients suffering from asthma and other allergic diseases.

Regulatory Status

The FDA has recently approved Celox, a proprietary product of chitosan, sold in the form of free granules or granules in soluble bag. It is used as a hemostat in emergencies for the temporary control of severe topical bleeding [Food and Drug Administration 2007].

Related Substances

Chitin has close structural similarity to cellulose, the most abundant polysaccharide in nature. However, it differs from cellulose by the presence of acetamide functionality instead of hydroxyl group at the C-2 position.

CHOLINE

Typical Properties and Description

Choline is biosynthesized in the human body by sequential phosphatidylethanolamine methylation pathway [Bremer and Greenberg 1961]. There are many studies that indicate choline as an essential nutrient in humans. Choline and its derivatives are an important source of methyl groups, essential in the synthesis of acetylcholine (cholinergic neurotransmitter) and membrane phospholipids [Zeisel and Blusztajn 1994]. Choline and its derivatives are present in rich quantities in egg yolks, chicken, beef, pork, turkey, veal, and lamb legumes [Zeisel et al. 2003].

Structural Formula

See Figure 4.9 for the structural formula.

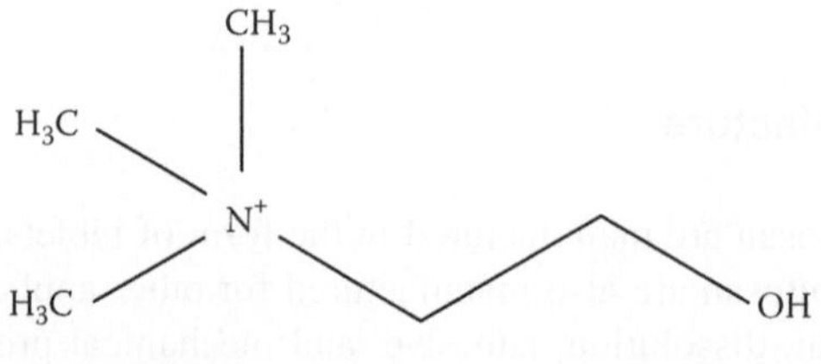

CH_3
H_3C
N^+
H_3C
OH

Figure 4.9 The structure of choline.

Functional Category

Choline is a saturated quaternary trimethyl ammonium compound. Choline is used as a precursor for the synthesis of phosphatidylcholine, a class of phospholipids.

Applications in Nutraceuticals

Alzheimer's disease (AD) is typically characterized by the low levels of acetylcholine. Therefore, scientists have studied the effects of supplementation choline on AD. In a multicenter, double-blind, randomized, placebo-controlled trial, it was found that the group receiving choline alfoscerate (glycerylphosphorylcholine), 400 mg capsules three times a day, had a cognitive improvement in mild-to-moderate AD when compared with the group receiving placebo [De Jesus Moreno Moreno 2003]. However, a Cochrane review of 12 randomized trials found that supplementation of lecithin, a major dietary source of choline, did not support its use in the treatment of dementia [Higgins and Flicker 2003].

In an open study, it was indicated that supplementation of choline may serve as a beneficial adjunct in the treatment of human complex partial seizures [McNamara et al. 1980].

In trained athletes, it was found that, after a marathon, the choline levels have dropped significantly [Conlay, Sabounjian, and Wurtman 1992]. This lead to the speculation that supplementation of choline may improve the performance of the athletes. However, in a double-blind crossover study, it was demonstrated that supplementation of choline did not improve physical or cognitive performance [Deuster et al. 2002].

Stability and Storage Conditions

Choline is stable at room temperature and should be stored away from light and atmosphere.

Interactions Known

No interactions have been reported.

Method of Manufacture

Choline is available at the market in the form of powders and tablets.

Safety

The Food and Nutrition Board of the Institute of Medicine recommends choline intake of 550 mg/day for men and 425 mg/day for women [United States Department of Agriculture 2008]. The tolerable upper intake level of choline for adults is 3.5 g/day. Excess consumption of choline may result in a strong fishy odor and nausea and may trigger existing epilepsy.

Regulatory Status

The FDA under the Food and Drug Administration Modernization Act has authorized that food or dietary supplemented products containing choline may provide nutrient claims on their labels. In Europe and Japan, CDP-choline has been approved to treat stroke, head injuries, and other neurological disorders [D'Orlando and Sandage 1995].

COENZYME Q_{10}

Typical Properties and Description

Coenzyme Q_{10} (CoQ_{10}) is a vitamin-like substance that is present in the majority of human cells. CoQ_{10} is biosynthesized in the human body from tyrosine or phenylalanine and mevalonate [Schultz and Clarke 1999]. CoQ_{10} transports electrons in the oxidation-reduction reaction that drives the adenosine triphosphate (ATP) synthesis in mitochondria. In addition, CoQ_{10} provides stability, fluidity, and regulates apoptosis in cell membranes [Lenaz et al. 1999; López-Lluch et al. 1999]. Despite that CoQ_{10} is synthesized in humans, supplementation of CoQ_{10} may be necessary in populations having CoQ_{10} deficiency. The deficiency of CoQ_{10} is linked to aging, certain type of diseases, genetic mutations, and 3-hydroxy-3-methyl-glutaryl (HMG)-CoA reductase inhibitors [Crane 2001]. CoQ_{10} is present in common dietary consumptions of meat, fish, vegetables, and fruits [Lester and Crane 1959; Weber, Bysted, and Holmer 1997].

Structural Formula

See Figure 4.10 for the structural formula.

Functional Category

CoQ_{10} is benzoquinone with a long isoprenoid side chain (10 units). This side chain makes the molecule highly lipophilic, which is readily diffusible across the

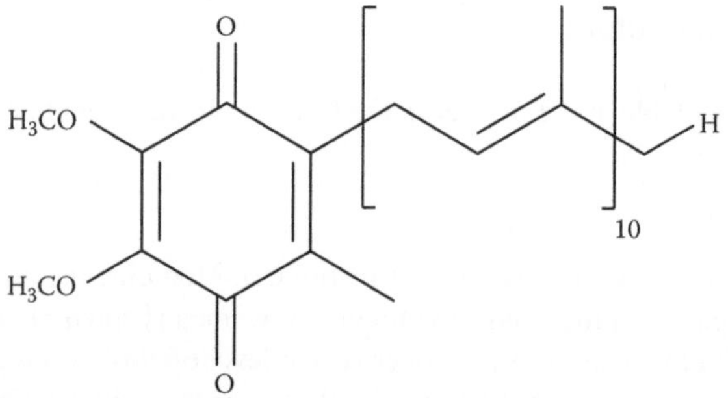

Figure 4.10 The structure of CoQ_{10}.

membranes. During the electron transport, CoQ_{10} is reduced to ubisemiquinone and ubiquinol by accepting one and two electrons, respectively.

Applications in Nutraceuticals

In a randomized, double-blind study, 22 aerobically trained and 19 untrained human subjects were supplemented with placebo or CoQ_{10} supplement. The results from this study suggest that acute (single dose) and/or chronic (14 days) supplementation of CoQ_{10} may enhance the exercise performance in both the trained and untrained subjects [Cooke et al. 2008].

In an 11.5-year-old patient suffering from mitochondrial myopathy, it was observed that CoQ_{10} concentration in the skeletal muscle decreased to 46% of normal average [Lalani et al. 2005]. The supplementation of CoQ_{10} resulted in complete recovery of myopathy. Furthermore, muscle biopsy specimens of 82 children showed CoQ_{10} deficiency to be the best indicator for electron transport chain abnormality. Therefore, the early identification of CoQ_{10} deficiencies in children and supplementation of this agent may cure certain mitochondrial disorders [Miles et al. 2008].

In a randomized, double-blind, placebo-controlled trial involving 42 patients, it was observed that supplementation of CoQ_{10} (three times at 100 mg/day) reduced headache frequency and nausea [Sandor et al. 2005]. In another study, Hershey et al. [2007] assessed CoQ_{10} levels in patients suffering from severe headaches. About 32.9% of 1,550 patients were below the reference range of CoQ_{10} levels. The CoQ_{10}-deficient patients were suggested to intake 1–3 mg/kg/day CoQ_{10}. In a subset of patients, supplementation with CoQ_{10} has resulted in improved CoQ_{10} levels, decreased headache frequency and disability, implying the potential application of CoQ_{10} for migraine.

In a multicenter, randomized, parallel-group, placebo-controlled, double-blind study, patients with Parkinson's disease were supplemented with placebo or CoQ_{10} at 300, 600, or 1200 mg/day. In this study, it was observed that CoQ_{10} supplementation is safe even at 1,200 mg/d, and the beneficial effects were observed in a dose-dependent manner [Shults et al. 2002].

It is observed that dietary supplementation of 0.07–0.7% CoQ_{10} for 26 weeks in a rat model of metabolic syndrome had a beneficial effect on increased oxidative and nitrative stress markers and inflammatory markers. In addition, CoQ_{10} has reduced elevated blood pressure and serum levels, implying that CoQ_{10} may have a beneficial effect in cardiovascular diseases in metabolic syndrome [Kunitomo et al. 2008].

Stability and Storage Conditions

CoQ_{10} should be stored at room temperature under atmosphere-, light-, and moisture-free conditions.

Interactions

HMG-CoA reductase inhibitors or statins are primarily used to treat hypercholesterolemia. It is evident that HMG-CoA reductase inhibitor reduces blood

CoQ_{10} concentrations. This is probably because CoQ_{10} and cholesterol share a similar biosynthetic pathway. In a mice model, it was shown that supplementation of CoQ_{10} during statin therapy reduces the oxidative stress caused by statin administration [Kettawan et al. 2007]. However, at this point in time, the coadministration of CoQ_{10} and statin to prevent myotoxicity remains questionable [Levy and Kohlhaas 2006].

There is an increasing amount of evidence showing that CoQ_{10} affects warfarin metabolism [Landbo and Almdal 1998]. Structurally CoQ_{10} is similar to vitamin K, which may explain its interaction with warfarin [Landbo and Almdal 1998]. It is speculated that concurrent administration of CoQ_{10} (100 mg) and warfarin would increase the total clearance of *S*-warfarin and *R*-warfarin by 32 and 17%, respectively [Zhou, Zhou, and Chan 2005]. Therefore, patients using this combination of medication must be kept under close medical supervision.

Method of Manufacture

CoQ_{10} is available in the market in the form of soft-gel capsules, tablets, and powder.

Safety

According to the OSL risk assessment method, CoQ_{10} is considered to be safe at intake of up to 1,200 mg/day. In fact, much higher levels of CoQ_{10} have been tested, but the data are not sufficient enough to provide a reasonable assurance of safety [Hathcock and Shao 2006].

Handling Precautions

CoQ_{10} should be handled at room temperature under inert conditions, away from light.

Regulatory Status

In the United States, CoQ_{10} holds GRAS status and is regulated as a dietary supplement. In the United Kingdom, CoQ_{10} is regulated as a dietary or food supplement. Japan approved CoQ_{10} in 1974 as a prescription drug for the treatment of congestive heart failure [Tran et al. 2001].

Related Substances

Coenzyme Q is closely related to other quinones such as vitamin K, plastoquinone (present only in plants), and menaquinone (present only in bacteria). Plastoquinone and menaquinone play the equivalent role of ubiquinone, i.e., they transport electrons in oxidation-reduction reactions.

CONCLUSION

See Table 4.1 for details and an overview of what this chapter has discussed regarding nutraceuticals obtained from animal origin.

Table 4.1 Nomenclature, Chemical Abstracts Service (CAS) Number, and Empirical Formula or Molecular Formula of Nutraceuticals Obtained from Animal Origin

Category	Synonyms	Chemical Name	CAS no.	Empirical Formula or Molecular Formula
Omega-3 fatty acids from fish	EPA; timnodonic acid	*cis*-5,8,11,14,17-EPA	10417-94-4	$C_{20}H_{30}O_2$
	DHA	*cis*-4,7,10,13,16,19-DHA	6217-54-5	$C_{22}H_{32}O_2$
Conjugated linoleic acids	9-*cis*,11-*trans*-Octadecadienoic acid solution; 9Z,11E-CLA; bovinic acid; conjugated linoleic acid (9Z,11E)	Conjugated (9Z,11E)-linoleic acid solution	2540-56-9	$C_{18}H_{32}O_2$
	(10E,12Z)-10,12-Octadecadienoic acid; 10E,Z12-CLA; linoleic acid (10-*trans*, 12-*cis*)	Conjugated (10E,12Z)-linoleic acid solution	2420-56-6	$C_{18}H_{32}O_2$
L-Carnitine	(−)-(*R*)-3-Hydroxy-4-(trimethylammonio) butyrate; vitamin B_T	L-Carnitine inner salt	541-15-1	$(CH_3)_3N^+CH_2CH(OH)CH_2COO$
Chondroitin	Chondroitin sulfate A sodium salt; CSA	Chondroitin 4-sulfate sodium salt from bovine trachea	39455-18-0	
Glucosamine	2-Deoxy-2-sulfamino-D-glucopyranose;	D-Glucosamine 2-sulfate sodium salt	38899-05-7	$C_6H_{12}NO_8SNa$
	2-amino-2-deoxy-D-glucose hydrochloride; chitosamine hydrochloride	D-(+)-Glucosamine hydrochloride	66-84-2	$C_6H_{13}NO_5{\bullet}HCl$
Chitosan	Chitosan from crab shells; deacetylated chitin; poly(D-glucosamine)	Chitosan low molecular weight	9012-76-4	Chitosan from crab shells; deacetylated chitin; poly(D-glucosamine)
	Poly(*N*-acetyl-1,4-β-D-glucopyranosamine)	Chitin from crab shells	1398-61-4	Poly(*N*-acetyl-1,4-β-D-glucopyranosamine)

(Continued)

Table 4.1 *(Continued)*

Category	Synonyms	Chemical Name	CAS no.	Empirical Formula or Molecular Formula
Choline	(2-Hydroxyethyl) trimethyl ammonium hydroxide; choline solution	Choline base solution	123-41-1	$HOCH_2CH_2N(CH_3)_3OH$
Coenzyme Q_{10}	Q-10; ubiquinone-50; ubiquinone-10	Coenzyme Q_{10}	303-98-0	$C_{59}H_{90}O_4$

REFERENCES

Albers, R., R. P. van der Wielen, E. J. Brink, H. F. Hendriks, V. N. Dorovska-Taran, and I. C. Mohede. 2003. Effects of cis-9, trans-11 and trans-10, cis-12 conjugated linoleic acid (CLA) isomers on immune function in healthy men. *Eur. J. Clin. Nutr.* 57:595–603.

Alibin, C. P., M. A. Kopilas, and H. D. Anderson. 2008. Suppression of cardiac myocyte hypertrophy by conjugated linoleic acid: Role of peroxisome proliferator-activated receptors alpha and gamma. *J. Biol. Chem.* 283:10707–10715.

Andersen, D. G. 1998a. Glucosamine. Part I. Basic science. http://www.chiroweb.com/mpacms/dc/article.php?id=37227. Accessed July 6, 2009.

Andersen, D. G. 1998b. Glucosamine. Part II. Forms. http://www.chiroweb.com/mpacms/dc/article.php?id=37263. Accessed July 6, 2009.

Banu, J., A. Bhattacharya, M. Rahman, M. O'Shea, and G. Fernandes. 2006. Effects of conjugated linoleic acid and exercise on bone mass in young male Balb/C mice. *Lipids Health Dis.* 5:7.

Benvenga, S., M. Lakshmanan, and F. Trimarchi. 2000. Carnitine is a naturally occurring inhibitor of thyroid hormone nuclear uptake. *Thyroid* 10:1043–1050.

Bhattacharya, A., J. Banu, M. Rahman, J. Causey, and G. Fernandes. 2006. Biological effects of conjugated linoleic acids in health and disease. *J. Nutr. Biochem.* 17:789–810.

Bohmer, T., H. Bergrem, and K. Eiklid. 1978. Carnitine deficiency induced during intermittent haemodialysis for renal failure. *Lancet* 1:126–128.

Borana, G., H. Karaçamb, and M. Boran. 2006. Changes in the quality of fish oils due to storage temperature and time. *Food Chem.* 98:693–698.

Bourgeois, P., G. Chales, J. Dehais, B. Delcambre, J. L. Kuntz, and S. Rozenberg. 1998. Efficacy and tolerability of chondroitin sulfate 1200 mg/day vs chondroitin sulfate 3 × 400 mg/day vs placebo. *Osteoarthritis Cartilage* 6 (Suppl. A):25–30.

Bourin, M. C. and U. Lindahl. 1993. Glycosaminoglycans and the regulation of blood coagulation. *Biochem. J.* 289:313–330.

Brass, E. P. and W. R. Hiatt. 1998. The role of carnitine and carnitine supplementation during exercise in man and in individuals with special needs. *J. Am. Coll. Nutr.* 17:207–215.

Bremer, J. and D. Greenberg. 1961. Methyltransferring enzyme system of microsomes in the biosynthesis of lecithin. *Biochem. Biophys. Acta* 46:205–216.

Bruyere, O. and J. Y. Reginster. 2007. Glucosamine and chondroitin sulfate as therapeutic agents for knee and hip osteoarthritis. *Drugs Aging* 24:573–580.

Buckley, M. S., A. D. Goff, and W. E. Knapp. 2004. Fish oil interaction with warfarin. *Ann. Pharmacother.* 38:50–52.

Bucsi, L. and G. Poor. 1998. Efficacy and tolerability of oral chondroitin sulfate as a symptomatic slow-acting drug for osteoarthritis (SYSADOA) in the treatment of knee osteoarthritis. *Osteoarthritis Cartilage* 6 (Suppl. A):31–36.

Carcione, A., E. Piro, S. Albano, G. Corsello, A. Benenati, M. Piccione, V. Verde, L. Giuffrè, and A. Albancse. 1991. Kabuki make-up (Niikawa-Kuroki) syndrome: Clinical and radiological observations in two Sicilian children. *Pediatr. Radiol.* 21:428–431.

Cargill Health and Food Technologies. 2004. Cargill Announces GRAS Status for OptaFlex (TM) Natural Chondroitin. http://www.npicenter.com/anm/templates/newsATemp.aspx?articleid=10704&zoneid=8. Accessed August 11, 2008.

Cargill Health and Food Technologies. 2007. Cargill announces GRAS status for non-animal glucosamine. http://www.nutraingredients-usa.com/Regulation/Cargill-announces-GRAS-status-for-non-animal-glucosamine. Accessed August 11, 2008.

Chan, Y. C., M. L. Tse, and F. L. Lau. 2007. Two cases of valproic acid poisoning treated with L-carnitine. *Hum. Exp. Toxicol.* 26:967–969.

Clegg, D. O., D. J. Reda, C. L. Harris, M. A. Klein, J. R. O'Dell, M. M. Hooper, J. D. Bradley, C. O. Bingham 3rd, M. H. Weisman, C. G. Jackson, N. E. Lane, J. J. Cush, L. W. Moreland, H. R. Schumacher Jr, C. V. Oddis, F. Wolfe, J. A. Molitor, D. E. Yocum, T. J. Schnitzer, D. E. Furst, A. D. Sawitzke, H. Shi, K. D. Brandt, R. W. Moskowitz, and H. J. Williams. 2006. Glucosamine, chondroitin sulfate, and the two in combination for painful knee osteoarthritis. *N. Engl. J. Med.* 354:795–808.

Conlay, L. A., L. A. Sabounjian, and R. J. Wurtman. 1992. Exercise and neuromodulators: choline and acetylcholine in marathon runners. *Int. J. Sports Med.* 13 (Suppl. 1):S141–S142.

Cooke, M., M. Iosia, T. Buford, B. Shelmadine, G. Hudson, C. Kerksick, C. Rasmussen, M. Greenwood, B. Leutholtz, D. Willoughby, and R. Kreider. 2008. Effects of acute and 14-day coenzyme Q10 supplementation on exercise performance in both trained and untrained individuals. *J. Int. Soc. Sports Nutr.* 5:8.

Crane, F. L. 2001. Biochemical functions of coenzyme Q10. *J. Am. Coll. Nutr.* 20:591–598.

D'Orlando, K. J. and B. W. Sandage Jr. 1995. Citicoline (CDP-choline): Mechanisms of action and effects in ischemic brain injury. *Neurol. Res.* 17:281–284.

De Jesus Moreno Moreno, M. 2003. Cognitive improvement in mild to moderate Alzheimer's dementia after treatment with the acetylcholine precursor choline alfoscerate: A multicenter, double-blind, randomized, placebo-controlled trial. *Clin. Ther.* 25:178–193.

Deuster, P. A., A. Singh, R. Coll, D. E. Hyde, and W. J. Becker. 2002. Choline ingestion does not modify physical or cognitive performance. *Mil. Med.* 167:1020–1025.

Gruppo Italiano per lo Studio della Sopravvivenza nell'Infarto-Prevenzione Investigators. 1999. Dietary supplementation with n-3 polyunsaturated fatty acids and vitamin E after myocardial infarction: results of the GISSI-Prevenzione trial. Gruppo Italiano per lo Studio della Sopravvivenza nell'Infarto miocardico. *Lancet* 354:447–455.

Erfle, J. D., L. J. Fisher, and F. Sauer. 1970. Carnitine and acetylcarnitine in the milk of normal and ketotic cows. *J. Dairy Sci.* 53:486–489.

Food and Drug Administration. 2001. Nutrient Content Claims Notification for Choline Containing Foods. http://vm.cfsan.fda.gov/~dms/flcholin.html. Accessed August 11, 2008.

Food and Drug Administration. 2004. OMACOR®. http://www.fda.gov/cder/foi/label/2004/21654lbl.pdf. Accessed August 11, 2008.

Food and Drug Administration. 2007. MedTrade CELOX Topical Hemostatic Granules in Soluble Bag. http://www.fda.gov/cdrh/pdf7/K072328.pdf. Accessed August 11, 2008.

Food and Drug Administration. 2008. FDA Media Briefing on Heparin. http://www.fda.gov/bbs/transcripts/2008/heparin_transcript_030508.pdf. Accessed August 11, 2008.

Geleijnse, J. M., E. J. Giltay, D. E. Grobbee, A. R. Donders, and F. J. Kok. 2002. Blood pressure response to fish oil supplementation: metaregression analysis of randomized trials. *J. Hypertens.* 20:1493–1499.

Ghosh, S., H. J. Blumenthal, E. Davidson, and S. Roseman. 1960. Glucosamine metabolism. V. Enzymatic synthesis of glucosamine 6-phosphate. *J. Biol. Chem.* 235:1265–1273.

Goldberg, H., A. Avins, and S. Bent. 2007. Chondroitin for osteoarthritis of the knee or hip. *Ann. Intern. Med.* 147:883; author reply 884–885.

Gómez-Amores, L., A. Mate, J. L. Miguel-Carrasco, L. Jiménez, A. Jos, A. M. Cameán, E. Revilla, C. Santa-María, and C. M. Vázquez. 2007. L-carnitine attenuates oxidative stress in hypertensive rats. *J. Nutr. Biochem.* 18:533–540.

Harris, W. S. 2007. Omega-3 fatty acids and cardiovascular disease: A case for omega-3 index as a new risk factor. *Pharmacol. Res.* 55:217–223.

Harris, W. S., H. N. Ginsberg, N. Arunakul, N. S. Shachter, S. L. Windsor, M. Adams, L. Berglund, and K. Osmundsen. 1997. Safety and efficacy of Omacor in severe hypertriglyceridemia. *J. Cardiovasc. Risk* 4:385–391.

Hathcock, J. N. and A. Shao. 2006. Risk assessment for carnitine. *Regul. Toxicol. Pharmacol.* 46:23–28.

Hathcock, J. N. and A. Shao. 2006. Risk assessment for coenzyme Q10 (Ubiquinone). *Regul. Toxicol. Pharmacol.* 45:282–288.

Hathcock, J. N. and A. Shao. 2007. Risk assessment for glucosamine and chondroitin sulfate. *Regul. Toxicol. Pharmacol.* 47:78–83.

Held, U. 2004. L-Carnitine: The natural choice for functional foods. http://www.lonza.com/group/en/company/news/publications_of_lonza.-ParSys-0002-ParSysdownloadlist-0011-DownloadFile.pdf/11_L- Carnitine_040831_NutraCos_2004_general.pdf. Accessed August 11, 2008.

Held, U. and S. Siebrecht 2003. L-carnitine: A biological route to a key nutraceutical. http://www.carnipure.com/carnipure/en/promotional_material/l_carnitine_in_the_press.-ParSys-0022-DownloadFile.tmp/LONZA%20MAKEUP%20(proof).pdf. Accessed August 11, 2008.

Herrera, J. A., M. Arevalo-Herrera, A. K. Shahabuddin, G. Ersheng, S. Herrera, R. G. Garcia, and P. Lopez-Jaramillo. 2006. Calcium and conjugated linoleic acid reduces pregnancy-induced hypertension and decreases intracellular calcium in lymphocytes. *Am. J. Hypertens.* 19:381–387.

Herrera, J. A., A. K. Shahabuddin, G. Ersheng, Y. Wei, R. G. Garcia, and P. Lopez-Jaramillo. 2005. Calcium plus linoleic acid therapy for pregnancy-induced hypertension. *Int. J. Gynaecol. Obstet.* 91:221–227.

Hershey, A. D., S. W. Powers, A. L. Vockell, S. L. Lecates, P. L. Ellinor, A. Segers, D. Burdine, P. Manning, and M. A. Kabbouche. 2007. Coenzyme Q10 deficiency and response to supplementation in pediatric and adolescent migraine. *Headache* 47:73–80.

Higgins, J. P. and L. Flicker. 2003. Lecithin for dementia and cognitive impairment. *Cochrane Database Syst. Rev.* 2003:CD001015.

Hirano, S., Y. Tanaka, M. Hasegawa, K. Tobetto, and A. Nishioka. 1985. Effect of sulfated derivatives of chitosan on some blood coagulant factors. *Carbohydr. Res.* 137:205–215.

Hug, G., C. A. McGraw, S. R. Bates, and E. A. Landrigan. 1991. Reduction of serum carnitine concentrations during anticonvulsant therapy with phenobarbital, valproic acid, phenytoin, and carbamazepine in children. *J. Pediatr.* 119:799–802.

Inoue, N., K. Nagao, J. Hirata, Y. M. Wang, and T. Yanagita. 2004. Conjugated linoleic acid prevents the development of essential hypertension in spontaneously hypertensive rats. *Biochem. Biophys. Res. Commun.* 323:679–684.

Jacobs, M. N., D. Santillo, P. A. Johnston, C. L. Wyatt, and M. C. French. 1998. Organochlorine residues in fish oil dietary supplements: comparison with industrial grade oils. *Chemosphere* 37:1709–1721.

Jaffer, S. and J. S. Sampalis. 2007. Efficacy and safety of chitosan HEP-40 in the management of hypercholesterolemia: A randomized, multicenter, placebo-controlled trial. *Altern. Med. Rev.* 12:265–273.

Kaats, G. R., J. E. Michalek, and H. G. Preuss. 2006. Evaluating efficacy of a chitosan product using a double-blinded, placebo-controlled protocol. *J. Am. Coll. Nutr.* 25:389–394.

Kawasaki, T., H. Kurosawa, H. Ikeda, S. G. Kim, A. Osawa, Y. Takazawa, M. Kubota, and M. Ishijima. 2008. Additive effects of glucosamine or risedronate for the treatment of osteoarthritis of the knee combined with home exercise: A prospective randomized 18-month trial. *J. Bone Miner. Metab.* 26:279–287.

Kelley, D. S., G. L. Bartolini, J. W. Newman, M. Vemuri, and B. E. Mackey. 2006. Fatty acid composition of liver, adipose tissue, spleen, and heart of mice fed diets containing t10, c12-, and c9, t11-conjugated linoleic acid. *Prostaglandins Leukot. Essent. Fatty Acids* 74:331–338.

Kennedy, A., D. Peterson, G. Eden, D. Argyres, S. George, and J. Barnhil. 2006. Shelf-Life Stability of the Nutraceutical Agents Glucosamine Hydrochloride and Chondroitin Sulfate. http://www.aapsj.org/abstracts/AM_2006/AAPS2006-002218.pdf. Accessed July 6, 2009.

Kettawan, A., T. Takahashi, R. Kongkachuichai, S. Charoenkiatkul, T. Kishi, and T. Okamoto. 2007. Protective effects of coenzyme q(10) on decreased oxidative stress resistance induced by simvastatin. *J. Clin. Biochem. Nutr.* 40:194–202.

Knudsen, J. F. and G. H. Sokol. 2008. Potential glucosamine-warfarin interaction resulting in increased international normalized ratio: case report and review of the literature and MedWatch database. *Pharmacotherapy* 28:540–548.

Kritchevsky, D., S. A. Tepper, S. Wright, P. Tso, and S. K. Czarnecki. 2000. Influence of conjugated linoleic acid (CLA) on establishment and progression of atherosclerosis in rabbits. *J. Am. Coll. Nutr.* 19:472S–477S.

Kunitomo, M., Y. Yamaguchi, S. Kagota, and K. Otsubo. 2008. Beneficial effect of coenzyme Q10 on increased oxidative and nitrative stress and inflammation and individual metabolic components developing in a rat model of metabolic syndrome. *J. Pharmacol. Sci.* 107:128–137.

Lalani, S. R., G. D. Vladutiu, K. Plunkett, T. E. Lotze, A. M. Adesina, and F. Scaglia. 2005. Isolated mitochondrial myopathy associated with muscle coenzyme Q10 deficiency. *Arch. Neurol.* 62:317–320.

Landbo, C. and T. P. Almdal. 1998. Interaction between warfarin and coenzyme Q10 (translated from Danish). *Ugeskr. Laeger.* 160:3226–3227.

Lau, C. S., K. D. Morley, and J. J. Belch. 1993. Effects of fish oil supplementation on non-steroidal anti-inflammatory drug requirement in patients with mild rheumatoid arthritis: A double-blind placebo controlled study. *Br. J. Rheumatol.* 32:982–989.

Lenaz, G., R. Fato, S. Di Bernardo, D. Jarreta, A. Costa, M. L. Genova, and G. Parenti Castelli. 1999. Localization and mobility of coenzyme Q in lipid bilayers and membranes. *Biofactors* 9:87–93.

Lester, R. L. and F. L. Crane. 1959. The natural occurrence of coenzyme Q and related compounds. *J. Biol. Chem.* 234:2169–2175.

Leung, Y. H. and R. H. Liu. 2000. trans-10,cis-12-conjugated linoleic acid isomer exhibits stronger oxyradical scavenging capacity than cis-9,trans-11-conjugated linoleic acid isomer. *J. Agric. Food Chem.* 48:5469–5475.

Levy, H. B. and H. K. Kohlhaas. 2006. Considerations for supplementing with coenzyme Q10 during statin therapy. *Ann. Pharmacother.* 40:290–294.

Lin, H., T. D. Boylston, M. J. Chang, L. O. Luedecke, and T. D. Shultz. 1995. Survey of the conjugated linoleic acid contents of dairy products. *J. Dairy Sci.* 78:2358–2365.

Longo, N., C. Amat di San Filippo, and M. Pasquali. 2006. Disorders of carnitine transport and the carnitine cycle. *Am. J. Med. Genet. C. Semin. Med. Genet.* 142C:77–85.

López-Lluch, G., M. P. Barroso, S. F. Martin, D. J. Fernández-Ayala, C. Gómez-Díaz, J. M. Villalba, and P. Navas. 1999. Role of plasma membrane coenzyme Q on the regulation of apoptosis. *Biofactors* 9:171–177.

Malaguarnera, M., M. P. Gargante, E. Cristaldi, M. Vacante, C. Risino, L. Cammalleri, G. Pennisi, and L. Rampello. 2008. Acetyl-L-carnitine treatment in minimal hepatic encephalopathy. *Dig. Dis. Sci.* 53:3018–3025.

Masters, N., M. A. McGuire, K. A. Beerman, N. Dasgupta, and M. K. McGuire. 2002. Maternal supplementation with CLA decreases milk fat in humans. *Lipids* 37:133–138.

McAlindon, T. E., M. P. LaValley, J. P. Gulin, and D. T. Felson. 2000. Glucosamine and chondroitin for treatment of osteoarthritis: A systematic quality assessment and meta-analysis. *JAMA* 283:1469–1475.

McNamara, J. O., S. Carwile, V. Hope, J. Luther, and P. Miller. 1980. Effects of oral choline on human complex partial seizures. *Neurology* 30:1334–1336.

Miles, M. V., L. Miles, P. H. Tang, P. S. Horn, P. E. Steele, A. J. DeGrauw, B. L. Wong, and K. E. Bove. 2008. Systematic evaluation of muscle coenzyme Q10 content in children with mitochondrial respiratory chain enzyme deficiencies. *Mitochondrion* 8:170–180.

Minemoto, Y., S. Adachi, Y. Shimada, T. Nagao, T. Iwata, Y. Yamauchi-Sato, T. Yamamoto, T. Kometani, and R. Matsuno. 2003. Oxidation kinetics for cis-9,trans-11 and trans-10,cis-12 Isomers of CLA. *J. Am. Oil Chem. Soc.* 80:675–678.

Mingrone, G., A. V. Greco, E. Capristo, G. Benedetti, A. Giancaterini, A. De Gaetano, and G. Gasbarrini. 1999. L-carnitine improves glucose disposal in type 2 diabetic patients. *J. Am. Coll. Nutr.* 18:77–82.

Nagao, K., N. Inoue, Y. M. Wang, and T. Yanagita. 2003. Conjugated linoleic acid enhances plasma adiponectin level and alleviates hyperinsulinemia and hypertension in Zucker diabetic fatty (fa/fa) rats. *Biochem. Biophys. Res. Commun.* 310:562–566.

Navarro, V., J. Miranda, I. Churruca, A. Fernandez-Quintela, V. M. Rodriguez, and M. P. Portillo. 2006. Effects of trans-10,cis-12 conjugated linoleic acid on body fat and serum lipids in young and adult hamsters. *J. Physiol. Biochem.* 62:81–87.

Pariza, M. W. 2004. Perspective on the safety and effectiveness of conjugated linoleic acid. *Am. J. Clin. Nutr.* 79 (6 Suppl):1132S–1136S.

Pariza, M. W. and W. A. Hargraves. 1985. A beef-derived mutagenesis modulator inhibits initiation of mouse epidermal tumors by 7,12-dimethylbenz[a]anthracene. *Carcinogenesis* 6:591–593.

Pariza, M. W., Y. Park, and M. E. Cook. 2001. The biologically active isomers of conjugated linoleic acid. *Prog. Lipid Res.* 40:283–298.

Park, Y., K. J. Albright, W. Liu, J. M. Storkson, M. E. Cook, and M. W. Pariza. 1997. Effect of conjugated linoleic acid on body composition in mice. *Lipids* 32:853–858.

Pavelka, K., J. Gatterova, M. Olejarova, S. Machacek, G. Giacovelli, and L. C. Rovati. 2002. Glucosamine sulfate use and delay of progression of knee osteoarthritis: A 3-year, randomized, placebo-controlled, double-blind study. *Arch. Intern. Med.* 162:2113–2123.

Pham, T., A. Cornea, K. E. Blick, A. Jenkins, and R. H. Scofield. 2007. Oral glucosamine in doses used to treat osteoarthritis worsens insulin resistance. *Am. J. Med. Sci.* 333:333–339.

Rapport, L. and B. Lockwood. 2000. Carnitine. http://www.pharmj.com/Editorial/20000819/articles/nutraceuticals4_carnitine.html. Accessed August 11, 2008.

Rebouche, C. J. 2004. Kinetics, pharmacokinetics, and regulation of L-carnitine and acetyl-L-carnitine metabolism. *Ann. NY Acad. Sci.* 1033:30–41.

Reese, T. A., H. E. Liang, A. M. Tager, A. D. Luster, N. Van Rooijen, D. Voehringer, and R. M. Locksley. 2007. Chitin induces accumulation in tissue of innate immune cells associated with allergy. *Nature* 447:92–96.

Reginster, J. Y., R. Deroisy, L. C. Rovati, R. L. Lee, E. Lejeune, O. Bruyere, G. Giacovelli, Y. Henrotin, J. E. Dacre, and C. Gossett. 2001. Long-term effects of glucosamine sulphate on osteoarthritis progression: A randomised, placebo-controlled clinical trial. *Lancet* 357:251–256.

Reichenbach, S., R. Sterchi, M. Scherer, S. Trelle, E. Bürgi, U, Bürgi, P. A. Dieppe, and P. Jüni. 2007. Meta-analysis: Chondroitin for osteoarthritis of the knee or hip. *Ann. Intern. Med.* 146:580–590.

Reliant Pharmaceuticals Inc. Changes Name of OMACOR® (omega-3-acid ethyl esters) to LOVAZA™ (omega-3-acid ethyl esters). http://www.lipid.org/display.php?n=43. Accessed July 6, 2009.

Riserus, U., S. Basu, S. Jovinge, G. N. Fredrikson, J. Arnlov, and B. Vessby. 2002. Supplementation with conjugated linoleic acid causes isomer-dependent oxidative stress and elevated C-reactive protein: a potential link to fatty acid-induced insulin resistance. *Circulation* 106:1925–1929.

Rozenfeld, V., J. L. Crain, and A. K. Callahan. 2004. Possible augmentation of warfarin effect by glucosamine-chondroitin. *Am. J. Health Syst. Pharm.* 61:306–307.

Rudman, D., C. W. Sewell, and J. D. Ansley. 1977. Deficiency of carnitine in cachectic cirrhotic patients. *J. Clin. Invest.* 60:716–723.

Sakurabayashi, T., S. Miyazaki, Y. Yuasa, S. Sakai, M. Suzuki, S. Takahashi, and Y. Hirasawa. 2008. L-carnitine supplementation decreases the left ventricular mass in patients undergoing hemodialysis. *Circ. J.* 72:926–931.

Sandor, P. S., L. Di Clemente, G. Coppola, U. Saenger, A. Fumal, D. Magis, L. Seidel, R. M. Agosti, and J. Schoenen. 2005. Efficacy of coenzyme Q10 in migraine prophylaxis: A randomized controlled trial. *Neurology* 64:713–715.

Schneider, A., L. Richert, G. Francius, J. C. Voegel, and C. Picart. 2007. Elasticity, biodegradability and cell adhesive properties of chitosan/hyaluronan multilayer films. *Biomed. Mater.* 2:S45–S51.

Shields, K. M., N. Smock, C. E. McQueen, and P. J. Bryant. 2003. Chitosan for weight loss and cholesterol management. *Am. J. Health Syst. Pharm.* 60:1310–1312, 1315–1316.

Shults, C. W., D. Oakes, K. Kieburtz, M. F. Beal, R. Haas, S. Plumb, J. L. Juncos, J. Nutt, I. Shoulson, J. Carter, K. Kompoliti, J. S. Perlmutter, S. Reich, M. Stern, R. L. Watts, R. Kurlan, E. Molho, M. Harrison, and M. Lew; Parkinson Study Group. 2002. Effects of coenzyme Q10 in early Parkinson disease: Evidence of slowing of the functional decline. *Arch. Neurol.* 59:1541–1550.

Schultz, J. R. and. C. F. Clarke. 1999. Functional roles of ubiquinone. In *Mitochondria, Oxidants and Ageing*. Eedited by Cardenas, E. and L. Packer. New York, NY: Marcel Dekker, pp. 95–118..

Silberberg, J. S., P. E. Barre, S. S. Prichard, and A. D. Sniderman. 1989. Impact of left ventricular hypertrophy on survival in end-stage renal disease. *Kidney Int.* 36:286–290.

Silveira, M. B., R. Carraro, S. Monereo, and J. Tebar. 2007. Conjugated linoleic acid (CLA) and obesity. *Public Health Nutr.* 10:1181–1186.

Song, H. J., I. Grant, D. Rotondo, I. Mohede, N. Sattar, S. D. Heys, and K. W. Wahle. 2005. Effect of CLA supplementation on immune function in young healthy volunteers. *Eur. J. Clin. Nutr.* 59:508–517.

Tallia, A. F. and D. A. Cardone. 2002. Asthma exacerbation associated with glucosamine-chondroitin supplement. *J. Am. Board Fam. Pract.* 15:481–484.

Tamblyn, R., L. Berkson, W. D. Dauphinee, D. Gayton, R. Grad, A. Huang, L. Isaac, P. McLeod, and L. Snell. 1997. Unnecessary prescribing of NSAIDs and the management of NSAID-related gastropathy in medical practice. *Ann. Intern. Med.* 127:429–438.

Tat, S. K., J. P. Pelletier, J. Verges, D. Lajeunesse, E. Montell, H. Fahmi, M. Lavigne, and J. Martel-Pelletier. 2007. Chondroitin and glucosamine sulfate in combination decrease the pro-resorptive properties of human osteoarthritis subchondral bone osteoblasts: A basic science study. *Arthritis Res. Ther.* 9:R117.

Teng, D., J. Hou, X. Zhang, X. Wang, Z. Wang, and C. Li. 2008. Glucosamine-carrying temperature- and pH-sensitive microgels: Preparation, characterization, and in vitro drug release studies. *J. Colloid Interface Sci.* 322:333–341.

Tran, M. T., T. M. Mitchell, D. T. Kennedy, and J. T. Giles. 2001. Role of coenzyme Q10 in chronic heart failure, angina, and hypertension. *Pharmacotherapy* 21:797–806.

Uebelhart, D., E. J. Thonar, P. D. Delmas, A. Chantraine, and E. Vignon. 1998. Effects of oral chondroitin sulfate on the progression of knee osteoarthritis: A pilot study. *Osteoarthritis Cartilage* 6 (Suppl. A):39–46.

United States Department of Agriculture. 2008. USDA Database for the Choline Content of Common Foods Release Two. http://www.nal.usda.gov/fnic/foodcomp/Data/Choline/Choln02.pdf. Accessed August 11, 2008.

Vesela, E., J. Racek, L. Trefil, V. Jankovy'ch, and M. Pojer. 2001. Effect of L-carnitine supplementation in hemodialysis patients. *Nephron* 88:218–223.

Wasserman, K. and B. J. Whipp. 1975. Excercise physiology in health and disease. *Am. Rev. Respir. Dis.* 112:219–249.

Weber, C., A. Bysted, and G. Holmer. 1997. Coenzyme Q10 in the diet: Daily intake and relative bioavailability. *Mol. Aspects Med.* 18 (Suppl.):S251–S254.

Whigham, L. D., A. C. Watras, and D. A. Schoeller. 2007. Efficacy of conjugated linoleic acid for reducing fat mass: A meta-analysis in humans. *Am. J. Clin. Nutr.* 85:1203–1211.

Yamasaki, M., K. Mansho, Y. Ogino, M. Kasai, H. Tachibana, and K. Yamada. 2000. Acute reduction of serum leptin level by dietary conjugated linoleic acid in Sprague-Dawley rats. *J. Nutr. Biochem.* 11:467–471.

Zeisel, S. H. and J. K. Blusztajn. 1994. Choline and human nutrition. *Annu. Rev. Nutr.* 14:269–296.

Zeisel, S. H., M. H. Mar, J. C. Howe, and J. M. Holden. 2003. Concentrations of choline-containing compounds and betaine in common foods. *J. Nutr.* 133:1302–1307.

Zhang, A. and Z. Chen. 1997. Oxidative stability of conjugated linoleic acids relative to other polyunsaturated fatty acids. *J. Am. Oil Chem. Soc.* 74:1611–1613.

Zhou, Q., S. Zhou, and E. Chan. 2005. Effect of coenzyme Q10 on warfarin hydroxylation in rat and human liver microsomes. *Curr. Drug Metab.* 6:67–81.

CHAPTER 5

Nutraceuticals with Mineral Origin

Miriam A. Ansong and Seema Y. Pathak

CONTENTS

INTRODUCTION

Minerals are simple inorganic elements. They are found in the body in the form of mostly salts. Four to 6% of the body weight is attributed to minerals. Almost half of it comes from calcium, and a quarter is phosphorus (phosphates); other essential minerals contribute to the whole picture. Calcium plays a major role in skeletal muscle formation and functions, bone and teeth building and formation, the heart and digestive systems, and blood cell formation. Potassium is important in metabolic processes, such as maintaining water balance in a cell unit. Iodine is a component of endocrine system. On their own, they are part of many organic molecules. Minerals are the basis of important digestive fluids, such as hydrochloric acid. The important minerals in the human body are calcium, phosphorus, potassium, chromium, sodium, chlorine, sulfur, copper, magnesium, manganese, molybdenum iron, iodine, zinc, and selenium [Otten, Helwig, and Meyers 2006].

The amount of mineral the body needs divides the spectrum of minerals into macro minerals, trace minerals, and ultra trace minerals. Calcium, phosphorus (phosphates), sulfur sodium, potassium, chloride, and magnesium are required in the amount of 100 mg and more. They are called macro minerals.

The minerals required in much smaller amounts, such as 15 mg/day, are iron, iodine, chromium, molybdenunm, copper, manganese, selenium, and zinc. These are called trace minerals. Ultra trace minerals are needed in extremely small quantities, such as micrograms; they are arsenic, boron, nickel, silicon, and vanadium. They have been shown to play a role in experimental animals, but sufficient data for humans are lacking. Many factors influence the absorption of minerals in the body. Fiber and phytates from plant source are a hindrance in absorption. The quantity of minerals present in food is dependent on the composition of soil and water, in which the food is grown.

CALCIUM

Calcium is an essential mineral that is the most common and abundant in the body. Calcium balance in humans is very important and turns out to be positive during growth, becomes steady in mature adults, and declines in the elderly [CDC 2008; Mason 2008; Jelin et al. 2008]

Mechanism of Action

Calcium plays a major role in skeletal muscle formation and functions, bone and teeth building and formation, the heart and digestive systems, and blood cell formation. It is highly concentrated in bones and teeth in an amount in excess of 99% of calcium that exists in the human body [Mason 2007; "Calcium, Copper" 2008; Jelin et al. 2008].

Absorption

Calcium is primarily absorbed in the small intestine of the gut involving an active process with the help of vitamin D. Absorption of calcium in general is low and can be increased with intake of food [Mason 2007]. Several factors affect the absorption of calcium, which may include age, environmental conditions, race, and dietary status. Aging decreases the ability to absorb calcium, resulting in negative calcium balance and bone loss [Pattanaungkul et al. 2000; Heaney 2001]. It is also known that Asians and Africans tend to have higher calcium absorption properties than Caucasians [Celotti and Bignamini 1999]. Weight loss has been linked to decrease in calcium absorption, which significantly affects bone loss. Absorption also tends to be optimal in conditions of high requirement of calcium, such as in childhood, adolescence, pregnancy, and breastfeeding [Mason 2007].

Distribution

Approximately 50% of the serum calcium is in a bound form to plasma proteins. Calcium is converted to an active free ionized form in the blood. The freely available calcium is used as an indicator for calcium levels in humans [Power et al. 1999; Jelin et al. 2008]. However, about 99% of body calcium is stored in bones and teeth [Mason 2007].

Elimination

Calcium is eliminated from the human body via various routes. These routes include feces, urine, sweat, skins cells, and breast milk. Feces serve as the main route of elimination for unabsorbed and secreted calcium [Mason 2007; Jelin et al. 2008].

Bioavailability

Bioavailability of calcium in the body depends highly on absorption. Absorption may be reduced by certain food such as high-fiber cereal, spinach, and cauliflower. It has also been documented that high sodium-containing food may reduce calcium retention in the body [Mason 2008].

Uses

There are many uses for calcium based on several studies conducted in humans and animals [Mason 2007; "Calcium, Copper" 2008]. Calcium may play a significant role in bone loss prevention, calcium deficiency, cardiopulmonary resuscitation, high blood phosphorus level, osteoporosis, high blood potassium level, and high blood pressure ["Calcium, Copper" 2008; Jelin et al. 2008].

Dietary Sources

Several dietary types have been documented to be rich in calcium. Selected examples of these food types include cereal products, milk and dairy products, fish, fruits, vegetables, and nuts. These groups of food represent the major sources of dietary calcium [Mason 2007].

Commercial Preparations

Calcium is available in various salt forms. The normal recommended daily dose for adults range from 400 to 3,000 mg as documented in various studies. However, different doses may be required for certain conditions; therefore, it is highly recommended to seek advice from a healthcare provider for dosing recommendations [Mason 2007; "Calcium, Copper" 2008]. To cite an example, the recommended dose for the prevention of osteoporosis is 1,000–1,200 mg daily [Mason 2007].

Deficiency

The major problem with calcium deficiency may lead to reduction of peak bone mass and mineral content [Mason 2007].

Adverse Effects, Contraindications, and Interactions

Common adverse effects associated with the use of calcium include abdominal pain, elevated calcium in the blood, confusion, dry mouth, frequent urination, kidney stones, nausea, thirst, and vomiting ["Calcium, Copper" 2008]. Calcium supplements should not be considered in certain conditions, such as high levels of calcium in the blood and in the urine as well as patients with chronic kidney diseases [Mason 2007]. Calcium can interact with various medicines, such as antacids, seizure medications, blood pressure medicines, cholesterol medications, diuretics, thyroid medicines, and weight loss products ["Calcium, Copper" 2008]

CHROMIUM

Chromium plays a role in the body's use of energy-providing carbohydrates, proteins, and fats. Short supply of chromium is associated with impaired glucose tolerance and diabetes-like symptoms [Glinsmann and Mertz 1966]. A chromium diabetes link was discovered when severe diabetic symptoms of a long-term tube-fed patient were alleviated by supplemental chromium [Jeejeebhoy et al. 1977].

Mechanism of Action

Chromium is an essential trace element. Metallic chromium has no biological activity. Chromium is referred to as a glucose tolerance factor. This glucose tolerance factor is a complex of molecules. Glycine, cysteine, glutamic acid, and nicotinic acid, along with chromium, form this complex [Jelin et al. 2008].

Absorption

The commonly available salts of chromium are chromium chloride, chromium picolinate, and chromium polynicotinate. Chromium picolinate at 4% is a more absorbed salt form than chromium chloride [Anderson et al. 1997]. The low absorption percentage, which decreases further when intake is increased, may be part of the reason chromium is not toxic. Chromium absorption can increase with exercise [Otten, Helwig, and Meyers 2006].

Distribution

Chromium is stored in the liver, spleen, soft tissues, and bone [Otten, Helwig, and Meyers 2006].

Elimination

Most absorbed chromium is excreted rapidly in the urine, whereas unabsorbed chromium is excreted through feces [Otten, Helwig, and Meyers 2006].

Bioavailability

The optimum solubility of chromium compounds is achieved at stomach pH [Otten, Helwig, and Meyers 2006]. Vitamin C enhances absorption of chromium [Jelin et al. 2008]. Urinary chromium excretion is related to the insulinogenic properties of the carbohydrates. Thus, diets high in simple sugars are likely to influence chromium absorption negatively [Otten, Helwig, and Meyers 2006].

Uses

Some diabetes studies have shown that chromium supplementation has a beneficial effect on plasma lipids, although study results have been far from uniform. In one study, total cholesterol was significantly reduced in the group receiving the higher dose of chromium [Anderson et al. 1997]. It has been documented in clinical studies that triglyceride levels in people with type II diabetes who took chromium supplements were lowered significantly [Lee and Reasner 1994].

Dietary Sources

Natural food supply is rich in chromium content. Geochemical factors determine the chromium content in food. Refining processes deplete chromium content from foods such as cereal and grains. Rich sources of chromium include beef, black strap molasses, brewer's yeast, brown rice, calves' liver, unrefined cereals, chicken, corn, dairy products, dried beans, and pulses [Otten, Helwig, and Meyers 2006].

Interactions and Side Effects

Vitamin C may enhance the absorption of chromium. Simple sugars and phytates decrease chromium absorption [Otten, Helwig, and Meyers 2006].

COPPER

Copper is a type of mineral that occurs in many foods and plays numerous roles in the human body [Mason 2007].

Mechanism of Action

Copper is known to play a significant role in bone formation, connective tissues integrity, iron absorption, synthesis of hemoglobin, and metabolic pathways, including glucose and cholesterol [Mason 2007].

Absorption

Absorption of copper occurs primarily in the small intestine and to a very small extent in the stomach. Normally, absorption occurs with a concentration gradient with higher intake of copper and an alternative mechanism at a lower level of intake [Mason 2007].

Distribution

The liver serves as major organ in uptake, transport, and storage of copper in the body [Mason 2007].

Elimination

Copper can be eliminated via various routes, with a major emphasis on bile and feces. Excretion via the urine and sweat serve as routes of elimination to a small extent [Mason 2007].

Bioavailability

High-fiber-containing food can reduce absorption of copper, which can ultimately lower bioavailability of copper in the body. This is not the case with the daily recommended dietary intake of fiber-containing foods [Mason 2007].

Uses

There is a claim that copper plays a protective role against high cholesterol levels. However, results of studies in humans are inconclusive [Mason 2007]. The primary use of copper is for copper deficiency [Pattanaungkul et al. 2000].

Deficiency

Copper deficiency is rare; however, when it occurs, it may lead to conditions such as anemia and immune system impairment [Mason 2007].

Dietary Sources

Food types such as cereal, meat, vegetables, fruits, nuts, and plain chocolate are known to provide major sources of copper in the body [Mason 2007].

Commercial Preparations

The benefits of using copper supplements have not been proven in studies [Mason 2007]. However, supplements do exist and may be available in dosage forms such as tablets and capsules. It is important to acknowledge that copper supplements are usually part of multivitamin and mineral preparations on the market. The most commonly available salts of copper are copper amino acid chelate, copper gluconate, and copper sulfate without any specific dosing requirements [Mason 2007]. However, for adults, a maximum dose of 10,000 μg daily has been recommended as the RDA ["Calcium, Copper" 2008].

Adverse Effects, Contraindications, and Interactions

Higher doses of copper may cause problems such as nausea, vomiting, diarrhea, low blood pressure, and back pain. Abdominal pain, fatigue, and infection are additional side effects that have been documented. Copper does interact with many drugs, including oral contraceptives, penicillamine, seizure drugs, and antacids [Mason 2007; "Calcium, Copper" 2008].

IODINE

Iodine is an essential element in the human body. Its primary function involves thyroid function. Deficiency has been associated with many health issues, including thyroid malfunctioning, skin problems, and neurological problems ["Calcium, Copper" 2008].

Mechanism of Action

Iodine is a major component of the thyroid hormones and is important for many functions, including enzyme activities and protein regulation [Otten, Helwig, and Meyers 2006].

Absorption

Iodine undergoes a process known as reduction to a reduced form known as iodide in the gut for easy absorption. Absorption can be greatly reduced by soya flour, which is found in some infant formula [Otten, Helwig, and Meyers 2006].

Distribution

Once iodine is absorbed, it is freely taken up by certain organs, such as the thyroid and the kidney. The thyroid gland uses the absorbed iodide for thyroid hormone synthesis, whereas the rest get eliminated via urine [Otten, Helwig, and Meyers 2006].

Elimination

The primary elimination route of iodine is through the urine [Otten, Helwig, and Meyers 2006].

Bioavailability

Bioavailability of iodine is enhanced by the complete absorption of the mineral [Otten, Helwig, and Meyers 2006; Mason 2007].

Uses

Iodine is useful in many conditions; however, the most important ones are goiter, iodine deficiency prevention, skin disinfectant, water purification, infection prevention, hearing loss, and Grave's disease ["Calcium, Copper" 2008].

Deficiency

Deficiency has been associated with numerous health conditions, such as intestinal problems, neurological issues, and skin problems. It can be very serious in pregnant and nursing mothers, resulting in neurocognitive disorders in infants. Severe deficiency has been associated with conditions such as goiter in adults and cognitive developmental issues in children [Otten, Helwig, and Meyers 2006; "Calcium, Copper" 2008].

Dietary Sources

Salt represents the most important source of iodine. Higher concentration of iodine usually comes from seafood [Otten, Helwig, and Meyers 2006].

Commercial Preparations

Lugol solution, saturated solution of potassium iodide, povidone-iodine, and sodium iodide are examples of iodine preparations ["Calcium, Copper" 2008].

Recommended daily dosing for adults ranges from 150 to 290 mcg (breastfeeding women) and ranges from 50 to 900 mcg depending on the age group ["Calcium, Copper" 2008].

Adverse Effects, Contraindications, and Interactions

Common side effects that have been associated with iodine use or exposure, including skin lesions, confusion, cough, depression, diarrhea, muscle aches, numbness, unpleasant taste, and weakness ["Calcium, Copper" 2008]. Iodine-based products should be avoided for patients with known hypersensitivity or allergy to those products. Iodine may interact with several medications, such as Amiodarone, antithyroid medications, lithium, diurectics, and high blood pressure drugs commonly described as angiotensin-converting enzyme inhibitors. Additionally, certain food types have been known to interact with iodine. These include cabbage, legumes, cassava, herbs, and supplements with identical effects ["Calcium, Copper" 2008].

IRON

Iron is an essential mineral that is relevant for transport of oxygen for metabolic pathways in the human body ["Calcium, Copper" 2008]. It is found in two different forms in the body: a reduced state (ferrous iron) and an oxidized form (ferric iron) [Jelin et al. 2008].

Mechanism of Action

Iron is involved in a number of metabolic pathways, and it is the main component of hemoglobin, myoglobin, and many enzymes. It also plays a major role in the transport and storage of oxygen and DNA synthesis [Mason 2007; "Calcium, Copper" 2008; Jelin et al. 2008].

Absorption

Iron is absorbed in the small intestine primarily in the duodenum and proximal jejunum, with hemoglobin iron-containing products more easily absorbed than the non-hemoglobin-containing products. Absorptions is the primarily determinant of how iron content is regulated in the human body [Whitney, Cataldo, and Rolfes 1998; Panel on Micronutrients et al. 2002; Mason 2007; National Collegiate Athletic Association 2007; Jelin et al. 2008].

Distribution

Iron is normally bound to blood proteins, referred to as transferring. Once absorbed and transported, it is normally stored in organs such as the spleen, liver, and bone marrow [Mason 2007].

Elimination

Elimination of iron in the body is very limited and can easily build up in the body to toxic levels. A very small fraction is eliminated via the feces, urine, skin, sweat, as well as hair and nails. Also, during the menstrual cycle, there is some loss of iron in the menstrual blood [Mason 2007; Jelin et al. 2008].

Bioavailability

The bioavailability of hemoglobin-containing iron is increased when taken together with certain foods, such as meat, poultry, and fish [Mason 2007].

Uses

The primary function of iron is to combat anemia. Iron deficiency has been associated with chronic diseases, pregnancy, and menstruation in the general population ["Calcium, Copper" 2008]. It has been found to be useful in minimizing attention deficit-hyperactivity disorder (ADHD), canker sores, depression, and fatigue [Jelin et al. 2008]. The usual dose for the treatment of iron deficiency anemia is up to 300 mg daily of elemental iron [Jelin et al. 2008].

Deficiency

Uncontrolled iron deficiency may lead to microcytic hypochromic anemia, which is a very serious condition clinically, the symptoms of which include fatigue and weakness. Mental retardation and growth or developmental abnormalities have been associated with iodine deficiency. Learning disability has also been linked to iodine deficiency [Otten, Helwig, and Meyers 2006; Mason 2007].

Dietary Sources

Certain types of food are documented to be rich in iron. These include cereal products, eggs, meat, nuts, fruits, and vegetables [Mason 2007].

Commercial Preparations

There are various salt forms of iron. These include ferrous fumurate, gluconate, glycine sulfate, orotate, succinate, and sulfate. The daily normal dose as a supplement is about 10–17 mg [Mason 2007]. For iron deficiency anemia in adults, the normal recommended dose is 50–100 mg of elemental iron given in divided doses three times daily [Jelin et al. 2008; McEvoy 2008].

Adverse Effects, Contraindications, and Interactions

The most significant adverse effects seen with intake of iron supplements are nausea and constipation, which can be serious in elderly people [Panel on Micronutrients et al. 2002; Mason 2007; Jelin et al. 2008]. The most common side effects associated with iron or iron intake include, but are not limited to, abdominal pain, joint pain, death, constipation, fatigue, shortness of breath, and vomiting ["Calcium, Copper" 2008]. Individuals with known hypersensitivity or allergy to iron should be cautious about iron-containing products and should avoid it if at all possible. Conditions such as kidney diseases, pancreatitis, and peptic ulcer diseases warrant avoidance of iron supplements ["Calcium, Copper" 2008]. Certain drugs, such as antacids, quinolones, and tetracyclines, among others, can interact with iron-containing products or supplements [Mason 2007]. Zinc can impede iron absorption when taken on an empty stomach [Jelin et al. 2008].

MAGNESIUM

Magnesium is another trace element found in the human body [Mason 2007].

Mechanism of Action

Magnesium is an important mineral required for RNA and DNA synthesis and calcium metabolism [Mason 2007].

Absorption

Magnesium is absorbed primarily in the small intestine by both active process and diffusion. Absorption decreases with increasing intake of magnesium [Mason 2007; Jelin et al. 2008].

Distribution

Distribution of magnesium is carried out in both the soft tissues and skeleton.

Elimination

The major elimination route of magnesium is via the kidney in the form of urine and, to some extent, via the stool. Saliva and breast milk also serve as elimination routes for a very small percentage of magnesium excretion [Mason 2007].

Bioavailability

Bioavailability of magnesium can be greatly increased by vitamin D. Fiber-containing products or foods decrease bioavailability of magnesium [Otten, Helwig, and Meyers 2006].

Uses

The main function or use of magnesium is for the prevention of low levels of magnesium and as a laxative. It has been used in the management of asthma and seasonal allergies and useful in the treatment of ADHD [Jelin et al. 2008]. Magnesium may play a role in certain diseases such as hypertension, diabetes, migraine headaches, osteoporosis, premenstrual syndrome, and normal bone structure [Mason 2007; Jelin et al. 2008].

Deficiency

The following conditions have been documented as clinical signs and symptoms of magnesium deficiency: low calcium and potassium levels, muscle spasm, tremor, lethargy, apathy, convulsions, coma, anorexia, nausea, vomiting, abdominal pain, intestinal paralysis, arrhythmias, tachycardia, and sudden death with cardiovascular origin [Mason 2007].

Dietary Sources

Foods rich in magnesium include cereals, milk, dairy products, meat, fish, vegetables, fruits, and nuts. High-fiber-containing foods provide high sources of magnesium [Mason 2007; Jelin et al. 2008].

Magnesium supplements are available in two main dosage forms, tablets and capsules. It can also be found in combination with calcium supplement or with vitamin D. The daily dose has not been decided yet in the literature [Mason 2007]. However, a dose of 350 mg/day can be taken without any safety issues [Jelin et al. 2008]. Examples of commercial preparations are magnesium chloride (Slo-Mag), magnesium lactate (Mag-Tab SR), and magnesium oxide (Magox) [Jelin et al. 2008]

Adverse Effects, Contraindications, and Interactions

Magnesium is fairly safe and may only pose a problem in individuals with renal or kidney insufficiency. Magnesium has been associated with nausea, vomiting, and diarrhea [Jelin et al. 2008]. It is known to cause cathartic effect at a certain dose, normally ranging between 3 and 5 g of magnesium [Mason 2007]. Alcohol, diuretics, quinolone antibiotics such as ciprofloxacin, and tetracyclines may interact with magnesium, and caution should be exercised when using these combinations [Mason 2007]. As an example, the diuretics may enhance excretion of the magnesium from the body, whereas the antibiotics may significantly reduce absorption of magnesium in the small intestine [Mason 2007]. It does interact with some herbal products and supplements, such as boron, calcium, malic acid, vitamin D, and zinc [Jelin et al. 2008].

MANGANESE

Manganese is another essential trace mineral [Mason 2007].

Mechanism of Action

Manganese is involved in several metabolic pathways involving several enzymes. It is also known to regulate glucose and calcium activities in the body [Mason 2007; Jelin et al. 2008].

Absorption

Absorption is known to take place in the entire small intestine. Normally, the absorption of manganese is not very efficient and is actually considered poor [Mason 2007].

Distribution

Manganese is distributed in the blood primarily bound to plasma proteins and distributed into several organs in the body, including the liver, kidney, pancreas, and the bones [Mason 2007].

Elimination

The main elimination route of manganese is through the feces [Mason 2007].

Bioavailability

Vitamin C and meat are known to increase the bioavailability of manganese. Conversely, iron or iron-containing products and a fiber diet decrease the bioavailability of manganese [Mason 2007].

Uses

It is documented with limited scientific evidence that manganese is useful in certain conditions, such as diabetes. Also, a manganese supplement is known to help treat arthritis conditions [Mason 2007]. It is used for the treatment and prevention of deficiency of manganese, anemia, osteoporosis, and premenstrual syndrome [Jelin et al. 2008].

Deficiency

Weight loss, high cholesterol levels, inflammation of the skin, hair and nail growth retardation, and hair discoloration can be caused by deficiency of manganese [Mason 2007].

Dietary Sources

Dietary sources of manganese come from the usual diets, such as cereal products, milk and dairy products, meat and fish, vegetables, and fruits [Mason 2007].

Adverse Effects, Contraindications, and Interactions

Manganese is very safe when taken orally and has not been associated with any side effects. Chronic inhalation of manganese has associated with toxic reactions in the body. However, this usually occurs in individuals that work in mines and close to industrial plants with a higher tendency to inhale large amounts of manganese [Mason 2007]. There are no interactions reported with manganese [Mason 2007]. However, it found that manganese can interact with calcium, iron, zinc, and certain antibiotics, such as quinolones and tetracyclines [Jelin et al. 2008].

MOLYBDENUM

Molybdenum is an essential trace element for virtually all life forms. It is an important cofactor for a number of enzymes that catalyze important chemical transformations in the carbon, nitrogen, and sulfur cycles [Wuebbens et al. 2000]. Thus, molybdenum-dependent enzymes are required for human health and very interrelated ecosystems. Molybdenum is a trace mineral that plays a role in some anemic conditions, dental caries prevention, and tumor restricting [Wuebbens et al. 2000].

Mechanism of Action

Molybdenum functions as a cofactor for several enzymes, such as sulfite oxidase, xanthine oxidase, and aldehyde oxidase [Otten, Helwig, and Meyers 2006].

Molybdenum functions as an electron carrier in those enzymes that catalyse the reduction of nitrogen and nitrate [Turnlund et al. 1995].

Absorption

In a highly efficient way, the dietary molybdenum is utilized in the body. The mechanism of the action is a passive diffusion process. The gastrointestinal tract readily absorbs soluble, but not insoluble, molybdenum compounds. Absorption rate of molybdenum from the diet of both patients and healthy volunteers averaged about 50% in one study and 88–93% in another study [Wester 1971].

Distribution

Molybdenum is transported in the blood, is loosely attached to erythrocytes, and binds specifically alpha macroglobulin. The highest concentrations are found in the liver and kidney [Mason 2007].

Elimination

Excretion is primarily achieved through the urine and is directly related to the dietary intake [Turnlund et al. 1999].

Dietary Sources

Legumes, grain products, and nuts grown in molybdenum-rich soil are primary sources of dietary molybdenum. Animal products, fruits, and vegetables do not contribute much toward the molybdenum content of diet [American Society 2007; Mason 2007].

Bioavailability

Like most minerals, molybdenum is less absorbed from soy than other food sources [CDC 2008].

Uses

Molybdenum has been found to prevent dental caries in children because of its cariostatic effect [International Molybdenum Association 2002].

Interactions and Side Effects

There are no reported side effects and interactions of molybdenum. One reason for this can be the rapid excretion from urine [Otten, Helwig, and Meyers 2006]. Molybdenum is present in most biological forms [Panel on Micronutrients et al. 2002]. During inborn, errors of metabolism of molybdenum, a genetic defect, prevents sulfite oxidase synthesis. Because of this, in nonconversion of sulfite to sulfate, severe neurological damage leading to early death occurs in infants [Otten, Helwig, and Meyers 2006].

PHOSPHORUS

The active form of phosphorus is found in bones and teeth as phosphates. Bones contain the highest amount of phosphorus, about 85%. It is involved in various metabolic pathways [Food and Nutrition 2002; Otten, Helwig, and Meyers 2006].

Mechanism of Action

The primary function of phosphorus is to maintain normal pH. In addition, it is responsible for storage and energy distribution in the body. It is also known to play a role by activating certain proteins through a mechanism referred to as phosphorylation [Otten, Helwig, and Meyers 2006].

Structurally, phosphorus occurs as phospholipids, which are a major component of most biological membranes [Otten, Helwig, and Meyers 2006].

Absorption

Phosphorus can be found in different food items, both organic and inorganic, in a form of phosphate. Inorganic phosphates therefore represent the most common form of phosphorus that is normally absorbed via concentration gradient [Otten, Helwig, and Meyers 2006].

Elimination

Excretion of endogenous phosphorus is mainly through the kidneys. In healthy adults, urine phosphorus is essentially equal to absorbed dietary phosphorus. Smaller amounts of phosphorus are lost in shedding of cells of skin and intestinal mucosa [Otten, Helwig, and Meyers 2006].

Bioavailability

Food such as beans, cereals, and nuts are rich sources of phosphorus. Certain whole-grain foods tend to have higher phosphorus bioavailability, as well as human milk for infants [Otten, Helwig, and Meyers 2006].

Uses

Athletes sometimes use phosphate supplements because phosphorus supports tissue growth [Healthnotes 2004].

Deficiency

Phosphorus deficiency is rare because of its abundance in various foods. Refeeding of energy-depleted individuals, either orally or parenterally, without adequate attention to supplying phosphorus can precipitate extreme, even fatal, hypophosphatemia. Such outcomes can occur on recovery from alcoholic bouts or from diabetic ketoacidosis [Panel on Micronutrients et al. 2002; Knochel et al. 2006].

Dietary Sources

Nuts, seeds, fish, and poultry are good sources of dietary phosphorus. Phosphates are found in naturally occurring foods and as food additives in the form of various phosphate salts. These salts are used in processed foods for moisture retention, smoothness, and binding [Otten, Helwig, and Meyers 2006].

Adverse Effects

Higher levels of phosphates can lead to calcification in nonskeletal tissues, leading to organ damage. Because kidneys are very efficient in excreting phosphorus, this is not seen in healthy individuals. Hyperphosphatemia has occurred as a result of increased intestinal absorption of phosphate salts taken by mouth as well as a result of colonic absorption of the phosphate salts in enemas [Knochel et al. 2006].

Interactions

Pharmacological doses of calcium carbonate may interfere with phosphorus absorption. When taken in large doses from antacids, aluminum may interfere with phosphorus absorption [Otten, Helwig, and Meyers 2006].

POTASSIUM

The mineral potassium is the main intracellular cation in the body. Potassium is an essential dietary mineral and electrolyte. Normal body function depends on tight regulation of potassium concentrations both inside and outside of cells [Peterson 1997]. It also plays a key role in cardiac, skeletal, and smooth muscle contraction, making it an important nutrient for normal heart, digestive, and muscular function. A diet high in potassium from fruits, vegetables, and legumes is generally recommended for optimum heart health [Otten, Helwig, and Meyers 2006].

Mechanism of Action

Potassium is the principal intracellular cation. It is fundamental to the regulation of acid base and water balance [Sheng 2000; Jelin et al. 2008].

Absorption

The small intestine is the main site of absorption of potassium [Mason 2007]. In a healthy population, about 80–85% of the dietary potassium is absorbed. As insulin stimulates the sodium-potassium ATPase pump, insulin concentration can affect extracellular as well as plasma concentration of potassium [Otten, Helwig, and Meyers 2006].

Distribution

Potassium is present in the form of a cation intracellular fluid, whereas sodium is the principal cation in extracellular fluid. Potassium concentrations are about 30 times higher inside than outside cells, whereas sodium concentrations are more than 10 times lower inside than outside cells [Sheng 2000].

Elimination

The relation between dietary potassium intake and urinary potassium content is high. Excretion is mainly via urine. Kidneys cannot conserve potassium. The rest of the unabsorbed potassium is excreted through feces and a smaller amount in sweat [Otten, Helwig, and Meyers 2006].

Bioavailability

Because of the high water solubility of potassium, its absorption is very efficient. There is limited information about the bioavailability from individual foods, because it is well absorbed from most of the foods [Otten, Helwig, and Meyers 2006].

Deficiency

Potassium deficiency is uncommon because of its natural abundance, unless there is restricted food intake. African Americans normally would benefit from increased potassium intake because of their sodium sensitivity [Otten, Helwig, and Meyers 2006].

Dietary Sources

Fruits and vegetables are good sources of potassium [Otten, Helwig, and Meyers 2006].

Adverse Effects and Interactions

In healthy individuals, food alone cannot cause potassium excess. However, supplements can be responsible for acute toxicity in healthy individuals. Cardiac arrhythmias and gastrointestinal discomfort are results of excess potassium intakes [Otten, Helwig, and Meyers 2006].

SELENIUM

Introduction

Selenium is an essential trace element, and it is also an antioxidant nutrient involved in the defense of the body from oxidative stress [Mason 2007]. Humans and animals require selenium for the function of a number of selenium-dependent enzymes that are essential in many metabolic functions in the body [Rayman 2000].

Mechanism of Action

During selenoprotein synthesis, selenocysteine is incorporated into a very specific location in the amino acid sequence to form a functioning protein. However, when selenium is present in the soil, plants incorporate it into compounds that contain sulfur [Rayman 2000].

Absorption

Most dietary selenium is in the form of selenomethionine or selenocysteine. Both these forms are well absorbed. Other forms of selenium include selenate and selenite, commonly found in supplements and fortified foods. Selenium is destroyed when foods are refined or processed. Selenium in general is known to be well absorbed after intake; however, a significant amount is known to be eliminated in the urine [Otten, Helwig, and Meyers 2006].

Distribution

Two pools of reserve selenium are present in the body. The first is selenomethionine, which may have a similar function as methionine. The second reserve pool is the selenium found in liver glutathione [Otten, Helwig, and Meyers 2006].

Elimination

Selenite, selenate, and selenocysteine are metabolized to selenide in the body. These selenides can be metabolized further or be converted to excretory metabolite. Selenium is excreted through urine [Otten, Helwig, and Meyers 2006].

Bioavailability

Bioavailability of selenium from animal sources, and particularly fish, is comparable with plant sources such as yeast. Bioavailability from fortified foods is lower than natural sources [Otten, Helwig, and Meyers 2006].

Uses

Selenium can possibly be used in cancer prevention. Selenium supplementation can increase sperm motility. Selenium has been found to have a positive effect on mood swings in some individuals. Patients with asthma, decreased immune function, and HIV infection may benefit from selenium supplements [Mason 2007].

Deficiency

Insufficient selenium intake results in decreased activity of certain enzymes that are responsible for antioxidant activities in the body, such as glutathione peroxidases, thioredoxin reductase, and thyroid deiodinases. Obvious clinical illness is rare with selenium deficiency, even in a more severe form. However, selenium-deficient individuals appear to be more susceptible to additional physiological stresses. Muscle pain and tenderness have both been associated with selenium deficiency [Burk and Levander 1999; Mason 2007].

Dietary Sources

Selenium content of the food greatly depends on the soil in which the animal was raised or the plant was grown. The richest food sources of selenium are organ meats and seafood, followed by muscle meats. Brazil nuts grown in areas of Brazil with selenium-rich soil may provide more than 100 mcg of selenium in one nut, whereas those grown in selenium-poor soil may provide 10 times less [Chang 1995].

Commercial Preparations

Selenium supplements are available in several forms. Sodium selenite and sodium selenate are inorganic forms of selenium, whereas selenomethionine represents the organic form, which is rich in most naturally occurring food [Panel on Dietary Antioxidants and Related Compounds et al. 2000].

Adverse Effects

High doses of selenium can be toxic. Acute and fatal toxicities have occurred with accidental or suicidal ingestion of gram quantities of selenium. Chronic selenium toxicity (selenosis) may occur with smaller doses of selenium over long periods of time. Hair and nail brittleness are known to be common in selenium toxicity. Gastrointestinal problems, skin rashes, garlic breath odor, fatigue, irritability, and nervous system problems have also been reported with selenium toxicity [Panel on Dietary Antioxidants and Related Compounds et al. 2000].

Interactions

There are no known food interactions affecting absorption and usage of selenium [Jelin et al. 2008]. Limited information is currently found on interaction between selenium and medications. Medications such as valproic acid have been found to decrease selenium levels [Flodin 1990]. Clozapine, an antipsychotic drug, has also been found to decrease selenium levels [Mason 2007].

The efficacy of cholesterol-lowering agents, such as simvastatin and niacin, has been reduced when used in conjunction with beta-carotene, vitamin C, vitamin E, and selenium [Jelin et al. 2008].

ZINC

Zinc is an essential trace element throughout the life process [Mason 2007].

Mechanism of Action

Zinc is the second most prevalent trace element in the body. It is known that the cell nucleus contains about 30% of zinc. It acts as a cofactor in many biological activities that take place in the body such as DNA, RNA, and protein synthesis in general [Jelin et al. 2008].

Absorption

The small intestine, particularly jejunum, acts as the most effective zinc absorption site. Zinc deficiency enhances absorption [Otten, Helwig, and Meyers 2006; Mason 2007; Jelin et al. 2008].

Distribution

Zinc is well distributed, and more than 85% is stored in skeletal muscle and bone. Other zinc storage tissues include the liver, kidney, pancreas, prostate gland, and retina [Otten, Helwig, and Meyers 2006; Mason 2007].

Elimination

Zinc is eliminated mainly through feces. Starvation and trauma tend to increase zinc losses through urine. Skin cell shedding, sweat, hair, semen, and menstruation are the other means of zinc loss through the body [Otten, Helwig, and Meyers 2006].

Bioavailability

Absorption of zinc can be impaired by phytate and calcium, whereas proteins have positive impact on zinc absorption [Lonnerdal 2000].

Uses

A zinc supplement is indicated for treatment and prevention of its deficiency. It can also be used for managing the common cold, recurrent ear infection, and respiratory infection [Jelin et al. 2008].

Deficiency

There is a broad range of physiological signs of zinc deficiency because of the multiple biological functions and multiple distribution sites in tissues. Clinical signs of frank zinc deficiency may be seen in skin problems such as inflammation [Van Wouwe 1989]. Diarrhea, impaired cognitive, behavioral problems, impaired memory, learning disability, and neuronal atrophy can be seen in infants as indications of zinc deficiency [Hambidge 1986]. Zinc deficiency in pregnancy can lead to growth retardation in the unborn baby in addition to conditions such as congenital abnormalities in fetus [Mason 2007].

Dietary Sources

Zinc occurs in a wide variety of foods but is found in highest concentrations in animal sources, particularly in red meat and seafood. Whole grains contain more zinc than refined grains [Otten, Helwig, and Meyers 2006].

Adverse Effects

High intakes of zinc are possible either through supplemental zinc or by contact with environmental zinc. Toxicity symptoms, such as nausea, vomiting, epigastric pain, diarrhea, and lethargy, may occur with acute high intakes [Fosmire 1990].

Interactions

Absorption of copper, iron, and folate can be decreased by zinc. Conversely, zinc absorption may be reduced by phytic acid, fiber, calcium, and phosphates. Protein intake positively affects zinc absorption [Otten, Helwig, and Meyers 2006]

REFERENCES

American Nutraceutical Association. http://www.ana-jana.org.

Anderson R. A., N. Cheng, N. A. Bryden, M. M. Polansky, N. Cheng, J. Chi, and J. Feng. 1997. Elevated intakes of supplemental chromium improve glucose and insulin variables with type 2 diabetes. *Diabetes* 46:1786–1791.

Burk, R. F. and O. A. Levander. 1999. Selenium. In *Modern Nutrition in Health and Disease.* 9th edition. Edited by Shils, M., J. A. Olson, M. Shike, and A. C. Ross. Baltimore, MD: Lippincott Williams and Wilkins, pp. 265–276, 600.

Calcium Monograph. http://kroger.naturalstandard.com. Accessed October 21, 2008.

Celotti, F. and A. Bignamini. 1999. Dietary calcium and mineral/vitamin supplementation: A controversial problem. *J. Int. Med. Res.* 27:1–14.

Centers for Disease Control. 2008. *Calcium and Bone Health.* http://www.cdc.gov/nutrition/everyone/basics/vitamins/calcium.html. Accessed November 3, 2008.

Chang, J. C. 1995. Selenium content of brazil nuts from two geographic locations in Brazil. *Chemosphere* 30:801–802.

Copper Monograph. http://kroger.naturalstandard.com. Accessed October 21, 2008.

Flodin, N. W. 1990. Micronutrient supplements: toxicity and drug interactions. *Prog. Food Nutr. Sci.* 14:277–331.

Fosmire, G. J. 1990. Zinc toxicity. *Am. J. Cin. Nutr.* 51:225.

Glinsmann, W. H. and W. Mertz. 1966. Effect of trivalent chromium on glucose tolerance. *Metabolism* 15:510–519.

Hambidge, K. M. 1986. Zinc deficiency in the weanling: How important? *Acta. Peadiatr. Scand. Suppl.* 323:52.

Healthnotes. 2004. Phosphorus for sports and fitness. http://www.evitamins.com/healthnotes.asp?ContentID=3892007#Reference-List. Accessed November 12, 2008.

Heaney, R. P. 2001. Calcium needs of the elderly to reduce fracture risk. *J. Am. Coll. Nutr.* 20 (2 Suppl.):192S–197S.

International Molybdenum Association. 2008. Molybdenum in Human Health. http://www.imoa.info/HSE/environmental_data/human_health/molybdenum_therapeutic_uses.html. Accessed October 11, 2008.

Iodine Monograph. http//kroger.naturalstandard.com. Accessed October 21, 2008.

Jeejeebhoy, K. N., R. C. Chu, E. B. Marliss, G. R. Greenberg, and A. Bruce-Robertson. 1977. Chromium deficiency, glucose intolerance, and neuropathy reversed by chromium supplementation in a patient receiving long-term total parenteral nutrition. *Am. J. Clin. Nutr.* 30:531–538.

Jelin, J. M., P. J. Gregory, F. Batz, et al. 2008. *Pharmacist's Letter/Prescriber's Letter Natural Medicines Comprehensive Database*. 10th ed. Stockton, CA: Therapeutic Research Faculty, pp. 278, 607, 844, 973, 990.

Knochel, J. P. 2006. Phosphorus. In *Modern Nutrition in Health and Disease.* 10th edition. Edited by Shils, M. E., M. Shike, A. C. Ross, B. Caballero, R. J. Cousins. Baltimore, MD: Lippincott Williams and Wilkins, pp. 211–222.

Lee, N. A. and C. A. Reasner. 1994. Beneficial effect of chromium supplementation on serum triglyceride levels in NIDDM. *Diabetes Care* 17:1449–1451.

Lonnerdal, B. 2000. Dietary factors influencing zinc absorption. *J. Nutr.* 130:1378.

Manganese Monograph. 2008. http://online.lexi.com/crlsql/servlet/crlonline. Accessed November 11, 2008.

Mason, P. 2007. *Dietary Supplements.* 3rd edition. Grayslake, IL: Pharmaceutical Press.

McEvoy, G. K., ed. 2008. *AHFS Drug Information 2008*. Bethesda, MD: American Society of Health-System Pharmacists.

National Collegiate Athletic Association. 2007. List of banned drug classes 2004–2005. http://www.ncaa.org/wps/wcm/connect/resources/file/ebab6d4a1276fc7/checklist.pdf?MOD=AJPERES. Accessed October 2008.

Otten, J. J., J. P. Helwig, and L. D. Meyers. 2006. *Dietary reference intakes: The essential guide to nutrient requirements.* Washington, D.C.: The National Academies Press, pp. 321–327.

Panel on Dietary Antioxidants and Related Compounds, et al. 2000. Selenium. In *Dietary Reference Intakes for Vitamin C, Vitamin E, Selenium, and Carotenoids.* Washington, D.C.: National Academy Press, 284–324. http://books.nap.edu/openbook.php?record_id=9810&page=R1.

Panel on Micronutrients, et al. 2002. Dietary reference intakes for vitamin A, K, arsenic, boron, chromium, copper, iodine, manganese, molybdenum, nickel, silicon, vanadium, and zinc. Washington, D.C.: Academy Press. http://books.nap.edu/catalog.php?record_id=10026. Accessed October 2008.

Pattanaungkul, S., B. L. Riggs, A. L. Yergey, N. E. Vieira, W. M. O'Fallon, S. Khosla. 2000. Relationship of intestinal calcium absorption to 1,25-dihydroxyvitamin D [1,25(OH)2D] levels in young versus elderly women: Evidence for age-related intestinal resistance to 1,25(OH)2D action. *J. Clin. Endocrinol. Metab.* 85:4023–4027.

Peterson, L. N. 1997. Potassium in nutrition. In *Handbook of Nutritionally Essential Minerals.* Edited by O'Dell, B. L. and R. A. Sunde. New York: Marcel Dekker, pp. 153–183.

Power, M. L., R. P. Heaney, H. J. Kalkwarf, R. M. Pitkin, J. T. Repke, R. C. Tsang, and J. Schulkin. 1999. The role of calcium in health and disease. *Am. J. Obstet. Gynecol.* 181:1560–1569.

Rayman, M. P. 2000. The importance of selenium to human health. Lancet 356:233–241.

Sheng, H.-W. 2000. Sodium, chloride and potassium. In *Physiological Aspects of Human Nutrition.* Edited by Stipanuk M. Philadelphia, PA: W. B. Saunders Company, pp. 686–710.

Turnlund, J. R., W. R. Keyes, G. L. Peiffer, and G. Chiang. 1995. Molybdenum absorption, excretion, and retention studied with stable isotopes in young men during depletion and repletion. *Am. J. Clin. Nutr.* 61:1102–1109.

Turnlund, J. R., C. M. Weaver, S. K. Kim, W. R. Keyes, Y. Gizaw, K. H. Thompson, and G. L. Peiffer. 1999. Molybdenum absorption and utilization in humans from soy and kale intrinsically labeled with stable isotopes of molybdenum. *Am. J. Clin. Nutr.* 69:1217–1223.

Van Wouwe, J. P. 1989. Clinical and laboratory diagnosis of acrodermatitis enteropathica. *Eur. J. Pediatr.* 149:2.

Wester, P. O. 1971. Trace element balances in two cases of pancreatic insufficiency. *Acta. Med. Scand.* 190:155–161.

Whitney, E., C. B. Cataldo, S. R. Rolfes, eds. 1998. Understanding normal and clinical nutrition. Belmont, CA: Walsworth.

Wuebbens, M. M., M. T. Liu, K. Rajagopalan, and H. Schindelin. 2000. Insights into molybdenum cofactor deficiency provided by the crystal structure of the molybdenum cofactor biosynthesis protein MoaC. *Struct. Fold. Des.* 8:709–718.

CHAPTER 6

Physiochemical Characterization of Nutraceuticals

Ajoy Koomer

CONTENTS

INTRODUCTION

According to the American Nutraceutical Association, the term "nutraceutical" was derived by condensing the terms "pharmaceutical and nutrition" in 1989 by Stephen DeFelice, who was founding chairman of the Foundation for Innovation in Medicine [American Nutraceutical 2008]. DeFelice suggests that, "A nutraceutical is any substance that is a food or a part of a food and provides medical or health benefits, including the prevention and treatment of disease. Such products may range from isolated nutrients, dietary supplements and specific diets to genetically engineered designer foods, herbal products, and processed foods such as cereals, soups and beverages" (2008). Nutraceuticals and dietary supplements can be broadly classified into five categories, namely vitamins, minerals, botanical substances, herbal extracts, and miscellaneous or specialty components [Krull and Swartz 2001]. The vitamins category includes fat and water-soluble vitamins and nutritional factors [Krull and Swartz 2001]. The minerals category comprises mineral chelates, salts, single and trace elements, and multiple minerals consisting of amino acids mixes [Krull and Swartz 2001]. Botanical substances include mixed and single whole herbs, essential oils, tea mixtures, and traditional formulas [Krull and Swartz 2001]. The specialty components include antioxidants, carotenoids, essential (omega-3) and omega-3 and -6 fatty acids, phytosterols, anthocyanins, flavonoids, probiotics, lecithins, glandular, diet acids, and digestive acids [Krull and Swartz 2001]. As noted by specialists, nutraceuticals are gaining public acceptance because of escalating consumer market share for "wellness products" [Krull and Swartz 2001; Dureja, Kaushik, and Kumar 2003; Metha et al. 2007]. Because of a lack of standardization of active ingredients in nutraceutical-related products coupled with increased market interest, the United States has had increased attention "in the marketing, claims substantiation, manufacturing, and FDA-based regulations of nutraceuticals" [Krull and Swartz 2001; Dureja, Kaushik, and Kumar 2003; Mehta et al. 2007]. The tightening of the regulation apparatus is the driving force for the implementation of reliable analytical techniques for the reliable detection of active gradients in nutraceutical complex matrices and their physicochemical characterization for enhanced formulation and quality-control studies [Krull and Swartz 2001; Dureja, Kaushik, and Kumar 2003]. The nutraceutical testing methods available in industry include stability testing, dissolution testing, *in vitro* release rate testing, content uniformity testing, high-performance liquid chromatography (HPLC) and thin-layer chromatography (TLC), mass spectrometric analysis, Fourier transform infrared spectroscopy, and ultraviolet/visible light (UV-vis) spectroscopy [Analytical Solutions 2008]. A major factor in physicochemical characterization is stress and stability testing. Stress testing is performed to identify all potential degradants. Generally, chromatographic techniques are used as methods of validation in stress testing, with detection being performed by UV-vis spectroscopy [Mehta et al. 2007]. In this chapter, we will focus on physicochemical characterization of phytosterols, fatty acids, caratenoids, amino acids, anthocyanins, and water-soluble vitamins.

PHYTOSTEROLS

Phytosterols are diphenolic compounds present in plant and animals that resemble the human sex hormone estrogen [Hurst 2002]. They can be classified into three categories, which are coumestans, lignans, and isoflavonoids [Hurst 2002]. They can be detected and quantified by TLC, HPLC, and gas chromatography-mass spectrometry (GC-MS) coupled with UV-vis spectroscopy [Hurst 2002]. However, before implementation of any methods for validation, sample preparation and extraction play a critical role. Usually, phytosterol matrix samples are spiked with 20% (w/v) of purified isoflavones, such as daidzein, genistein, and biochanin A [Hurst 2002]. The spiked solution consists of 1.3% (w/v) of tertiary butyl hydroquinone in methanol [Hurst 2002]. The use of isoflavones as internal standards stem from the pros that it allows for accurate "determination of analyte recoveries and account for weight loss in the extraction procedure" [Hurst 2002]. After sample preparation, they are extracted with 80% methyl alcohol for 0.5 h, although other solvents, such as ethanol, acetonitrile, and acetone, can be used [Hurst 2002]. TLC has been used routinely in the identification of isoflavonoids [Hurst 2002]. Scientists have used precoated polyamide TLC plates for the separation of isoflavones and other diphenolic compounds from soya bean extracts [Hurst 2002]. As noted by scientists, the methanolic extracts spotted on 20 × 20 plates were developed with methanol/acetic acid/water mixture in proportions of 90:5:5 (weight/volume) [Hurst 2002]. After eluting the bands with ethanol, they were "rechromatographed on polyamide using chloroform-methanol-methyl ethyl ketone solvent system in the ratio 12:2:1" [Hurst 2002]. When viewed under a UV lamp at 366 nm, the analyzed fractions produced characteristics Rf values corresponding to daidzein, genistein, formononetin, and biochanin A [Hurst 2002]. It is to be noted that TLC is a qualitative procedure that cannot be used for quantification of individual phytosterol fractions from complex plant and food matrices [Hurst 2002]. HPLC is the method of choice here. Although MS techniques can be used with improved sensitivity, they are not popular because of cumbersome sample preparation time and costs [Hurst 2002]. Generally, phytosterols can be purified and quantified by mixtures of "methanol or acetonitrile and aqueous acids or buffers" by the use of reverse-phase (RP) C18 separation columns [Hurst 2002]. The most frequently used HPLC techniques use either linear or nonlinear gradient elutions for isoflavonoid separation and "quantitative estimation from legume or soybean matrices" [Hurst 2002]. Generally, in gradient elution HPLC, "acetonitrile increases by 2.25% min," enabling "the separation of isoflavone-β-glucoside conjugates and aglucones" in a single experiment [Hurst 2002]. The experimental runtime usually does not exceed 60 min, with equilibration between the cycles [Hurst 2002]. Recently, RP-HPLC has been used to separate isoflavones such as daidzein and genistein [Hurst 2002]. Isocratic elution conditions have been reported by researchers in the separation of phytosterols, although the method has been unsuccessful, except for genistein [Hurst 2002]. Because of the high concentration of isoflavonoids in soy or legume products, UV-vis spectroscopy coupled with RP-HPLC is used for detection [Hurst 2002]. However, the method suffers from challenges in quantifying isoflavonoids from sources with low concentration or if phytosterols other than isoflavonoids are involved [Hurst 2002]. In those cases, "fluorometric detection, amperometric methods, or thermospray MS with SIMI" can be considered to increase sensitivity of common UV detection systems [Hurst 2002].

FATTY ACIDS

This section focuses on quantitative functional analysis of essential omega-3 fatty acids, namely EPA, DHA, and ALA [Hurst 2002]. For the quantitative estimation of fatty acid profile in complex matrix (functional foods), the usual steps are extraction to release the free fatty acids, derivation of the released fatty acids, and chromatogram analysis with GC with flame ionization detection, although other techniques such as GC-MS and HPLC have been used with success [Hurst 2002]. The Folch method of lipid extraction, using chloroform and methanol as solvents in the volume ratio of 2:1, is most common [Hurst 2002]. To eliminate the nonlipid contaminants such as carbohydrates and amino acids, the components chloroform-methanol-water/0.88% potassium chloride must be adjusted to 8:4:3 by volume, yielding a biphasic system

[Hurst 2002]. The free fatty acids are then usually derivatized to fatty acid methyl esters (FAMEs) [Hurst 2002]. These derivatives are volatile and render themselves to excellent gas chromatographic analysis [Hurst 2002]. In GC, the area percentage of each FAME peak is proportional to the weight percentage of each FAME [Hurst 2002]. However, this method overestimates long-chain fatty acids and underestimates short-chain ones. Neglecting this caveat, this is still the most commonly used procedure [Hurst 2002]. The quantification of individual FAME in milligrams per gram of the total sample is facilitated by the use of an internal standard, which can be added during either extraction or derivatization process [Hurst 2002]. Recently, GC-MS and HLPC techniques to characterize DHA and ALA have been reported [Hurst 2002].

CAROTENOIDS

Carotenoids are lipophilic compounds present in both plant and animal species. The unique chromophore for these compounds containing polyene structures make amenable to light absorption in the visible range; hence they are identified by UV-vis spectroscopy [Hurst 2002]. Colors usually vary from yellow to red via different shades of orange. The most common techniques used for quantitative identification of carotenoids from plant matrices involve TLC/UV-vis spectroscopy or HPLC/UV-vis spectroscopy. Different carotene components can be separated using TLC following the method of Gross [Hurst 2002]. Silica gel is conventionally used to prepare the stationary phase in chromatographic separation with low polarity solvents used as elutents. After separation in TLC plates, the bands corresponding to each pigment are recovered from the chromatographic support and eluted with low polarity solvents, such as acetone. Filtration is used to remove the adsorption materials, and the absorption spectrum of the pigments is quantified using the Lambert–Beer law. The actual quantification of carotenes in the sample is then performed, knowing the following parameters in the hand: "weight of sample extracted, final volume of extract, volume of extract chromatographed and volume of elution of pigments chromatographed" using a modified equation based on the Lambert–Beer law [Hurst 2002]. Both normal-phase and RP-HPLC coupled with visible spectroscopy have been reported for the quantitative estimation of carotenes. The former uses a polar compound in the stationary phase, whereas the latter uses nonpolar stationary phases such as octa silane or octadecyl silane. If fixed wavelength detector is used, the selected wavelength is 450 nm; for diode arrays or multiple wavelength detectors in UV-vis, spectra are obtained at the absorption maxima of each pigment [Hurst 2002].

ANTHOCYANINS

Anthocyanins are a diverse group of water-soluble pigments responsible for the red, blue, and purple colors of plants. They are all characterized by the cyanidin aromatic ring structure and are classified by the number of sugars and their position in the aglycon chain [Hurst 2002]. Andersen and his colleagues at the University of Bergen in Norway developed analytical techniques for the isolation, separation, and characterization of anthocyanins. In the initial step, materials were extracted in methanolic solution containing 1% trifluoroacetic acid, followed by partitioning the extract against ethyl acetate to eliminate the flavonoid contaminants [Hurst 2002]. The partitioned mixture was then subjected to column chromatography using an Amberlite XAD-7 column, eluted with 50% methanol and 100% methanol containing trifluoroacetic acid, subjected to fractionation in a Sephadex LH 20 column, and further eluted with methanol, water, and trifluoroacetic acid mixture; it is then subjected to additional purification by preparative HPLC using a RP C18 column. The purified compounds are detected by UV-vis adsorption spectroscopy based on absorption maxima of individual aglycon units. For example, peonidin-3-glucoside exhibits absorption maximum between 520 and 526 mn whereas delphinidin shows the same between 532 and 537 nm. Recently ^{1}H and ^{13}C nuclear magnetic resonance spectroscopy have been used to detect anthocyanins [Hurst 2002].

AMINO ACIDS

Of more than 200 amino acids found in nature, about 20 are components of proteins including enzymes. Of the 20 amino acids, nearly half of them are essential amino acids. These amino acids cannot be synthesized by humans and had to be obtained as nutritional supplements [Hurst 2002]. A nutritionist generally focuses on obtaining amino acid profile for the essential amino acids. The physicochemical characterization of amino acids involves acid or alkaline hydrolysis of the samples and extraction, followed usually by HPLC purification and detection. However, detection of amino acids poses challenges because only three of them, phenylalanine, tyrosine, and tryptophan, have significant UV absorption. Also, only tyrosine and tryptophan fluoresce. Thus, the universal detection techniques stress "on refractive index and light scattering or derivatizing the amino acid to an intermediate product that shows absorption in the UV-visible range of the spectra" [Hurst 2002]. This derivatization can be performed precolumn or postcolumn. Precolumn indicates before HPLC separation whereas postcolumn refers to after HPLC purification. It should be noted that each procedure has from pros and cons. The groups commonly used in precolumn derivation include 9-fluorenylmethyl chloroformate and *o*-phthalaldehyde. In postcolumn derivatization, typically ninhydrin or and *o*-phthalaldehyde is preferred. When using postcolumn derivation, ones uses ion exchange chromatography, which separates amino acids based on charges [Hurst 2002]. Also, in this case, tedious sample preparation is not required. In cation-exchange chromatography used for amino acid purification, the stationary phase is negatively charged and traps the positively charged amino acid at a low pH. On the contrary, when precolumn derivatization is preferred, RP-chromatography is used with either C8 or C18 on silica. The added advantage stems from the fact that silica supports can stand high pressures compared with polymeric supports used in exchange chromatography with increased flow rates and reduced runtimes. The major drawback with precolumn derivatization is the chance of sample manipulation before HPLC separation and purification. The cons of the postcolumn method include the fact that adding reagents to the eluant can adversely interfere with the chromatogram resolution [Hurst 2002]. Also, the postcolumn reactor is costly and derivatization is restricted by reaction kinetics and the chemical compatibility of the introduced group with mobile phases of the chromatograph [Hurst 2002]. Thus, choosing the best method will depend on the sample properties and reaction conditions.

WATER-SOLUBLE VITAMINS

As the name suggests, vitamins are "vital amines" that play an important role in human growth and development. Since the past decade, vitamins have drawn attention because of their nutritional implications and the need to develop standardized quality-control procedures [Hurst 2002]. However, development of physicochemical methods for characterization is difficult because these compounds are diverse in nature, present in low quantities in complex matrices, such as functional foods and vegetables, mimicking of their activity by other compounds, and the special stability considerations. Most of the analytical methods described in literature focus on extraction, purification, and detection [Hurst 2002]. The extraction procedure from the complex matrices may involve heat, acid or alkali or enzymes, followed by cleanup procedures, and then quantitative estimation by RP-HPLC coupled with UV, fluorescence, or protein binding. The water-soluble B vitamins, such as thiamin, riboflavin, and niacin, may be estimated by RP-HPLC coupled with fluorescence or UV absorption. For thiamine, after extraction, it is convenient to convert it to fluorescent thiocrome by oxidation and then do quantitative separation and identification by RP-HPLC coupled with fluorometry using an excitation wavelength of 360–365 nm and an emission wavelength of 460–480 nm [Hurst 2002]. For RP-HPLC, usually C8 or C18 columns are used with the mobile phase ranging from organic solvents and ion pairs to organic-aqueous buffer mixtures. It should be noted that ion exchange and normal HPLC purification has also been reported for thiamine and other B vitamins. Cobalamin or vitamin B_{12} can be characterized by HPLC coupled with specific protein binding assays involving radio isotopes or enzymes. Its diversity and low propensity in food matrices make it an unsuitable candidate for UV detection; the same applies to biotin or folic acid [Hurst 2002].

REFERENCES

American Nutraceutical Association. 2008. Nutraceutical Information. http://www.ana-jana.org/nut_info_details.cfm?NutInfoID=4.

Analytical Solutions. 2008. Nutraceuticals. http://www.asi-rtp.com.

Dureja, S., D. Kaushik, and V. Kumar. 2003. Developments in nutraceuticals. *Indian J. Pharmacol.* 35:363–372.

Hurst, W. J. 2002. *Methods of Analysis for Functional Foods and Nutraceuticals*. 2nd edition. Boca Raton, FL: CRC Press.

Krull, I. and M. Swartz. 2001. Striving to validate nutraceuticals. *LCGC* 19:1142–1149.

Mehta, J., R. L. Chapman, W. G. Chambliss, M. A. Murray, and W. Obermeyer. 2007. Formulation and Stability of Nutraceuticals. http://www.aapspharmaceutica.com/meetings/annualmeet/am07/ProgramInfo/Prelim/program/pt_form.asp.

CHAPTER 7

Development of Techniques for Analysis of Nutraceuticals with Specific Reference to Glucosamine and Coenzyme Q_{10}

John Adams and Brian Lockwood

CONTENTS

INTRODUCTION

Over the past two decades, the public has gained an increasing interest in the role of diet, nutrition, and dietary supplements in the prevention and treatment of disease [Childs and Poryzees 1998]. This has resulted in a dramatic increase in the use of commercially available nutraceuticals and their foodstuffs [G. Lockwood 2007].

This widespread use of nutraceuticals has stimulated the development of a number of analytical techniques to monitor their identification and levels in available raw materials, simple extracts, formulated products, biological fluids, and a number of other matrices.

This review covers the wide range of techniques developed for the analysis of two of the most commonly studied nutraceuticals. There are a number of scientific reasons for the development of improved analytical methods, predominantly in the area of additional understanding of disease states and their treatment and particularly the modes of action of specific nutraceuticals. Other important reasons include prediction of disease states and compliance.

The analytical procedures used for identification and quantification of nutraceuticals are becoming increasingly sophisticated, reflecting current analytical advances. Manufacturers, independent researchers, and consumer organizations such as ConsumerLab.com often produce detailed information about levels of active constituents, in both the natural materials and formulated products. Clinical researchers are increasingly investigating biological fluids, in clinical trials and *in vivo* research in an attempt to determine the biological fate of nutraceuticals.

There are few published reviews on the analytical techniques used for nutraceuticals. There has been a limited overview of 26 examples published in 2007 [B. Lockwood 2007], chapters in two edited texts published in 2002 [Ho and Zheng 2002; Hurst 2002], but, for the majority, there are few available methods. Monographs, such as those (being) produced by the FDA and the Office of Dietary Supplements of the National Institutes of Health of the United States, list techniques for sample preparation and extraction, chromatographic separation, detection methods, and quantitation methods [B. Lockwood 2007]. In addition, these publications highlight the wide range of techniques increasingly being used for their identification and quantification in both manufactured products and biological fluids. HPLC, using a range of different detectors, has been used for the majority of nutraceuticals, but newer techniques are also becoming widespread.

The phytoestrogens from soy and flax, n-3 PUFAs and CLA, carotenoids, lycopene, zeaxanthin, astaxanthin, and lutein have been the subject of a recent edited book on the subject [Ho and Zheng 2002]. The other text by Ho and colleagues reviews the techniques available for cocoa, chocolate, cranberry, and guggul (*Commiphora wightii*) constituents [Hurst 2002]. Two recent publications have reviewed the analysis of soy [Dentith and Lockwood 2008] and tea polyphenols [Kirrane and Lockwood 2008].

This chapter will focus on the techniques used for two important and widely used nutraceuticals whose analysis has not been reviewed previously, namely glucosamine and CoQ_{10}. The chapter will also explain the development of techniques used in analysis of such nutraceuticals.

Glucosamine is widely used to improve joint health, specifically osteoarthritis and rheumatoid arthritis, and skin health. It is also widely used for joint problems in domestic animals. CoQ_{10} has been reported to have activities in cardiovascular health, cancer prevention, respiratory and skin diseases, and animal ailments.

TECHNIQUES FOR ANALYSIS OF GLUCOSAMINE

There are many challenges involved in the analysis of glucosamine. First, the hydrophilic nature of the molecule makes the extraction with organic solvents from plasma ineffectual. Along with the drug, a variety of endogenous compounds with a chemical structure similar to glucosamine, such as glucose, galactose, other sugars and amino sugars [Huang et al. 2006b], would also be extracted from a biological sample [Zhang et al. 2006]. To circumvent this, often lengthy sample preparation techniques are used. Second, several ingredients are often found within a commercial formulation of glucosamine. Chondroitin, also a glycoaminoglycan but with molecular weight of 5,000–55,000, is commonly found in such formulations, and this compound has the potential for interference with the quantitative analysis of glucosamine [Shen, Yang, and Tomellini 2007]. Analysis of chondroitin is not nearly as highly developed as that for glucosamine.

Radiolabeling

Previous analytical methods for biological samples depended on radiolabeling a compound; however, using radioactivity to quantify glucosamine may potentially confound results because the parent drug molecule cannot be differentiated from its degradation products and metabolites [Huang et al. 2006a; Zhang et al. 2006; Shen, Yang, and Tomellini 2007]. Therefore, there is a need for specific analytical techniques to analyze the content of glucosamine in both commercial products and biological fluids. Both areas present different problems in terms of analysis. A variety of methods have been suggested in an attempt to overcome such challenges, with each method having its own applications, limitations, and advantages, which will be discussed below.

HPLC Separation

It has been suggested that, of all the methods, HPLC is the most extensively used and highly sensitive technique for the analysis of glucosamine [Huang et al. 2006b]. Glucosamine poses several challenges for analysis using liquid chromatographic methods, and several techniques have been used and studied. The first challenges are created by the nature of the glucosamine molecule (Figure 7.1).

Because sugars and amino sugars (such as glucosamine) are highly polar molecules, they are not retained on common hydrophobic HPLC column packing materials (e.g., RP C18), which makes separation difficult [Roda et al. 2006]. A possible solution to this problem is the use of more expensive amino columns [Shao et al. 2004] or the complexation of glucosamine through its amino group with a hydrophobic compound. The chemical structure of glucosamine also lacks a suitable chromophore or fluorophores, which absorbs in the wavelength range useful for HPLC with either UV or fluorescence detection [Nemati et al. 2007; Shen, Yang, and Tomellini 2007].

Precolumn Derivatization Agents for UV/Fluorescent Visualization

To overcome this problem, glucosamine is often derivatized to incorporate a suitable chromophore or fluorophore into its structure to improve its detection by UV methods [Shen Yang, and Tomellini 2007], and such agents can be bound to glucosamine through its amino group.

It has been reported that precolumn derivatization steps can often be lengthy, making the time of analysis for such techniques unsuitable for routine use [Shen, Yang, and Tomellini 2007].

Issues with the stability of glucosamine-derivative complexes have been raised. Shen, Yang, and Tomellini [2007] monitored the UV absorption of a glucosamine derivative σ-phthalaldehyde-3-mercaptopropionic acid (OPA-MPA) over time to illustrate stability issues. The derivatization reaction was performed with an OPA-MPA molar ratio of 1:50, according to a published method in the U.S. Pharmacopeia. The 1:50 ratio was shown to increase stability and was therefore the chosen method. The maximum absorption of the glucosamine-OPA-MPA derivative in a borate buffer (80 mM, pH 9.5) was found to occur at 335 nm. The absorbance was measured for 4.5 h after mixing the reagents. Shen, Yang, and Tomellini [2007] showed that the UV absorbance decreased to approximately half the maximum over a period of 4 h. This instability could lead to inaccuracy in analysis if the time period varies between performing the derivatization reaction and injection of the sample into the HPLC system.

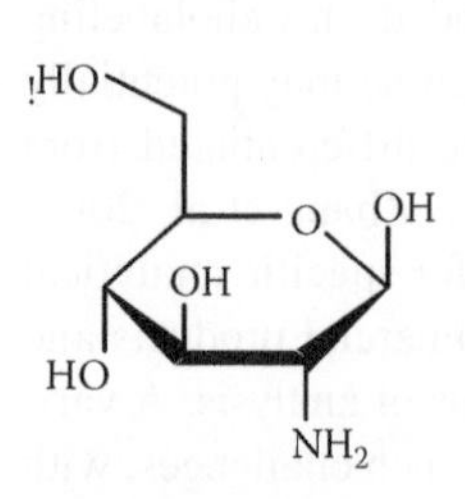

Figure 7.1 The structure of glucosamine.

Several agents have been suggested as suitable to make detectable complexes with glucosamine. Studies investigated the usefulness of each agent for a variety of applications, such as glucosamine content in a commercial product, or biological fluid analysis. Possible precolumn derivatization agents include the following: OPA [Nemati et al. 2007], phenylisothiocyanate (PITC) [Liang et al. 1999; B. Lockwood 2007], *N*-(9-fluorenyl-methoxycarbonyloxy) succinimide, and 9-fluorenyl-methyl chloroformate (FMOC-Cl) [Huang et al. 2006b; Zhang et al. 2006].

All of these agents can be complexed with glucosamine precolumn before separation. Nemati et al. [2007] investigated the use of OPA-MPA as a precolumn derivatization agent for analysis of glucosamine. The derivatization reagent was made by dissolving 5 mg of OPA in 900 μl of methanol, 100 μl of borate buffers, and 10 μl of MPA solutions. Separation was achieved using a 250 × 4 mm RP C18 column. The mobile phase was isocratic and consisted of methanol-sodium phosphate (pH 6.5, 12.5 mM; 10+90, volume/volume [v/v]) and methanol-tetrahydrofuran (97+3, v/v) for 20 min at a flow rate of 1 ml/min in proportions of 85+15. Detection was performed using a fluorescence detector. A calibration curve was constructed using standards ranging from 5 to 200% of the nominal assay concentration of 10 μg/ml glucosamine HCl, which showed excellent linearity with a correlation coefficient of 0.9980. The lower limit of detection (LOD) and quantification (LOQ) were 0.009 and 0.027 μg/ml, respectively [Nemati et al. 2007]. Nemati et al. concluded that derivatization with OPA-MPA was extremely simple and robust and showed greater sensitivity than other derivatization methods. The derivatization reaction took less than 5 min. The process would also be suitable for automation because there is no need to remove excess derivatization agent, and the reaction does not require an evaporation step, which would reduce human error. OPA also has the advantage that both UV and fluorescence detectors can be used, giving it a wider applicability [Nemati et al. 2007]. Stability issues with OPA have been noted, because glucosamine-OPA complexes decompose over time [Huang et al. 2006a]. The analysis of glucosamine in human fluids requires different analytical techniques and is essential for understanding the physiological role of glucosamine, metabolism, pharmacokinetics, fate, and mechanism of action. Huang et al. investigated the use of FMOC-Cl as a derivatization agent for the analysis of glucosamine content in human plasma. The levels of glucosamine in human plasma are relatively low, which presents an additional analytical challenge. A suitable analytical technique therefore needs to be sensitive, simple, and fast. Derivatization reagent consisted of 0.8 mM FMOC-Cl in acetonitrile. Separation was achieved using a C18 analytical column, 150 × 4.6 mm inner diameter, 5 μm. Elution was obtained by using gradient steps of solvents A (acetonitrile) and B (water): 30:70 (A/B) for 10 min and then 98:2 for 15 min at a flow rate of 1 ml/min. The LOD was found to be 15 ng/ml (signal to noise ratio ≥3), which is well below the expected drug concentration in plasma samples from a patient given therapeutic doses of glucosamine sulphate. Huang concluded that the use of FMOC-Cl as a derivatization agent allowed for sensitive, rapid (attributable to faster sample preparation times), simple, and reliable analytical results. It has also been shown that there should be no stability issues with the FMOC-Cl-glucosamine complexes, because samples have been shown to be stable for 24 h at 4°C [Zhang et al. 2006]. Thus, the technique should be appropriate for the routine analysis of glucosamine for pharmacokinetic, bioavailability, or bioequivalence studies [Huang et al. 2006b]

Postcolumn Agents Used for Indirect Fluorescent Detection

Another method used to overcome the issue of the limited UV or fluorescence detection attributed to glucosamine is to attach a visualizing agent postcolumn separation. Shen, Yang, and Tomellini [2007] exploited an indirect fluorescence detection method for the analysis of glucosamine in dietary supplements. This method claimed to avoid the stability and time issues associated with precolumn derivatization and thus could be advantageous. The method is based on the fluorescent signal of either L-tryptophan (L-TRP) or DL-5-methoxytryptophan (5-MTP). Either compound is added postcolumn as a copper complex; when bound to the copper (II) ion, the fluorescent signal of these compounds is quenched, thus giving no absorbance. Glucosamine is capable of complexing with the copper (II) ion and thus displaces some fraction of the L-TRP or 5-MTP recovering their fluorescent signal. The amount of glucosamine present can be calculated indirectly from the intensity of signal produced by the displaced L-TRP or 5-MTP. Shen, Yang, and Tomellini used the following chromatographic conditions: mobile phase, 1.6 mM sodium borate, pH 9.0; flow rate, 1 ml/min; postcolumn interaction component 2×10^{-5} M CU $(L\text{-}TRP)_2$ in 40 mM sodium borate at pH 9.0, or 2×10^{-5} M Cu $(5\text{-}MTP)_2$ in 40 mM sodium borate at pH 8.4; flow rate, 1 ml/min; column, strong anion-exchange column, PRP-X100 (250 × 4.1 mm, 10 μm) with a fluorescence detector [Shen, Yang, and Tomellini 2007]. The main disadvantage reported was the lower sensitivity of the method compared with precolumn derivatization. The detection limit found using postcolumn derivatization corresponded to a concentration of glucosamine of 3.2 μg/ml compared with detection limits of 0.009 μg/ml [Nemati et al. 2007] and 0.075 μg/ml [Ji et al. 2005] for precolumn derivatization with OPA-MPA and PITC, respectively. However, the technique would still be an acceptable alternative if detection limits were not an issue (e.g., in the analysis of glucosamine in commercial products) because it avoids the time-consuming derivatization reactions and avoids the possibility of stability issues [Shen, Yang, and Tomellini 2007].

HPLC Using Electrospray Ionization-MS Detection

Because of the possible limitations in sensitivity of the methods involving precolumn and postcolumn derivatization followed by UV or fluorescence detection, researchers investigated the possibility of HPLC separation followed by MS detection [Huang et al. 2006a; Roda et al. 2006].

A method developed by Huang et al. looked at using precolumn derivatization with PITC, followed by electrospray ionization (ESI)-MS detection. Huang used an RP C18 column, and, because of the high polarity of glucosamine, the elution was too fast; thus, suitable derivatization was used to help facilitate HPLC component separation. PITC was chosen specifically over other agents because it gave selective mass spectra. Elution was performed with 0.2% glacial acetic acid (A) and methanol (B) at a flow rate of 0.3 ml/min. Gradient HPLC was used with a changing solvent ratio from 80:20 to 10:90 over 8 min and then 80:20 for 13 min for A/B, respectively. Quantification was achieved by MS in the positive ionization mode, with ESI as an interface, a drying gas flow of 10 L/min, a drying gas temperature of 350°C, a nebulizer pressure of 50 psi, a capillary voltage of 4,000 V, and a fragmentation energy of 130 V. This method obtained good precision, accuracy, and speed with an LOQ of 0.1 μg/ml and an LOD of 35 ng/ml (signal to noise ratio of 3) with a time of 8 min per analysis. It was concluded that this method could be used for the analysis of glucosamine in human plasma, including the analysis of basal glucosamine plasma levels [Huang et al. 2006a]. This technique allows for information to be gathered on the physiological role of endogenous glucosamine and its involvement in disease processes.

Another method studied by Roda et al. [2006] used HPLC, followed by ESI-MS/MS detection. In this study a polymer-based amino column was used to increase the retention of glucosamine, helping to prevent the need for a derivatization step. This is beneficial for several reasons. First, the speed of analysis is increased. Second, derivatization reactions are aspecific and could derivate other plasma components, resulting in an increased "noise" in the separation and detection steps. Separation of components on the column was achieved using a gradient composed of Milli-Q water (A) and acetonitrile (B) at a flow rate of 0.3 ml/min. The gradient elution program used was as follows: 20% A for 7 min, 7–8 min a linear increase from 20 to 50% A, 50% A for 8 min, 16–17 min a linear decrease from 50 to 20% A, 10 min 20% A. Detection was effected using a triple quadrapole mass spectrometer that was set on the positive ionization mode, with quantification performed in the multiple reaction monitoring mode. This method achieved an LOQ and LOD of 10 and 5 ng/ml, respectively, without the need for a derivatization step. Roda et al. concluded that this method was suitable for the measurement of endogenous glucosamine plasma levels attributable to its high sensitivity. The method is said to be advantageous over other techniques because no preanalytical derivatization step is required, resulting in analysis that can be performed quickly and reduce analytical variability [Roda et al. 2006].

Both methods using HPLC coupled to ESI-MS detection give high sensitivity and accuracy. However, criticisms exist around the use of HPLC ESI-MS methods for the routine analysis of glucosamine. Some claimed that LC-MS technology is not yet widely available in laboratories, thus making the method nonsuitable for widespread analysis. It is also claimed that precolumn derivatization with FMOC-Cl with UV detection gives the same sensitivity at less cost than the LC-MS technique [Huang et al. 2006b].

The Use of HPLC with Alternative Detection Methods

HPLC with precolumn derivatization methods are cheap and effective. They offer a suitable level of accuracy for analysis of glucosamine products, and derivatization with FMOC-Cl may offer suitable sensitivity for analysis of biological samples. The derivatization step is involved with many criticisms of the method. The initial derivatization reaction may be time consuming, increasing the time of analysis rendering the methods unsuitable for routine use. Once made, the derivative-glucosamine complexes could have stability issues that may confound the results. Finally, the derivative compounds themselves could negatively affect the performance of the assay. To avoid these issues, HPLC ESI-MS/MS is available using an amino column that circumnavigates the use of derivatization agents. However, this method is more expensive and less readily available but offers good sensitivity. The last HPLC option available is the use of indirect fluorescence detection with L-TRP or 5-MTP. This method avoids the derivatization step and is cheap and accessible; however, it lacks suitable accuracy for analysis of biological samples.

Overall, the HPLC method of choice depends on the analytical function need that is to be performed.

Gas Chromatography

Glucosamine has been determined in soil using gas chromatographic methods [Zhang 1996]. The technique is said to suitably sensitive, but the need to make solutes volatile by complicated sample preparation procedures can be time consuming and can result in multiple peaks for a single component [Qui et al. 2006]. This makes the method unsuitable for routine use. Zhang et al. [1996] looked at the determination of glucosamine in soil samples, and, thus, its relevance to the quantification of glucosamine in commercial products or plasma samples may be questionable.

High-Performance Thin-Layer Chromatography

A quantitative densitometric high-performance thin-layer chromatographic method was developed by Ester et al. [2006] for determination of glucosamine in dietary supplements. The method was sought because no rapid, simple, or selective HPLC method was reasonably available for the quantitative determination of glucosamine from its complex matrix. Ester et al. achieved separation on 20 × 10 cm silica gel 60 F_{254} high-performance thin-layer chromatographic plates, with a mobile phase consisting of 2-propanol/ethyl acetate/ammonia solution (10:10:10 v/v/v). Because glucosamine lacks a suitable chromophore, the plates were immersed in a visualization reagent solution consisting of diethyl-ether/glacial acetic acid/anisaldehyde/sulphuric acid (136:91:1.2:20 v/v/v/v). The plates were then processed, and glucosamine appeared as brownish-red zones on a colorless background. Densitometric determination of glucosamine was performed at 415 nm by reflectance scanning. The amounts of glucosamine were found from the intensity of diffusely reflected light. The method is advantageous because it circumvents the tedious and time-consuming sample preparation steps necessary for the HPLC techniques discussed above. Ester et al. [2006] concluded the method to be reliable, repeatable, and accurate.

Capillary Electrophoresis Determination

Both HPLC with UV/fluorescence or ESI-MS detection include particular disadvantages that are discussed above, including a relatively high cost. Capillary electrophoresis (CE) has been considered as a suitable alternative because of its high efficiency, fast speed, small sample requirement, simplicity, and flexibility [Liang et al. 1999]. For such reasons, CE has been widely used for the rapid analysis of biomolecules, such as carbohydrates, amines, and amino acids, giving high resolution and short separation times of approximately 3 min [Skelley et al. 2005]. Similar to the HPLC method with precolumn derivatization, the main drawback of CE is its low detection sensitivity to glucosamine, which has to be improved by incorporation of a suitable UV absorbing group onto the glucosamine molecule [Qi et al. 2006; Skelley et al. 2006].

Capillary Electrophoresis with Dansylation of Glucosamine under Microwave Irradiation

Qi et al. [2006] appreciated that a major drawback of derivatization is its increase in the time of analysis. To obviate this problem, they investigated the use of dansyl chloride as a suitable labeling agent. To accelerate the labeling process of glucosamine, they performed the dansylation reaction in a microwave oven, which gave labeling speeds up to 50 times faster than common methods. Derivatization was achieved by mixing 10 mg of dansyl chloride with 10 ml of acetone and the glucosamine solution (tablets dissolved in 80 mM borate buffer at pH 9.5). The solution was placed in a microwave oven and irradiated at 385 W for 6 min. CE was performed using a bare fused-silica capillary of 75 μm inner diameter × 57 cm. Samples were injected at 0.5 psi for 2 s and separated at +18 kV at 20°C. The separated bands were detected by UV absorption at 214 nm. This technique gave reasonable sensitivity with an LOD of 1 μg/ml. Qi et al. concluded that the use of CE using accelerated labeling of glucosamine with dansyl chloride under microwave irradiation was a reliable, accurate, quantitative, and highly applicable method for determination of glucosamine content within commercially available glucosamine tablets but is not sufficiently sensitive for determination of glucosamine in biological samples [Esters et al. 2006].

Microchip Capillary Electrophoresis with Fluorescamine Labeling for Anomeric Composition Determination

Glucosamine can exist as either an alpha or beta anomer, an anomer being an epimer that is a stereoisomer of a saccharide differing only at the reducing carbon atom. Because of this, there is a need to assess the anomeric composition of glucosamine products and to assess the interconversion rates between the two anomers. Skelley and Mathies [2006] developed a microchip capillary electrophoretic method able to resolve both the alpha and beta anomers of glucosamine, allowing for determination of the anomeric composition of a sample. Previous chromatographic methods reported timescales of analysis between 10 and 20 min, which led to poor resolution between the alpha and beta anomers; however, because of the inherent speed of CE, this did not occur. The blurring in resolution occurs as a result of on-column mutarotation between the two anomers of glucosamine; thus, the faster the analytical technique the less time mutarotation has to occur and less blurring of resolution is seen. Skelley et al. labeled glucosamine with fluorescamine by mixing 2 μl of sample with 20 mM fluorescamine in dimethylsulfoxide. The samples were analyzed at room temperature using all-glass microfabricated devices made in-house. The CE separations were performed at 700 V/cm on a portable CE instrument with a 100-μm-diameter fiber optic-coupled photomultiplier tube for fluorescence detection. The inherent speed of microchip CE enabled Skelley et al. to observe the alpha and beta anomers of glucosamine and their interconversion rates in real time. It was noted that the interconversion between the anomers may have been effected by the fluorescamine labels. Furthermore, if interconversion was allowed to take place before labeling, more accurate results could be acquired. However, this was at the expense of time, and, if labeling took place after mutarotation, the time of overall analysis rose from approximately 2 min up to 2 h. Skelley and Mathies [2006] concluded that this method was fast, portable, and suitable for glucosamine determination.

The Pros and Cons of Capillary Electrophoresis

The use of CE for the analysis of glucosamine offers a suitable low-cost alternative to HPLC methods. CE has been shown to be a fast, quantitative, and efficient method of analysis. The latter method discussed has the unique advantage that it enables the real-time observation of the mutarotation of the glucosamine anomers, which no other analytical technique is able to accomplish. Disadvantages include a limited sensitivity when compared with other methods, and also the apparatus is less commonly available in laboratories [Huang et al. 2006b].

Infrared Spectroscopy with Chemometrics

The chemical structure of glucosamine makes it suitable for analysis with infrared (IR) spectroscopy. Previously, IR spectroscopy has been used to gather information about glucosamine and other GAGs, how GAGs absorb water (relates to physical properties of cartilage), and the breakdown products of cartilage after enzyme degradation. This technique can therefore give useful information about arthritic disease processes.

In this method, IR is combined with chemometric pattern recognition techniques for the analysis of glucosamine and similar compounds. Foot and Mulholland [2005] analyzed all samples by transmission Fourier transform infrared spectroscopy on a Nicolet Magna IR 760 spectrometer. Data analysis was then undertaken using computer programs. Microsoft Excel was used to produce the first derivative spectra. Then both the original and derivative data were transposed to SIRIUS to perform principal components analysis as a data reduction technique linking that with hierarchical cluster analysis and soft independent modeling of class analogies. These mathematical techniques are used to distinguish between the IR spectra produced by similar compounds, such as glucosamine, galactosamine, and chondroitin. This allows for fast and simple evaluation of the spectra of such compounds. Foot and Mulholland concluded that samples were best classified using the first derivative spectra. However, this technique has the disadvantage of only being qualitative and thus cannot be used to quantify the active ingredients within products [Foot and Mulholland 2005].

In conclusion, this technique maybe less useful than other methods because it can only be used for qualification of components, but the authors have suggested that additional work be undertaken to enable the technique to be used for quantification.

TECHNIQUES FOR ANALYSIS OF COENZYME Q_{10}

CoQ_{10} (see Figure 7.2) has many physiological roles; therefore, being able to analyze its levels in plasma could possibly aid in the treatment and prediction of diseases. Not only this, the new HMG-CoA reductase inhibitors (cholesterol lowering drugs) affect the synthesis of CoQ_{10} and, therefore, have the potential to cause adverse effects such as cardiac myopathy. This is thought to be attributable to a decrease in ubiquinone plasma levels, meaning that a determination of CoQ_{10} is important in early drug development to predict and avoid adverse effects [Teshima and Kondo 2005].

Finally, with the ever-increasing use of nutraceutical products, it is important to be able to determine levels of these compounds within commercially available products to ensure public safety.

The quantification of CoQ_{10} plasma levels is, therefore, important from a medical and epidemiological point of view [Karpińska et al. 2006]. The analysis of CoQ_{10} presents several challenges, and numerous techniques have been investigated for potential use [Jiang et al. 2004; Zu et al. 2006].

The average level of coenzyme in a healthy subject is 0.8 ± 0.2 mg/L [Grossi et al. 1992]. Thus, the main analytical challenge was to develop a method selective enough and sensitive enough to quantify levels in human fluids and commercial products, which was fast enough to be used routinely in quality-control laboratories. This would allow for useful information on the action, bioavailability, and fate of CoQ_{10} products to be collected.

The challenge is made more difficult by the chemical nature of CoQ_{10}. First, CoQ_{10} contains isopentenyl, which makes the compound susceptible to photo-oxidation. This means that all experiments should be undertaken in particular conditions to

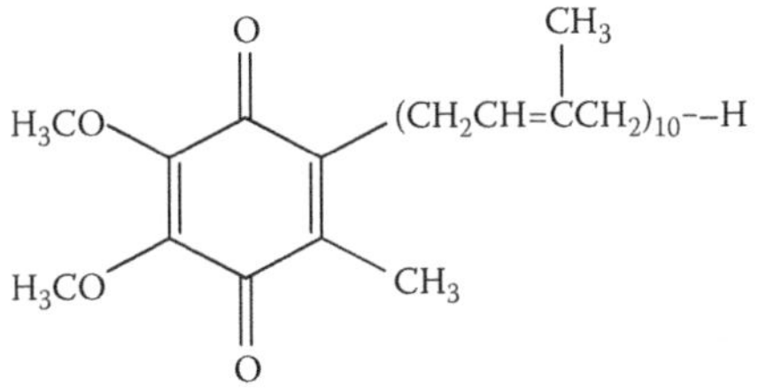

Figure 7.2 The structure of coenzyme Q_{10}.

avoid the photodegradation, specifically in the absence of light. Second, biological samples of CoQ_{10} will be accompanied by a complex matrix. This means that any sample will require intensive and time-consuming sample preparation before an assay can be undertaken [Karpińska et al. 2006]. This is often a factor that limits the value of a technique if it is to be widely applied.

HPLC

HPLC is probably the most popular technique used in the analysis of CoQ_{10} [Li et al. 2006]. Several detection methods have been used, including MS, UV, and electrochemical detection. UV and electrochemical detection are perhaps the most popular techniques used because of the wider accessibility of the technology [Karpińska et al. 1998]. Each method is associated with certain disadvantages and advantages that will be discussed below.

UV or Electrochemical Detection?

It has been shown that determination of CoQ_{10} in human plasma can be achieved by HPLC with UV detection after TLC, liquid/liquid [Katayama et al. 1980], or solid-phase extraction [Jiang et al. 2004]. These techniques are used to remove complex components that are found within a biological sample. The usual method used is an alcohol-hexane extraction, in which the extract is dried and redissolved before injection. This precolumn manipulation of the sample is time consuming, which makes the methods unsuitable for routine use in clinical chemistry laboratories [Jiang et al. 2004]. Karpińska et al. [2006] developed such a method. Before sample injection, a lengthy preparation took place in which 0.25 ml of sample was deproteinized using 0.5 ml of methanol. After that, 0.75 ml of hexane was added. The mixture was then vortexed for 5 min and centrifuged at 5000 rpm for 15 min. Next, the clear hexane layer was transferred to another tube, and extraction of plasma vitamins was repeated with a new 0.75 ml portion of *n*-hexane. The plasma extracts were combined and evaporated to dryness under a stream of nitrogen. Finally, the dry residue was dissolved in 0.25 ml of mobile phase and chromatographed by HPLC. Separation took place on a RP C18 125 × 4 mm (5 μm) column with a guard column, 4 × 4 mm (5 μm), with a mobile phase of methanol and hexane 72:28 (v/v). The flow rate used was 1 ml/min, with detection achieved by a UV detector at 276 nm. This gave a retention time of 7.47 ± 0.12 min and an LOD and LOQ of 0.87 and 2.98 μM, respectively, for CoQ_{10}.

The use of an electrochemical detector has been shown to be 10–20 times more sensitive than a UV detector. For this reason, the method was used by Tang et al. [2006] for detection of CoQ_{10} in human breast milk, in which the levels are even lower than that of the plasma. The analytical column used was a Microsorb-MV RP C18 column (150 × 4.6 mm; 5 μm bead size). The mobile phase used consisted of a mixture of sodium acetate anhydrous (4.2 g), 15 ml of glacial acetic acid, 15 ml of 2-propanol, 695 ml of methanol, and 275 ml of hexane. The mobile phase was filtered through a 0.2 μm pore-sized nylon filter, and the flow rate was 1 ml/min. An electrochemical detector was used and was shown to give an LOQ of 60 nmol/L, with a time of analysis of approximately 8 min per sample. With the basal concentration of CoQ_{10} in breast milk being 0.166 ± 0.002 μmol/L, this method achieves high sensitivity [Tang et al. 2006]. The other main advantage of electrochemical detection over UV methods is that it allows for the detection of the reduced form of ubiquinone, ubiquinol. This allows for estimation of the total plasma CoQ_{10}. To accomplish this, all CoQ_{10} in a sample is converted into the reduced form by precolumn treatment and then detected using such methods [Grossi et al. 1992]. However, as with UV detection, this method still requires manual sample preparation before chromatographing, which has the disadvantage of being time consuming.

Column Switching HPLC

Because of the complexity of biological fluid specimens, most samples normally require some sort of purification before injection into the HPLC system, such as solid-phase extraction, liquid/liquid, or TLC. Column switching avoids this by allowing online sample purification. This results in the method being suitable for automation and decreased handling time, which can result in artificial error and poor reproducibility. Jiang et al. [2004] used such a method for the determination of CoQ_{10} in human plasma. In this technique, a short precolumn (20 × 4.6 mm inner diameter) was packed with 5 μm Hypersil C18. This was used to eliminate both polar compounds, including reagents and strongly retained solutes. The mobile phase used for the precolumn was pure methanol at a flow rate of 0.5 ml/min. The analyte was then transferred to the analytical column (150 × 4.6 mm inner diameter packed with 5 μm of Hypersil ODS_2). The mobile phase through the analytical column contained 10% (v/v) isopropanol in methanol at a flow rate of 1.5 ml/min. The detection was performed with UV set at 275 nm. The lower limit of detection achieved was approximately 1 μg/ml in human plasma, which is comparable with methods using solid-phase extraction to purify samples [Jiang et al. 2004]. The disadvantage noted by Grossi et al. [1992] is that column-switching technology may not be available in many laboratories. Whether this comment still applies 15 years on is debatable.

Mass Spectrometry Detection

As technology develops, researchers are adopting the use of HPLC coupled with MS as a useful method for the identification and determination of compounds [Teshima and Kondo 2005; Zu et al. 2006]. The use of LC-MS for the detection of CoQ10 in complex samples is effective because of the high sensitivity and selectivity. Strazisar et al. [2005] demonstrated that an LC-MS method was more sensitive than LC-UV. Hansen et al. [2004] showed that HPLC-MS has comparable or better sensitivity than HPLC-electrochemical detection.

Zu et al. [2006] used an LC-MS/MS method for the determination of CoQ_{10} in tobacco leaves. Chromatographic analysis was performed on an RP C18 column, with a mobile phase consisting of a mixture of acetonitrile and isopropanol (8:7 v/v) containing 0.5% formic acid at a flow rate of 0.3 ml/min. ESI-MS, operating in the positive ion mode, was used for detection and analysis. CoQ_{10} was monitored using the multiple reaction monitoring mode. Under these conditions, a retention time of 2.91 ± 0.1 min and an LOQ and LOD of 4.0 and 1.2 ng/ml, respectively, were established for CoQ_{10}. This method was therefore able to give an LOD 205.8, 71.9, 41.7, and 208.6 times lower than the LOD obtained by spectrophotometric, voltammetric, and chromatographic methods with UV and fluorescence detection. It has been noted in literature that matrix components do have the potential to cause erroneous results unless optimization of sample preparation and the chromatographic system takes place.

Another LC-MS/MS technique using turbo-spray ionization was investigated by Teshima and Kondo [2005] for analysis of CoQ_{10} levels in rat heart and thigh muscle. Teshima and Kondo looked at existing methods using LC-MS/MS with a turbo-spray ionization source for the analysis of ubiquinones and looked to improve their sensitivity. Separation was achieved on a YMC Pack Pro (75 × 2 mm inner diameter) analytical column, and a mixture of methanol/2-propanol/formic acid (45:55:0.5) containing 5 mmol/L methylamine was selected as the mobile phase. The addition of the methylamine to the mobile phase led to 12.5-fold higher signal intensity than without. A triple-stage quadrapole mass spectrometer with a turbo-spray ionization source was used for the detection of CoQ_{10}. Under these conditions, CoQ_{10} had the observed retention time of 4 min, and the method was able to quantify CoQ_{10} over a wide concentration range in rat thigh and heart muscle, 1–500 and 10–10,000 μg/g, respectively [Teshima and Kondo 2005].

All of the above HPLC methodologies are suitable for analysis of CoQ_{10} in products and biological samples and are routinely used. Intuitively, the column-switching method seems the most sensible for use in clinical investigations of CoQ_{10} because of its fast, accurate, and automated properties.

Derivative Spectrophotometry

Although HPLC methods are highly sensitive, it has been noted that they can be time consuming and complicated. An alternative method was therefore sought by Karpińska et al. [2006]. They investigated the use of spectrophotometry because it is a rapid, specific, simple, and reliable analytical technique. Unfortunately, classical UV spectrophotmetric methods could not be used because of spectra interference from the complex matrix of plasma samples of CoQ_{10}. This made determination of CoQ_{10} impossible. To avoid this problem, Karpińska et al. used a mathematical technique to eliminate the influences of the matrix/background, namely derivative spectrophotometry (derivative of the spectra is taken using the Savitzky–Golay algorithm). This technique enhances the sensitivity and selectivity of the analysis. Karpińska used a HP-8452A diode array spectrophotometer, and the spectrophotometric measurements were performed at 284 nm. This technique was able to determine the presence of CoQ_{10} in pharmaceuticals in the range of 0.5–10 ppm. Because levels of CoQ_{10} do not exceed 1.5 ppm, the method was applicable. Karpińska et al. concluded that the technique could be used as an alternative to HPLC analysis of

CoQ_{10} in commercial formulations; however, the method has limited application for analysis of biological samples, and a relatively large sample volume is required.

Voltammetric Determination

Voltammetry is an electroanalytical technique that is suitable for use with CoQ_{10} because of its redox and adsorptive behavior. Two methods for the analysis of CoQ_{10} have been used in the literature.

Square-Wave Voltammetry with Glassy Carbon Electrode

Litescu et al. [2001] analyzed commercially available CoQ_{10} formulations using a standard addition voltammetry method on a glassy carbon electrode. This is made possible because CoQ_{10} is electroactive at the glassy carbon electrode. This technique is said to be a simple, rapid, sensitive, and accurate means for CoQ_{10} analysis via a reduction process. Litescu et al. recommended the technique as a means of analysis for quality-control laboratories because it does not require any lengthy sample preparation steps except for previous extraction with hexane.

Mercury Hanging Electrode

Emons et al. [1992] developed a method for determination of ubiquinones with differential pulse voltammetry on a mercury hanging electrode. They concluded that this method was only suitable for determination of CoQ_{10} in samples with a less complex matrix, meaning that it has limited application for routine analysis of biological samples.

Both of the above methods have been said to lack analytical sensitivity and have a disadvantage of needing to be operated in dim light because of the sensitivity of CoQ_{10} to photodegradation. The application of these methods above HPLC, or spectrophotometry, would be debatable.

Electron Paramagnetic Spectroelectrochemistry

Electron paramagnetic spectroscopy is a powerful tool for characterization of radical ion intermediates in reactions. If these intermediates are sufficiently stable, it is possible to determine relative concentrations from the relative resonance intensities. The absolute concentrations can then be found by comparing the relative resonance intensities with those of a standard. Long [1999] used this methodology for the determination of CoQ_{10} in commercial products. In this case, the intermediate radical ubisemiquinone is measured that is formed as part of the reduction reaction of CoQ_{10} at a silver electrode by in situ electron paramagnetic spectroelectrochemical techniques. Long concluded that this method showed good agreement to data produced from HPLC methods and with that of the manufacturers' labeling. The technique was stated as useful for enhancing selectivity of other methods when other chemicals present in the sample do not interfere with the assay [Long 1999].

CONCLUSIONS

This review describes a number of techniques used for the analysis of glucosamine and CoQ_{10}, including radiolabeling, HPLC with both precolumn and postcolumn derivatization reagents, using UV, fluorescence, and ESI-MS detection, GC, high-performance thin-layer chromatography, CE with derivatization reagents, IR with chemometrics, derivative spectrophotometry, voltammetric determination, and electron parametric spectroelectrochemistry. In addition to the techniques described here, high-speed counter-current chromatography, micellar electrokinetic capillary chromatography, microemulsion electrokinetic chromatography, ^1H nuclear magnetic resonance, and near IR spectroscopy have also been used recently for analysis of other nutraceuticals. A number of novel alternative techniques have also been devised, such as the use of biosensors [Kirrane and Lockwood 2008]. The plethora of methods available include almost the whole gamut of techniques available for organic molecules.

With the ever-increasing use of nutraceuticals, the need to have analytical techniques capable of attaining accurate and reliable information is of growing importance, gaining more insight as to their distribution. If the traditional paradigm for drug development continues to fail, pharmaceutical companies will pay an ever-increasing amount of attention to the potential use of nutraceuticals. Central to this are reliable, reproducible, sensitive, selective, and quantitative analytical techniques.

With respect to glucosamine and CoQ_{10}, many challenges are presented in terms of analysis, and every technique available has its reported drawbacks.

REFERENCES

Childs, N. M. and G. H. Poryzees. 1998. Foods that help prevent disease: Consumer attitudes and public policy implications. *Br. Food J.* 100:419–426.

Dentith, S. and B. Lockwood. 2008. Development of techniques for the analysis of isoflavones in soy foods and nutraceuticals. *Curr. Opin. Clin. Nutr. Metab. Care* 11:242–247.

Emons, H., G. Wittstock, B. Voigt, and H. Seidel. 1992. Voltammetric trace determination of ubiquinones at mercury electrodes. *Fresenius J. Anal. Chem.* 342:737–739.

Esters, V., L. Angenot, V. Brandt, M. Frédérich, M. Tits, C. Van Nerum, J. N. Wauters, and P. Hubert. 2006. Validation of a high-performance thin-layer chromatography/densitometry method for the quantitative determination of glucosamine in a herbal dietary supplement. *J. Chromatogr. A.* 1112:156–164.

Foot, M. and M. Mulholland. 2005. Classification of chondroitin sulfate A, chondroitin sulfate C, glucosamine hydrochloride and glucosamine 6 sulfate using chemometric techniques. *J. Pharm. Biomed. Anal.* 38:397–407.

Grossi, G., A. Bargossi, P. Fiorella, S. Piazzi, M. Battino, and G. Bianchi. 1992. Improved high-performance liquid chromatographic method for the determination of coenzyme Q10 in plasma. *J. Chromatogr.* 593:217–226.

Hansen, G., P. Christensen, E. Tüchsen, and T. Lund. 2004. Sensitive and selective analysis of coenzyme Q10 in human serum by negative APCI LC-MS. *Analyst* 129:45–50.

Ho, C.-T. and Q. Y. Zheng, eds. 2002. *Quality Management of Nutraceuticals*. Washington, D.C: American Chemical Society.

Huang, T., L. Cai, B. Yang, M. Zhou, Y. Shen, and G. Duan. 2006a. Liquid chromatography with electrospray ionization mass spectrometry method for the assay of glucosamine sulfate in human plasma: Validation and application to a pharmacokinetic study. *Biomed. Chromatogr.* 20:251–256.

Huang, T., C. Deng,, N. Chen, Z. Liu, and G. Duan. 2006b. High performance liquid chromatography for the determination of glucosamine sulfate in human plasma after derivatization with 9-fluorenylmethyl chloroformate. *J. Sep. Sci.* 29:2296–2302.

Hurst, W. J. 2002. *Methods of Analysis for Functional Foods and Nutraceuticals*. Boca Raton, FL: CRC Press.

Ji, D., L. Zhang, J. Chen, and E. Peng. 2005. Precolumn derivatization liquid chromatography method for analysis of dietary supplements for glucosamine: Single laboratory validation study. *J. AOAC Int.* 88:413–417.

Jiang, P., M. Wu, Y. Zheng, C. Wang, Y. Li, J. Xin, and G. Xu. 2004. Analysis of coenzyme Q(10) in human plasma by column-switching liquid chromatography. *J. Chromatogr. B. Analyt. Technol. Biomed. Life Sci.* 805:297–301.

Karpińska, J., B. Mikołuć, and J. Piotrowska-Jastrzebska. 1998. Application of derivative spectrophotometry for determination of coenzyme Q10 in pharmaceuticals and plasma. *J. Pharm. Biomed. Anal.* 17:1345–1350.

Karpińska, J., B. Mikołuć, R. Motkowski, and J. Piotrowska-Jastrzebska. 2006. HPLC method for simultaneous determination of retinol, alpha-tocopherol and coenzyme Q10 in human plasma. *J. Pharm. Biomed. Anal.* 42:232–236.

Katayama, K., M. Takada, T. Yuzuriha, K. Abe, and S. Ikenoya. 1980. Simultaneous determination of ubiquinone-10 and ubiquinol-10 in tissues and mitochondria by high performance liquid chromatography. *Biochem. Biophys. Res. Commun.* 95:971–977.

Kirrane, M. and B. Lockwood. 2008. Developments in the analysis of tea. *Nutrafoods* 7:2008.

Li, K., Y. Shi, S. Chen, W. Li., X.Shang, and Y. Huang. 2006. Determination of coenzyme Q10 in human seminal plasma by high-performance liquid chromatography and its clinical application. *Biomed. Chromatogr.* 20:1082–1086.

Liang, Z., J. Leslie, A. Adebowale, M. Ashraf, and N. Eddington. 1999. Determination of the nutraceutical, glucosamine hydrochloride, in raw materials, dosage forms and plasma using pre-column derivatization with ultraviolet HPLC. *J. Pharm. Biomed. Anal.* 20:807–814.

Litescu, S.-C., I.-G. David, G.-L. Radu, and H. Y. Aboul-Enein. 2001. Voltammetric determination of coenzyme Q10 at a solid glassy carbon electrode. *Instrument. Sci. Technol.* 29:109–116.

Lockwood, B. 2007. *Nutraceuticals: A Balanced View for Healthcare Professionals*. London, UK: Pharmaceutical Press.

Lockwood, G. 2007. The hype surrounding nutraceutical supplements: Do consumers get what they deserve? *Nutrition* 23:771–772.

Long, Y.-T. 1999. Determination of coenzyme Q10 by in situ spectroelectrochemistry. *Electrochem. Commun.* 1:194–196.

Nemati, M., H. Valizadeh, M. Ansarin, and F. Ghaderi. 2007. Development of a simple and sensitive high-performance liquid chromatography method for determination of glucosamine in pharmaceutical formulations. *J. AOAC Int.* 90:354–357.

Qi, L., S. Zhang, M. Zuo, and Y. Chen. 2006. Capillary electrophoretic determination of glucosamine in osteoarthritis tablets via microwave-accelerated dansylation. *J. Pharm. Biomed. Anal.* 41:1620–1624.

Roda, A., L. Sabatini, A. Barbieri, M. Guardigli, M. Locatelli, F.S. Violante, L.C. Rovati, and S. Persiani. 2006. Development and validation of a sensitive HPLC-ESI-MS/MS method for the direct determination of glucosamine in human plasma. *J. Chromatogr. B. Analyt. Technol. Biomed. Life Sci.* 844:119–126.

Shao, Y., R. Alluri, M. Mummert, U. Koetter, and S. Lech. 2004. A stability-indicating HPLC method for the determination of glucosamine in pharmaceutical formulations. *J. Pharm. Biomed. Anal.* 35:625–631.

Shen, X., M. Yang, and S. Tomellini. 2007. Liquid chromatographic analysis of glucosamine in commercial dietary supplements using indirect fluorescence detection. *J. Chromatogr. Sci.* 45:70–75.

Skelley, A. M., J. R. Scherer, A. D. Aubrey, W. H. Grover, R. H. Ivester, P. Ehrenfreund, F. J. Grunthaner, J. L. Bada, and R. A. Mathies. 2005. Development and evaluation of a microdevice for amino acid biomarker detection and analysis on Mars. *Proc. Natl. Acad. Sci. USA* 102:1041–1046.

Skelley, A. and R. Mathies. 2006. Rapid on-column analysis of glucosamine and its mutarotation by microchip capillary electrophoresis. *J. Chromatogr. A.* 1132:304–309.

Strazisar, M., M. Fir, A. Golc-Wondra, L. Milivojevic, M. Prosek, and V. Abram. 2005. Quantitative determination of coenyzme Q10 by liquid chromatography and liquid chromatography/mass spectrometry in dairy products. *J. AOAC Int.* 88:1020–1027.

Tang, P., M. Miles, P. Steele, B. Davidson, S. Geraghty, and A. Morrow. 2006. Determination of coenzyme Q10 in human breast milk by high-performance liquid chromatography. *Biomed. Chromatogr.* 20:1336–1343.

Teshima, K. and T. Kondo. 2005. Analytical method for ubiquinone-9 and ubiquinone-10 in rat tissues by liquid chromatography/turbo ion spray tandem mass spectrometry with 1-alkylamine as an additive to the mobile phase. *Anal. Biochem.* 338:12–19.

Zhang, L., T. Huang, X. N. Fang, X. N. Li, Q. S. Wang, Z. W. Zhang, and X. Y. Sha. 2006. Determination of glucosamine sulfate in human plasma by precolumn derivatization using high performance liquid chromatography with fluorescence detection: Its application to a bioequivalence study. *J. Chromatogr. B. Analyt. Thecnol. Biomed. Life Sci.* 842:8–12.

Zhang, X. Gas chromatographic determination of muramic acid, glucosamine, mannosamine, and galactosamine in soils. *Soil Biol. Biochem.* 28:1201–1206.

Zu, Y., C. Zhao, C. Li, and L. Zhang. 2006. A rapid and sensitive LC-MS/MS method for determination of coenzyme Q10 in tobacco (*Nicotiana tabacum* L.) leaves. *J. Sep. Sci.* 29:1607–1612.

CHAPTER 8

Pharmacological Characterization of Nutraceuticals

Charles Preuss

CONTENTS

DEFINITION OF PHARMACOLOGICAL CHARACTERIZATION

The pharmacological characterization of a nutraceutical is simply the determination of its efficacy and safety. Currently, many nutraceuticals (e.g., botanicals) do not require efficacy and safety testing before marketing. They are regulated by the Dietary Supplement Health and Education Act of 1994, which considers many nutraceuticals as food items [Dietary Supplement Health and Education Act 1994]. However, there is a concern that many nutraceuticals have pharmacological activity that can endanger the public health and that certain nutraceuticals (e.g., botanicals) should be regulated similarly to prescription drugs [Morrow et al. 2005]. Therefore, future marketing of nutraceuticals may require more rigorous testing of safety and efficacy before marketing.

In fact, as of June 22, 2007, the FDA developed a current good manufacturing practice requirement for dietary supplements that obligates manufacturers to evaluate the composition, identity, quality, and strength of their marketed products [Food and Drug Administration 2007]. With future increased regulation of nutraceuticals on the horizon, the current pharmacological characterization of drugs will be discussed. At this point, it would be helpful to briefly review the drug development and testing process in the United States. The following is a simplified drug review process [Food and Drug Administration 2002; Berkowitz 2007]:

1. Preclinical testing: *in vitro* studies, animal models
2. Phase I: 20–80 human subjects, safety, pharmacokinetics
3. Phase II: 36–300 human subjects, efficacy
4. Phase III: 300–3,000 human subjects, efficacy, double-blind studies
5. Phase IV: post-marketing surveillance

The drug review process is roughly divided into preclinical and clinical testing, and, as noted above, preclinical is primarily *in vitro* and animal studies, whereas clinical are human studies.

PRECLINICAL

Pharmacological Profile Tests

Preclinical testing involves pharmacological profile tests for drugs, and it can be further divided into the following [Berkowitz 2007]:

1. Molecular: receptor binding, enzyme inhibition
2. Cellular: cell cultures, isolated tissues
3. Disease models: pain, seizures

Initial pharmacological characterization of a nutraceutical would take place preclinically in *in vitro* studies and animals to determine efficacy and safety. Table 8.1 provides some examples of preclinical pharmacological characterizations of nutraceuticals.

Table 8.1 Preclinical Pharmacological Characterization

Nutraceutical	*In Vitro/In Vivo*	Pharmacology	Reference
St. John's Wort	Rat brain homogenates	Inhibition of monoamine oxidase	Bladt and Wagner 1994
St. John's Wort	Rat	Antidepressant	De Vry et al. 1999
Cat's claw	*Salmonella typhimurium*	Antimutagenic	Rizzi et al. 1993
Devil's claw	Rat	Anti-inflammatory	Andersen et al. 2004
Echinacea	Rat	Immunostimulation	Cundell et al. 2003
Feverfew	Rat leukocytes	COX inhibition	Capasso 1986
Ginkgo	Rat	Cognition Improved	Winter 1998
Kava	Chick	Anxiolytic	Feltenstein et al. 2003
Glucosamine and chondroitin	Horse	Stride Improvement	Forsyth, Brigden, and Northrop 2006
Lycopene	Rat	Antioxidant	Augusti et al. 2007

Safety Tests and Toxicology Tests

Preclinical safety testing assesses the potential toxicity of a drug in *in vitro* and animal studies [Food and Drug Administration 1985; Berkowitz 2007]. Below is a listing and a short commentary on the type of safety tests required by the FDA:

1. Pharmacology studies: determine ED_{50}
2. Acute toxicity studies: determine LD_{50}
3. Multidose toxicity studies
   a. Subchronic toxicity: duration of one to three months
   b. Chronic toxicity: duration of six months
   c. Carcinogenicity: duration of two years
4. Special toxicity studies: route of administration
5. Reproduction studies: birth defects
6. Mutagenicity studies: Ames test
7. Pharmacokinetics studies: ADME

EVALUATION IN HUMANS

Clinical studies involve human subjects, and they are divided into four phases: Phase I, Phase II, Phase III, and Phase IV. The goal of these clinical trials is to validate that the drug demonstrates efficacy and safety before it is marketed. As discussed previously, the preclinical testing would have rejected those chemical entities that did not demonstrate efficacy and/or caused unacceptable toxicity, such as a poor safety profile.

Because the test subjects for clinical studies are human beings, the highest ethical and moral regard for their safety must be paramount. Therefore, two important safeguards are inherent to clinical trials: institutional review boards (IRB) and informed consent. The primary function of institutional review boards is to review proposed clinical research procedures, ensure that the proposed research is going to be conducted according to proper procedures, including institutional, local, state, and federal regulations. They are made up of individuals who have no conflict of interest with the clinical research that is being conducted by the institution (i.e., a drug company). It is usually made up of five or more members of the institution with different backgrounds and at least one outside person. Each participant in the clinical study must be given informed consent, and essential components include the following: a description of the research to be conducted, risks/benefits, and the ability to withdraw from the trial for any reason [Food and Drug Administration 1997].

There are several important parameters to be carefully considered when conducting clinical research to ensure the highest scientific standards:

1. Design and analysis considerations
2. Selection of subjects

3. Number of patients
4. Randomization of patients
5. Study control
6. Patient compliance
7. Dose considerations
8. Pharmacokinetics
9. Tests for safety

Important design and analysis considerations include the following: the appropriate use of statistics, careful planning of the clinical trials, and rationale for the length of the clinical trials. The selection of human subjects should involve a wide variety of parameters, such as age, sex, and ethnicity. The numbers of patients in the clinical trials are important, especially with regard to statistical considerations. Randomization of patients increases the confidence in the conclusions drawn from the study [Edwards 2001]. Study control is the use of a placebo or, because of the nature of the disease, the use of a positive control is warranted. To enhance the validity of the study, patient compliance must be diligently documented. An important dose consideration is the effective range of the drug (i.e., lowest and highest doses). The pharmacokinetics of the investigational drug must be ascertained (e.g., ADME). This topic will be discussed in additional detail in the next chapter. Tests for safety involve the appropriate laboratory tests (e.g., blood urea nitrogen) to monitor the health of the patient [Food and Drug Administration 1997].

Before a Phase I clinical trial can start, the drug company must submit a Notice of Claimed Investigational Exemption for a New Drug (IND) to the FDA. The IND must include the following information:

1. Drug source and composition
2. Information on manufacturing and chemistry
3. Animal studies data
4. Clinical trial plans and protocols
5. Credentials of the physicians conducting the trials
6. Key drug information given to investigators and their institutional review boards

It can take four to six years to collect enough preclinical data to submit an IND.

Phase I

In Phase I, about 20–80 healthy human subjects are involved. An important goal of this phase is to determine the maximum tolerated dose with minimal toxicity. An exception would be the use of certain drugs (e.g., antineoplastics), which are very toxic; patients with the disease being studied would be involved at this point rather than healthy volunteers. Many pharmacokinetic parameters are determined in this phase, such as absorption and metabolites. These trials are nonblinded, which means that both the investigator and patient know what is being given.

Phase II

In Phase II, about 36–300 human subjects with the disease are studied for efficacy of the investigational drug. These trials are often single blinded, meaning that the investigator knows which treatment is used, with a placebo control and a positive control (i.e., an approved active drug).

Phase III

In Phase III, 300–3,000 patients with the disease being studied are given the investigational drug. With data from Phases I and II, this trial is able to minimize errors from placebo effects, disease variability, etc. This trial is double blinded, i.e., both the investigator and patient do not know which treatment is being given, with placebo, positive control, and crossover techniques. At this point, a new drug application can be submitted to the FDA if the data from Phase III demonstrates safety and efficacy. Vast amounts of preclinical and clinical data are submitted to the FDA.

Phase IV

Phase IV can begin once the drug is approved for marketing. This phase is primarily concerned with capturing toxicities not observed in the previous phases because of the lower number of human subjects and chronic dosing [FDA 1997; Berkowitz 2007]. Table 8.2 provides some examples of nutraceuticals that have undergone clinical trials.

Table 8.2 Clinical Trials with Nutraceuticals

Nutraceutical	Purpose	Reference
St. John's Wort	Treatment of depression	Gastpar, Singer, and Zeller 2006
Black cohosh	Treatment of vasomotor symptoms	Newton et al. 2006
Chondroitin	Treatment of knee osteoarthritis	Mazieres et al. 2007
Lycopene	Reduce polysialylic acid	Bunker et al. 2007
Ginkgo	Treatment of dementia	Scripnikov, Khomenko, and Napryeyenko 2007
Cat's claw	Treatment of rheumatoid arthritis	Mur et al. 2002
Devil's claw	Treatment of back pain	Laudahn and Walper 2001
Echinacea	Bioavailability	Woelkart et al. 2006
CoQ_{10}	Bioavailability	Nuku et al. 2007
Resveratrol	Pharmacokinetics	Boocock et al. 2007

REFERENCES

Andersen M., E. Santos, M. Seabra, A. da Silva, and S. Tufik. 2004. Evaluation of acute and chronic treatments with *Harpagophytum procumbens* on Freund's adjuvant-induced arthritis in rats. *J. Ethnopharmacol.* 91:325–330.

Augusti P., G. Conterato, S. Somacal, L. Einsfeld, A. T. Ramos, F. Y. Hosomi, D. L. Graça, and T. Emanuelli. 2007. Effect of lycopene on nephrotoxicity induced by mercuric chloride in rats. *Basic Clin. Pharmacol. Toxicol.* 100:398–402.

Berkowitz, B. A. 2007. Development and Regulation of Drugs. In *Basic and Clinical Pharmacology.* 10th edition. Edited by Katzung, B. G. New York, NY: McGraw-Hill, pp. 64–73.

Bladt, S. and H. Wagner. 1994. Inhibition of MAO by fractions and constituents of hypericum extract. *J. Geriatr. Psychiatry Neurol.* 7 (Suppl. 1):S57–S59.

Boocock, D., G. Faust, K. Patel, A. M. Schinas, V. A. Brown, M. P. Ducharme, T. D. Booth, J. A. Crowell, M. Perloff, A. J. Gescher, W. P. Steward, and D. E. Brenner. 2007. Phase I dose escalation pharmacokinetic study in healthy volunteers of resveratrol, a potential cancer chemopreventative agent. *Cancer Epidemiol. Biomarkers Prev.* 16:1246–1252.

Bunker, C., A. McDonald, R. Evans, N. de la Rosa, J. Boumosleh, and A. Patrick. 2007. A rodomized trial of lycopene supplementation in Tabago men with high prostate cancer risk. *Nutr. Cancer* 57:130–137.

Capasso, F. 1986. The effect of an aqueous extract of *Tanacetum parthenium* L. on arachidonic acid metabolism by rat peritoneal leucocytes. *J. Pharm. Pharmacol.* 38:71–72.

Cundell, D., M. Matrone, P. Ratajczak, and J. Pierce. 2003. The effect of aerial parts of Echinacea on the circulating white cell levels and selected immune functions of the aging Sprague-Dawley rat. *Int. Immunopharmacol.* 3:1041–1048.

De Vry, J., S. Maurel, R. Schreiber, R. deBeun, and K. Jentzsch. 1999. Comparison of hypericum extracts with imipramine and fluoxetine in animal models of depression and alcoholism. *Eur. Neuropsychopharmacol.* 9:461–468.

Dietary Supplement Health and Education Act (DSHEA). 1994. Public Law 103–417.

Edwards, D. J. 2001. Classification of Study Design-Observational Research. In *Evaluating Drug Literature: A Statistical Approach.* Edited by Slaughter, R. L. and D. J. Edwards. New York, NY:McGraw-Hill, pp. 63–81.

Food and Drug Administration. 1985. Guidance for industry: Guideline for the format and content of the nonclinical pharmacology/toxicology section of an application. http://www.fda.gov/cder/guidance/old032fn.pdf. Accessed December 10, 2007.

Food and Drug Administration. 1997. Guidance for industry: General consideration for the clinical evaluation of drugs. http://www.fda.gov/cder/guidance/old034fn.pdf. Accessed December 11, 2007.

Food and Drug Administration. 2002. The FDA's drug review process: Ensuring drugs are safe and effective. http://www.fda.gov/fdac/features/2002/402_drug.html. Accessed December 6, 2007.

Food and Drug Administration. 2007. Final rule promotes safe use of dietary supplements. http://www.fda.gov/consumer/updates/dietarysupps062207.html. Accessed December 17, 2007.

Feltenstein M., L. Lambdin, M. Ganzera, H. Ranjith, W. Dharmaratne, N. P. Nanayakkara, I. A. Khan, and K. J. Sufka. 2003. Anxiolytic properties of *Piper methysticum* extract samples and fractions in the chick social separation-stress procedure. *Phytother. Res.* 17:210–216.

Forsyth, R., C. Brigden, and A. Northrop. 2006. Double blind investigation of the effects of oral supplementation of combined glucosamine hydrochloride (GHCL) and chondroitin sulphate (CS) on stride characteristics of veteran horses. *Equine Vet. J. Suppl* 36:622–625.

Gastpar, M., A.Singer, and K. Zeller. 2006. Comparative efficacy and safety of a once-daily dosage of hypricum extract STW3-VI and citalopram in patients with moderate depression: a double-blind, randomized, multicentre, placebo controlled study. *Pharmacopsychiatry* 39:66–75.

Laudahn, D. and A. Walper. 2001. Efficacy and tolerance of Harpagophytum extract LI 174 in patients with chronic non-radicular back pain. *Phytother. Res.* 15:621–624.

Mazieres, B., M. Hucher, M. Zaim, and P. Garnero. 2007. Effect of chondroitin sulphate in symptomatic knee osteoarthritis: A multicentre, randomized, double-blind, placebo-controlled study. *Ann. Rheu. Dis.* 66:639–645.

Morrow, J., T. Edeki, M. El Mouelhi, R. Galinsky, R. Kovelesky, and C. Preuss. 2005. American Society for Clinical Pharmacology and Therapeutics position statement on dietary supplement safety and regulation. *Clin. Pharmacol. Ther.* 77:113–122.

Mur, E., F. Hartig, G. Eibl, and M. Schirmer. 2002. Radomized double blind trial of an extract from the pentacyclic alkaloid-chemotype of uncaria tomentosa for the treatment of rheumatoid arthritis. *J. Rheumatol.* 29:678–681.

Newton, K., S. Reed, A. LaCroix, L. Grothaus, K. Ehrlich, and J. Guiltinan. 2006. Treatment of vasomotor symptoms of menopause with black cohosh, multibotanicals, soy, hormone therapy or placebo: A randomized trial. *Ann. Intern. Med.* 145:869–879.

Nuku, K., Y. Matsuoka, T. Yamagishi, H. Miyaki, and K. Sato. 2007. Safety assessment of PureSorb-Q40 in healthy subjects and serum coenzyme Q10 level in excessive dosing. *J. Nutr. Vitaminol. (Tokyo)* 53:198–206.

Rizzi, R, F. Re, A. Bianchi, V. De Feo, F. de Simone, L. Bianchi, and L. Stivala. 1993. Mutagenic and antimutagenic activities of *Uncaria tomentosa* and its extracts. *J. Ethnopharmacol.* 38:63–77.

Scripnikov, A., A. Khomenko, and O. Napryeyenko. 2007. Effects of *Ginkgo biloba* extract EGb 761 on neuropsychiatric symptoms of dementia: Findings from a randomized controlled trial. *Wien Med. Wochenschr.* 157:295–300.

Winter, J. 1998. The effects of an extract of *Ginko biloba*, EGb 761, on cognitive behavior and longevity in the rat. *Physiol. Behav.* 63:425–433.

Woelkart, K., E. Marth, A. Suter, R. Schoop, R. B. Raggam, C. Koidl, B. Kleinhappl, and R. Bauer. 2006. Bioavailability and pharmacokinetics of *Echinacea purpurea* preparations and their interaction with the immune system. *Int. J. Clin. Pharmacol. Ther.* 44:401–408.

CHAPTER 9

Biopharmaceutical and Pharmacokinetic Characterization of Nutraceuticals

Charles Preuss

CONTENTS

INTRODUCTION

Some very important components of the drug approval process along with demonstrating safety and efficacy are biopharmaceutical and pharmacokinetic characterization. This is simply documenting the degree of absorption of the active ingredient(s) from the dosage form, for example, tablet absorption into the systemic circulation in which the active ingredient is assumed to be in equilibrium with the biophase, i.e., site of action. The following discussion will focus on guidelines for the pharmaceutical industry for biopharmaceutical and pharmacokinetic characterization of drugs. This brief summary will provide some insights into biopharmaceutical and pharmacokinetic characterization of nutraceuticals if it is eventually mandated by the FDA.

BIOPHARMACEUTICS: BIOAVAILABILITY AND BIOEQUIVALENCE

Oral Drug Dosage Forms

A commonly used dosage form for the delivery of medicines is the oral dosage form, such as tablets, capsules, suspensions, and solutions. Two important parameters that are related to the degree of absorption of the active ingredient of the dosage form are bioavailability and bioequivalence. Bioavailability is the measurement of the rate and extent of the active ingredient that reaches the systemic circulation and bioequivalence is when two different drug products with the same active ingredient have similar bioavailability [Shargel 1993]. For example, a generic company wants to market its own version of a brand name drug because the patent has expired.

Bioavailability and bioequivalence studies are important components of the drug approval process. Bioavailability for oral dosage forms can be documented via systemic exposure profile. This can be determined by measuring the active ingredient and, in some cases, its metabolite(s) from the system circulation, i.e., blood. An important regulatory concern is that the data from the dosage form used in clinical trials provide documentation of efficacy and safety. The bioavailability data can be used as a gold standard for subsequent bioequivalence studies [Food and Drug Administration 2003].

Four important types of studies to document bioavailability and bioequivalence in descending order of preference are as follows:

1. Pharmacokinetic
   a. Plasma
      i. t_{max}
      ii. C_{max}
      iii. AUC
   b. Urine
      i. D^{∞}_{u}
      ii. dD_u/dt
      iii. t^{∞}
2. Pharmacodynamic
3. Clinical
4. Dissolution

The active ingredient can be accurately quantitated pharmacokinetically in the plasma and urine, which gives the most objective data on bioavailability. Plasma data can determine three important parameters: t_{max}, C_{max}, and AUC. t_{max} is the time to reach maximum drug concentration in the plasma after administration. C_{max} is the maximum drug concentration in the plasma after administration. C_{max} is often a good marker for safety and efficacy for the drug because drug concentration in the plasma often corresponds to drug action and drug toxicity. AUC is the area under the drug concentration-time curve, and this is a measure of the degree of bioavailability. The AUC is the amount of drug that reaches the systemic circulation. Often, the AUC is proportional to drug dose, unless there is saturation of drug metabolizing enzymes [Shargel 1993; Food and Drug Administration 2003].

The drug excreted in the urine can be used to estimate bioavailability. However, for this to be valid, the drug must be excreted significantly. Also, from a clinical trial participant perspective, peeing in a cup is less traumatic than a venopuncture. D^{∞}_u is the total amount of drug excreted in the urine and this is related to the amount of drug absorbed. dD_u/dt is the rate of drug excreted in the urine, and it is graphically similar to the drug plasma concentrations. t^{∞} is the time for maximum urinary excretion, and this can be a useful parameter for bioequivalence studies [Shargel 1993; Food and Drug Administration 2003].

Pharmacodynamic studies or acute pharmacological effect studies measure drug action, for example, a reduction in blood pressure. Pharmacokinetic studies are preferred over pharmacodynamic studies. However, in some instances, a pharmacokinetic study is not possible, for example, no available bioassay or a bioassay with unacceptable accuracy or reproducibility. Thus, pharmacodynamic studies can be used to estimate the bioavailability. When conducting a pharmacodynamic study, measurements of drug action should be evenly divided over at least three half-lives of the drug to be within the estimated AUC [Shargel 1993; Food and Drug Administration 2003].

When a pharmacokinetic or pharmacodynamical study is not possible, then a clinical study can be used. The measurement of a clinical study is simply that the therapeutic treatment of the drug is a success, for example, complete cure from cancer. The assumption is that a therapeutic success occurred because there was enough bioavailability when the drug was administered. However, there can be a lot of pharmacodynamic and/or pharmacokinetic variability, such as diet, disease, or genetics, which can make it difficult to determine the cause of patient differences to drug action, such as therapeutic success or failure [Shargel 1993; Food and Drug Administration 2003].

In certain cases, dissolution studies, i.e., *in vitro* studies, can be used instead of the previously mentioned studies. This is especially true for drugs that are very soluble, permeable, and quickly dissolving. In addition, food-effect studies should be conducted because food can affect the bioavailability and bioequivalence of a drug dosage form, and this important topic will be discussed in more detail. Also, certain drug dosage forms contain multiple active ingredients that may be natural or synthetic, and it is unreasonable to quantify each active ingredient. Therefore, it is acceptable to conduct bioavailability or bioequivalence studies on a small number of markers of rate and extent of absorption. This can be an important point to remember when conducting bioavailability and bioequivalence studies with nutraceuticals because some, such as botanicals, potentially contain many active ingredients [Shargel 1993; Food and Drug Administration 2003].

Food Effect

Food-effect bioavailability and fed bioequivalence studies for oral drug dosage forms are part of IND applications and abbreviated new drug applications, respectively.

This includes both immediate-release and modified-release drug dosage forms. These studies are important because food can affect the rate and extent of absorption, which can impact the safety and efficacy of drugs and the focus of this chapter, namely nutraceuticals. Food can alter the bioavailability by various mechanisms, such as delay gastric emptying, alter gastrointestinal (GI) pH, and alter GI metabolism. High-calorie and high-fat meals, such as bacon, eggs, toast, hash browns, and a glass of milk, are used for food-effect bioavailability and fed bioequivalence studies because these are more likely to affect GI physiology, which can cause the largest effect when compared against fasting controls. It is important to keep in mind that food can affect the pharmacokinetics of the active ingredient and/or excipients, which can lead to the alteration of the bioavailability, and this is most likely to occur when the drug dosage form is taken right after the meal. The following are study considerations:

1. General design
2. Subject selection
3. Dosage strength
4. Test meal
5. Administration
6. Sample collection

The general design of the food-effect bioavailability study is fed versus fasting, single-dose, randomized, crossover design with an appropriate washout period. The fed bioequivalence study would include a comparison of the test and reference drug dosage forms. At least 12 healthy subjects should be included in bioavailability and bioequivalence studies unless safety concerns warrant the inclusion of patients with the disease state that is being studied, for example, antineoplastics for cancer treatment. The dosage strength to be studied is usually the highest strength that will be marketed. A high-fat, high-calorie meal is used as a test meal for food-effect bioavailability and bioequivalence studies. This type of meal is most likely to affect the GI tract, leading to the maximal affect on systemic availability. Subjects of fasted treatments should have fasted overnight for at least 10 h, and the drug dosage form, such as tablet or capsule, should be swallowed with 240 ml of water. No food should be eaten for at least 4 h after the dose was given. Subjects of fed treatments should have fasted for at least 10 h overnight, and they should have had 30 min to consume the high-calorie, high-fat meal, after which the oral dosage form with 240 ml of water is swallowed. No food should be eaten for at least 4 h after the dose was given. For both treatment arms, subjects receive standardized meals for lunch and dinner at the same time each day. Sample collection is usually from the plasma, and the types of pharmacokinetic parameters that are examined are AUC, t_{max}, C_{max}, and t_{lag} (for modified release dosage forms) [Food and Drug Administration 2002].

Clozapine Tablets: Dissolution Testing

An example of *in vivo* bioequivalence and *in vitro* dissolution testing guidance is with the clozapine tablet dosage form, which is required for abbreviated new drug applications. Clozapine (Clozaril®) is used for the treatment of schizophrenia. Some important side effects of clozapine are agranulocytosis, seizures, and syncope. Because of the side effects, it is recommended that healthy subjects not be used for bioequivalence studies. The goal of the bioequivalence and dissolution studies is to compare the generic clozapine with the brand Clozaril® with regard to the rate and extent of absorption at equal doses. This is important because the generic form must demonstrate safety and efficacy just like the brand drug dosage form. If company X has marketed a popular nutraceutical in a tablet form and now company Y wants to market its tablet form of the identical nutraceutical, future regulations might require bioequivalence studies. An equal number of patients in a randomized schedule will receive either the generic or brand tablet dosage form at the same dose with 240 ml of water every 12 h for 10 days. Then the patients should be switched to the other tablet for a second period of 10 days. The highest tablet strength is used, which is 100 mg. The study must be first approved by an IRB and the names, titles, and curriculum vitae of medical and scientific directors should be documented. Patients who have been stable while taking clozapine for three months are eligible for the study. The patients white blood cell counts, blood pressure, heart rate, and body temperature need to be carefully monitored. Food does not appear to affect the bioavailability of clozapine. Although on day 10, patients should fast for 8 h before and 4 h after the administration of the morning dose because 14 venous blood samples will be collected from times 0.25–12 h. The following pharmacokinetic data should be collected and evaluated for bioequivalence:

1. Patient and mean drug concentrations
2. Patient and mean trough concentrations ($C_{min\ SS}$)
3. Patient and mean peak concentrations ($C_{max\ SS}$)
4. Patient and mean steady-state $AUC_{interdose}$
5. Patient and mean percentage drug concentration fluctuation
6. Patient and mean t_{max}

Two additional factors to consider are batch size and potency. The test batch should be no less than 10% of the largest batch planned for full production or a minimum of 100,000 units. The assayed potency of the generic should not differ from that of the brand by more than 5%. Dissolution testing should be conducted for all strengths of both generic and brand, with an n = 12 for each. Note that the same batch as the bioequivalence study is used [Food and Drug Administration 2005].

PHARMACOKINETICS

Before drugs can be approved for marketing, their pharmacokinetic profile must be studied. In particular, the investigational drug's biotransformation or metabolism must be studied because the metabolism can affect the safety and efficacy of a drug. Inhibition of drug metabolism can lead to high concentrations, which can lead to toxicity, and induction of drug metabolism can lead to low concentrations, which can lead to lack of efficacy. Most drug dosage forms contain one active ingredient and many inactive ingredients or excipients. However, nutraceuticals may contain many active and inactive ingredients, especially with botanical or herbal nutraceuticals, which could potentially affect its safety and efficacy and therefore potentially complicate active ingredient studies.

Drug Metabolism

The following are studies of *in vitro* drug metabolism that can lead to a better understanding of potential drug interactions [Food and Drug Administration 1997]:

1. Hepatic cytochrome P450
2. Additional hepatic enzymes
3. GI tract enzymes
4. Animal studies

Cytochrome P450 is a large family of enzymes that metabolize the majority of drugs. Some clinically important examples are CYP3A4, CYP2D6, and CYP2C9. The liver is an important site for drug metabolism, and the human liver microsomes are a common *in vitro* technique to study drug metabolism. Isolated hepatocytes and precision cut liver slices can give a more complete understanding of drug metabolism. Unfortunately, these two techniques have limited enzyme stability. The investigational drug is incubated with human liver microsomes, and different inducers and inhibitors of cytochrome P450 can be tested. A positive result might require additional *in vivo* drug metabolism and drug interaction studies [Food and Drug Administration 1997].

There are many additional hepatic enzymes that can influence drug metabolism. Some clinically important examples are glucuronidation, sulfation, and acetylation. The *in vitro* tests to evaluate their drug metabolism are not as well developed as with the cytochrome P450s.

The GI tract enzymes are especially important for oral drug dosage forms because of gut metabolism and subsequent absorption into the system circulation. Two clinically important enzymes that can affect drug absorption are CYP3A4 and P-glycoprotein. Therefore, compounds that inhibit CYP3A4 and P-glycoprotein can lead to increased concentrations of the investigational drug, and, conversely, compounds that induce CYP3A4 and P-glycoprotein can lead to decreased concentrations of the investigational drug [Food and Drug Administration 1997].

In vitro or *in vivo* animal studies during preclinical drug development can help elucidate the safety and efficacy of potential drugs, including nutraceuticals. Of special interest are potentially active metabolites found during human *in vitro* studies. These potentially active metabolites can be studied in animals to determine their safety and efficacy, which can help determine future *in vitro* or *in vivo* human studies. Furthermore, animal studies can be used to examine drug interactions with regard to parent drug and its metabolites. A difficulty with animal studies is how they relate to humans. For example, a metabolite that is toxic to the selected experimental animal species might not be toxic to humans [Food and Drug Administration 1997].

In vitro tests can help elucidate the metabolic pathway of the investigational compound, i.e., nutraceutical and this can determine the type of human *in vivo* tests that need to be conducted. For example, if the *in vitro* tests indicate that the primary route of metabolism for the nutraceutical is via CYP3A4, then one would want to follow up with human *in vivo* tests on CYP3A4 inducers and inhibitors. However, if the *in vitro* tests concluded that the nutraceutical is not a substrate for CYP2C9, then it is not necessary to conduct human *in vivo* studies with CYP2C9 inhibitors and inducers [Food and Drug Administration 1997].

An important consideration is the timing of the *in vitro* metabolic studies. There are two important goals for the *in vitro* metabolic studies:

1. Identify important routes of metabolism and metabolites
2. Identify potential drug interactions

The earlier this information can be determined the better, but this must be balanced with capital expenditure. A company does not want to invest its precious capital if the nutraceutical encounters early problems and it is dropped from advancing. However, regulatory agencies are not going to allow a pharmacologically active compound to the market until there are *in vitro* and *in vivo* drug metabolism studies. Therefore, it is suggested that *in vitro* metabolic studies be conducted during Phase II preclinical drug development [Food and Drug Administration 1997].

If during *in vitro* drug metabolism studies the nutraceutical of interest is found to have activity as either an inhibitor or inducer, then human *in vivo* drug interaction studies should be conducted.

Drug Interactions

The following study design is adapted from drug-drug interaction studies as recommended by the FDA with its key components [Food and Drug Administration 1999]:

1. Design
2. Subjects
3. Interacting drug selection
4. Route of administration
5. Dose
6. Pharmacokinetic endpoints
7. Statistics

There are many different study designs that can be conducted, such as one-sequence crossover or randomized crossover. The selection of the study design can really depend on several factors, such as length of drug use or toxicity of the drug. These studies in general examine the concentration of the compound of interest, i.e., the nutraceutical, with the interacting compound, i.e., the inducer or inhibitor. An important consideration with the interacting compound is that inhibitors usually exert their activity much sooner than inducers, because inducers usually work by increasing the protein synthesis of the drug-metabolizing enzyme [Food and Drug Administration 1999].

Subjects for drug-drug interaction studies or nutraceutical-drug interaction studies are usually conducted in healthy volunteers unless safety concerns warrant the use of volunteers with the disease state of interest. In addition, pharmacogenomic analysis of the subject's drug-metabolizing enzymes is encouraged because polymorphisms of the enzymes can dramatically affect safety and efficacy. This is especially true for CYP2D6, UGT1A1, etc. [Food and Drug Administration 1999].

The selection of the interacting drug with the nutraceutical is a very important consideration. Two general types of experiments should be considered:

1. Nutraceutical as the interacting compound to the drug
2. Drug as the interacting compound to the nutraceutical

The nutraceutical could cause either inhibition or induction of drug-metabolizing enzymes, which could affect drug concentrations. This information is important because drugs such as cyclosporine with a narrow therapeutic window, i.e., a specific concentration range, could lead to toxicity or lack of efficacy if the CYP3A4 enzyme is inhibited or induced. Table 9.1 lists several nutraceuticals that inhibit or induce a cytochrome P450 and the affected drug. Another possibility to explore is that a drug could affect the concentrations of the nutraceutical via inhibition or induction of drug-metabolizing enzymes. These enzymes can also biotransform endobiotics and xenobiotics, such as testosterone and benzo[a]pyrene. Therefore, these enzymes are not restricted to biotransform drugs only. Table 9.2 lists several drugs that can induce or inhibit drug-metabolizing enzymes that could be used to design the appropriate *in vitro* tests with the follow up *in vivo* tests if deemed necessary. Potentially, the FDA would like to know if a drug could affect the concentration of the nutraceutical if taken together [Food and Drug Administration 1999].

The route of administration for the *in vivo* drug-nutraceutical interaction study would depend on the dosage form that will be marketed. For example, if the nutraceutical will be formulated via a capsule only, then studies with an intravenous solution or oral suspension would not be necessary [Food and Drug Administration 1999].

The dose selection for the drug-nutraceutical interaction should use the maximum approved dose for the nutraceutical and drug in the shortest dosing interval to

Table 9.1 Nutraceutical and Drug Interactions

Nutraceutical	Enzyme	Drug
Grapefruit	CYP3A4	Felodipine [Baily et al. 1991]
Seville Orange	CYP3A4	Felodipine [Malhotra et al. 2001]
St. John's Wort	CYP3A4	Cyclosporine [Bauer et al. 2003]
Watercress	CYP2E1	Chlorzoxazone [Leclercq, Desager, and Horsmans 1998]
Pomelo	CYP3A4	Tacrolimus [Egashira et al. 2004]
Lime	CYP3A4	Felodipine [Bailey et al. 2003]

Table 9.2 Inducers and Inhibitors of Drug Metabolism

Enzyme	Inducer	Inhibitor
CYP1A2	Omeprazole	Fluvoxamine
CYP2C9	Rifampin	Sulfinpyrazone
CYP2C19	Rifampin	Fluconazole
CYP2D6	Rifampin	Quinidine
CYP2E1	Isoniazid	Disulfirim
CYP3A4	Rifampin	Ketoconazole

Source: Adapted from Correia, M. 2007. Drug Biotransformation. In *Basic and Clinical Pharmacology.* 10th edition. Edited by Katzung, B.G. New York, NY: McGraw-Hill, pp. 64–73.

detect the interaction. The dose of both might need to be further reduced because of safety concerns [Food and Drug Administration 1999].

The following pharmacokinetic endpoints could be recommended to be measured for drug-nutraceutical interactions: AUC, C_{max}, t_{max} and sometimes CL, V_d, and $T_{½}$. For example, if nutraceutical A induces the metabolism of drug B, then the pharmacokinetic endpoints to be measured in a drug-nutraceutical interaction study would be AUC, C_{max}, and t_{max} [Food and Drug Administration 1999].

An important statistical consideration for drug-nutraceutical interaction studies would be to capture clinically significant interactions, and this can be determined as 90% confidence intervals about the geometric mean ratio of the pharmacokinetic endpoints with and without the interacting compound [Food and Drug Administration 1999].

REFERENCES

Bailey, D., G. Dresser, and J. Bend. 2003. Bergamottin, lime juice and red wine as inhibitors of cytochrome P450 3A4 activity: Comparison with grapefruit juice. *Clin. Pharmacol. Ther.* 73:529–537.

Bailey, D., J. Spence, C. Munoz, and J. Arnold. 1991. Interaction of citrus fruit juices with felodipine and nifedipine. *Lancet* 337:268–269.

Barnes, J., L. Anderson, and J. Phillipson. 2007. *Herbal Medicines.* 3rd edition. Grayslake, IL: Pharmaceutical Press.

Bauer, S., E. Störmer, A. Johne, H. Krüger, K. Budde, H. H. Neumayer, I. Roots, and I. Mai. 2003. Alterations in cyclosporine A pharmacokinetics and metabolism during treatment with St. John's wort in renal transplant patients. *Br. J. Clin. Pharmacol.* 55:203–211.

Correia, M. 2007. Drug Biotransformation. In *Basic and Clinical Pharmacology.* 10th edition. Edited by Katzung, B.G. New York, NY: McGraw-Hill, pp. 64–73.

Egashira, K., H. Ohtani, S. Itoh, N. Koyabu, M. Tsujimoto, H. Murakami, and Y. Sawada. 2004. Inhibitory effects of pomelo on the metabolism of tacrolimus and the activites of CYP3A4 and P-glycoprotein. *Drug. Metab. Disp.* 32:828–833.

Food and Drug Administration. 1997. Guidance for industry. Drug metabolism/drug interaction studies in the drug development process: Studies in vitro. http://www.fda.gov/cder/guidance/clin3.pdf. Accessed December 31, 2007.

Food and Drug Administration. 1999. Guidance for industry. *In vivo* drug metabolism/drug interaction studies: Study design, data analysis, and recommendations for dosing and labeling. http://www.fda.gov/cder/guidance/2635fnl.pdf. Accessed December 28, 2007.

Food and Drug Administration. 2002. Guidance for industry. Food-effect bioavailability and fed bioequivalence studies. http://www.fda.gov/cder/guidance/5194fnl.pdf. Accessed December 26, 2007.

Food and Drug Administration. 2003. Guidance for industry. Bioavailability and bioequivalence for orally administered drug products: General considerations. http://www.fda.gov/cder/guidance/5356fnl.pdf. Accessed December 18, 2007.

Food and Drug Administration. 2005. Guidance for industry. Clozapine tablets: In vivo bioequivalence and in vitro dissolution testing. http://www.fda.gov/cder/guidance/6077fnl.pdf. Accessed December 26, 2007.

Leclercq, I., J. Desager, and Y. Horsmans. 1998. Inhibition of chlorzoxazone metabolism, a clinical probe for CYP2E1, by a single ingestion of watercress. *Clin. Pharmacol. Ther.* 64:144–149.

Malhotra, S., D. Bailey, M. Paine, and P. Watkins. 2001. Seville orange juice-felodipine interactions: Comparison with dilute grapefruit juice and involvement of furanocoumarins. *Clin. Pharmacol. Ther.* 69:14–23.

Shargel, L. and A. Yu. 1993. Bioavailability and Bioequivalence. In *Applied Biopharmaceutics and Pharmacokinetics.* 3rd edition. Norwalk, CT: Appleton and Lange, pp. 193–223.

CHAPTER 10

Regulatory Considerations for Dietary Supplements and Functional cGMPs

Mike Witt and Yashwant Pathak

CONTENTS

INTRODUCTION

Current and proposed regulations to the dietary supplements industry are changing the way that companies approach manufacturing and labeling of products. Understanding the regulations and avoiding pitfalls will be a major focus of companies of all sizes. The size of the company determines how fast the proposed GMPs are implemented within a company. No matter what size the company may be, all manufactures need to be compliant as of June 2010.

The cGMP regulations proposed by the FDA are designed to bring the minimum manufacturing and labeling standards from the pharmaceutical industry into the dietary supplements industry. These standards establish basic standards of operation at the manufacturing level to ensure that finished dietary supplement products meet established specifications for identity, purity, strength, and composition and that they limit possible contamination. Manufacturers must adhere to standards regarding production and process controls, manufacturing and testing equipment, quality-control approvals, detailed specifications, master manufacturing records, and batch production records. Procedures must be written and followed to investigate and document product complaints and returns. Manufacturers must include company information on the label to allow consumers to report problems or adverse reactions. All of these regulations move the dietary supplement manufacturers in line with the consumers' expectations of quality and reliability associated with and expected of pharmaceutical manufacturers.

DEFINITION OF A DIETARY SUPPLEMENT

Dietary supplements are defined, in part, as products (other than tobacco) intended to supplement the diet that bear or contain one or more of the following dietary ingredients: (1) a vitamin, (2) a mineral, (3) an herb or other botanical, (4) an amino acid, (5) a dietary substance for use by man to supplement the diet by increasing the total dietary intake, or (6) a concentrate, metabolite, constituent, extract, or a combination of any ingredient mentioned above.

Furthermore, dietary supplements are products intended for ingestion, are not represented for use as a conventional food or as a sole item of a meal or the diet, and are labeled as dietary supplements. The complete statutory definition is found in § 201(ff) of the Federal Food, Drug, and Cosmetic Act (The Act) (21 U.S.C. 321).

A dietary supplement must be identified by use of the term "dietary supplement" as part of the statement of identity, except that the word "dietary" may be deleted and replaced with the name of the dietary ingredient(s) in the product (e.g., calcium supplement) or an appropriately descriptive term indicating the type of dietary ingredient(s) in the dietary supplement product (e.g., herbal supplement with vitamins) (21 CFR 101.3(g)).

LABELING REQUIREMENTS FOR DIETARY SUPPLEMENTS

The following five statements are required for dietary supplements:

1. Statement of identity (name of the dietary supplement)
2. Net quantity of contents statement (amount of the dietary supplement)
3. Nutrition labeling
4. Ingredient list
5. Name and place of business of the manufacturer, packer, or distributor. (21 CFR 101.3(a), 21 CFR 101.105(a), 21 CFR 101.36, 21 CFR 101.4(a)(1), and 21 CFR 101.5)

All required label statements must be placed on either the front label panel (the principal display panel) or the information panel (usually the label panel immediately to the right of the principal display panel, as seen by the consumer when facing the product), unless otherwise specified by regulation (i.e., exemptions) (21 CFR 101.2(b) and (d), 21 CFR 101.9(j)(13) and (j)(17), 21 CFR 101.36(g), (i)(2) and (i)(5)).

The statement of identity and the net quantity of contents statement must be placed on the principal display panel. When packages bear alternate principal display panels, this information must be placed on each alternate principal display panel (21 CFR 101.1, 21 CFR 101.3(a) and 21 CFR 101.105(a)).

The principal display panel of the label is the portion of the package that is most likely to be seen by the consumer at the time of display for retail purchase. Many containers are designed with two or more different surfaces that are suitable for use as the principal display panel. These are alternate principal display panels (21 CFR 101.1).

The "Supplement Facts" panel, the ingredient list, and the name and place of business of the manufacturer, packer, or distributor must be included on the information panel; if such information does not appear on the principal display panel, except that if space is insufficient, then the special provisions on the Supplement Facts panel in 21 CFR 101.36(i)(2)(iii) and (i)(5) may be used (21 CFR 101.2(b) and (d), 101.36(i)(2)(iii) and (i)(5), 101.5, 101.9(j)(13)(i)(A) and (j)(17)).

The information panel is located immediately to the right of the principal display panel as the product is displayed to the consumer. If this panel is not usable, because of package design and construction (e.g., folded flaps), then the panel immediately contiguous and to the right of this part may be used for the information panel. The information panel may be any adjacent panel when the top of a container is the principal display panel (21 CFR 101.2(a)).

The street address must be listed if it is not listed in a current city directory or telephone book, including the city or town, the state, and zip code. The address of the principal place of business may be used in lieu of the actual address (21 CFR 101.5).

Intervening material, which is defined as label information that is not required (e.g., universal product bar code), may not be placed between label information that is required on the information panel (21 CFR 101.2(e)).

Unless excepted by law, the Tariff Act requires that every article of foreign origin (or its container) imported into the United States conspicuously indicate the English name of the country of origin of the article (§ 304, Tariff Act of 1930, as amended (19 U.S.C. 304)).

Expiration dating does not need to be included on the label of a dietary supplement. However, a firm may include this information if it is supported by valid data demonstrating that it is not false or misleading.

NUTRITIONAL LABELING FOR A DIETARY SUPPLEMENT

The nutrition label for a dietary supplement is called a Supplement Facts panel (see example labels below) (21 CFR 101.36(b)(1)(i)).

The major differences between Supplement Facts panel and "Nutrition Facts" panel are as follows:

1. A company must list dietary ingredients without reference daily intakes (RDIs) or daily reference value (DRVs) in the Supplement Facts panel for dietary supplements. A company is not permitted to list these ingredients in the Nutrition Facts panel for foods.
2. A company may list the source of a dietary ingredient in the Supplement Facts panel for dietary supplements. A company cannot list the source of a dietary ingredient in the Nutrition Facts panel for foods.
3. A company is not required to list the source of a dietary ingredient in the ingredient statement for dietary supplements if it is listed in the Supplement Facts panel.
4. A company must include the part of the plant from which a dietary ingredient is derived in the Supplement Facts panel for dietary supplements. A company is not permitted to list the part of a plant in the Nutrition Facts panel for foods.
5. A company is not permitted to list "zero" amounts of nutrients in the Supplement Facts panel for dietary supplements. A company is required to list zero amounts of nutrients in the Nutrition Facts panel for food (21 CFR 101.36(b)(3) and (b)(2)(i), 21 CFR 101.4(h), 21 CFR 101.36(d) and (d)(1), and 21 CFR 101.9).

The names and quantities of dietary ingredients present in the product, the "Serving Size," and the "Servings Per Container" must be listed. However, the listing of Servings Per Container is not required when it is the same information as in the net quantity of contents statement. For example, when the net quantity of contents statement is 100 tablets and the Serving Size is one tablet, the Serving Per Container also would be 100 tablets and would not need to be listed (21 CFR 101.36(b)).

Definition of Serving Size

One serving of a dietary supplement equals the maximum amount recommended, as appropriate, on the label for consumption per eating occasion or, in the absence of recommendations, one unit (e.g., tablet, capsule, packet, teaspoonful, etc). For example, if the directions on the label say to take one to three tablets with breakfast, the serving size would be three tablets (21 CFR 101.12(b)). The term Serving Size must be used on the label (21 CFR 101.36(b)(1)).

Required Nutrient Declaration Required in the Supplement Facts Panel

Total calories, calories from fat, total fat, saturated fat, cholesterol, sodium, total carbohydrate, dietary fiber, sugars, protein, vitamin A, vitamin C, calcium, and iron must be listed when they are present in measurable amounts. A measurable amount is an amount that exceeds the amount that can be declared as zero in the nutrition label of conventional foods, as specified in 21 CFR 101.9(c). If present in a measurable amount, *trans* fat must be listed on a separate line underneath the listing of saturated fat, as of January 1, 2006. Calories from saturated fat and the amount of polyunsaturated fat, monounsaturated fat, soluble fiber, insoluble fiber, sugar alcohol, and other carbohydrate may be declared, but they must be declared when a claim is made about them (21 CFR 101.36(b)(2)(i)) (see 68 FR 41434 at 41505, July 11, 3003).

Declaring vitamins and minerals (other than vitamin A, vitamin C, calcium, and iron) is only required when they are added to the product for purposes of supplementation or if making a claim about them (21 CFR 101.36(b)(2)(i)). When vitamin E occurs naturally, it does not need to be declared. This is because vitamin E is not one of the 14 mandatory dietary ingredients (21 CFR 101.36(b)(2)(i)).

It is required to list any other nutrients used in manufacturing when making a claim about calories from saturated fat, insoluble fiber, polyunsaturated fat, sugar alcohol, monounsaturated fat, other carbohydrate, and soluble fiber (21 CFR 101.36(b)(2)(i)).

Dietary ingredients for which no "Daily Values" (DVs) have been established must be listed by their common or usual names when they are present in a dietary supplement. They must be identified as having no DVs by use of a symbol in the column for "% Daily Value" (% DV) that refers to the footnote "Daily Value Not Established" (21 CFR 101.36(b)(2)(iii)(F) and (b)(3)).

Ingredients in dietary supplements that are not dietary ingredients, such as binders, excipients, and fillers, must be included in the ingredient statement (21 CFR 101.4(g)). Products that contain only amino acids may not declare protein for the product (21 CFR 101.36(b)(2)(i)).

Dietary ingredients that have DVs must be listed in the same order as for the labels of conventional foods, except that vitamins, minerals and electrolytes are grouped together. This results in the following order for vitamins and minerals: vitamin A, vitamin C, vitamin D, vitamin E, vitamin K, thiamin, riboflavin, niacin, vitamin B_6, folate, vitamin B_{12}, biotin, pantothenic acid, calcium, iron, phosphorus, iodine, magnesium, zinc, selenium, copper, manganese, chromium, molybdenum, chloride, sodium, and potassium (21 CFR 101.36(b)(2)(i)(B)).

The label may use the following synonyms in parentheses after dietary ingredients: vitamin C (ascorbic acid), thiamin (vitamin B_1), riboflavin (vitamin B_2), folate (folacin or folic acid), and calories (energy). Alternatively, "folic acid" or "folacin" may be listed without parentheses in place of "folate." Energy content may also be expressed parenthetically in kilojoules immediately after the caloric content (21 CFR 101.36(b)(2)(i)(B)(2)).

Reporting Amounts

When using materials such as calcium carbonate as the source of calcium in the product, the list must include the weight of the calcium rather than the weight of the calcium carbonate in the Supplement Facts panel (21 CFR 101.36(b)(2)(ii)). Also, the amount of the dietary ingredient may be placed in a separate column or immediately after the name of the dietary ingredient (21 CFR 101.36(b)(2)(ii)). When using a separate column for amounts of dietary ingredients, the heading "Amount per Serving" may be placed over the column of amounts (21 CFR 101.36(b)(2)(i)(A)).

Language consistent with the declaration of the serving size, such as "Each Tablet Contains" or "Amount per 2 Tablets," may be used in place of the heading "Amount per Serving." Other terms may be used as well, such as capsule, packet, or teaspoonful (21 CFR 101.36(b)(2)(i)(A)). It is also acceptable to declare information on a "per unit" basis in addition to the required "per serving" basis (21 CFR 101.36(b)(2)(iv)).

If the product has different servings, such as one tablet in the morning and two at night, additional columns may be used. The columns must be labeled appropriately, e.g., "Amount per 1 Tablet" and "Amount per 2 Tablets" (21 CFR 101.36(b)(2)(i)(A)).

It is required to use the units of measurement specified for use in the Nutrition Facts panel. For example, the amount of fat would be listed in terms of grams in both the Nutrition Facts and Supplement Facts panels. However, units of measurement for amounts of vitamins and minerals are not specified for use in the Nutrition Facts panel because they must be listed by % DV and not by weight. The units of measurement given should be used in 21 CFR 101.9(c)(8)(iv) for the DVs of vitamins and minerals when listing these nutrients in Supplement Facts (e.g., the amount of vitamin C must be listed in terms of milligrams because its DV is stated in milligrams) (21 CFR 101.36(b)(2)(ii)(B) and 101.9(c)).

Percent of Daily Value

The % DV is the percentage of the DV (i.e., RDIs or DRVs) of a dietary ingredient that is in a serving of the product (21 CFR 101.36(b)(2)(ii)(B) and 21 CFR 101.9(c)(8) and (9)). The % DV must be declared for all dietary ingredients for which the FDA has established DVs, except that (1) the percentage for protein may be omitted, and (2) labels of dietary supplements to be used by infants, children less than four years of age, or pregnant or lactating women must not list any percent for total fat, saturated fat, cholesterol, total carbohydrate, dietary fiber, vitamin K, selenium, manganese, chromium, molybdenum, chloride, sodium, or potassium. See FDA's proposed labeling guides for the DVs to be used for adults and children four or more years of age and the DVs to be used for infants, children less than four years of age, or pregnant or lactating women (21 CFR 101.36(b)(2)(iii)).

The % DV is calculated by dividing the quantitative amount weight by the established DV for the specified dietary ingredient and multiplying by 100 (except that the % DV for protein must be calculated in accordance with 21 CFR 101.9(c)(7)(iii)). In this calculation, the unrounded amount must be used as the quantitative amount, except that for total fat, saturated fat, cholesterol, sodium, potassium, total carbohydrate, and dietary fiber, the quantitative amount by weight declared on the label may be used (i.e., the rounded amount). For example, the % DV for 60 mg of vitamin C is 100 (60 mg divided by the DV for vitamin C, multiplied by 100) (21 CFR 101.36(b)(2)(iii)(B) and 21 CFR 101.9(c)(7)(iii)). The % DV must be expressed to the nearest whole percent, except that "Less than 1%" or "<1 %" must be used when the amount present is big enough to be listed but so small that the % DV when rounded to the nearest percent is zero. For example, a product containing 1 g of total carbohydrate would list the % DV as Less than 1% or <1 % (21 CFR 101.36(b)(2)(iii)(C)).

If the amount of a dietary ingredient in the product in high enough to declare but so low that the % DV rounds to zero, Less than 1% or <1% must be declared because the label might confuse consumers if 5 mg is declared and the listed DV is 0%. For example, if a product contains 5 mg of potassium, the % DV calculates to 0.14% (5 mg divided by 3,500 mg), which would round to zero. In this case, Less than 1% or <1% would be declared for the % DV. Note that this does not pertain to dietary ingredients having RDIs because they may not be listed when present at less than 2% of the RDI (21 CFR 101.36(b)(2)(iii)(C) and 101.36(b)(2)(i)).

Other Dietary Ingredients

"Other Dietary Ingredients" are those dietary ingredients that do not have DVs (i.e., RDIs or DRVs), such as phosphatidylserine (21 CFR 101.36(b)(3)(i)). The statement Other Dietary Ingredients can be listed in the Supplement Facts panel after the listing of dietary ingredients having DVs (21 CFR 101.36(b)(3)(i)).

Furthermore, Other Dietary Ingredients must be listed by common or usual name in a column or linear display. The FDA has not specified an order that must be followed. The quantitative amount needs to be listed by weight per serving immediately after the name of the dietary ingredient or in a separate column. Furthermore, a symbol in the column for % DV that refers to the footnote Daily Value Not Established must be included, except that the symbol must follow the weight when not using the column format (21 CFR 101.36(b)(3)).

All liquid extracts are to be listed using the volume or weight of the total extract and the condition of the starting material before extraction when it was fresh. Also included may be information on the concentration of the dietary ingredient and the solvent used, e.g., "fresh dandelion root extract, x (y:z) in 70% ethanol," where x is the number of milliliters or milligrams of the entire extract, y is the weight of the starting material, and z is the volume (milliliters) of solvent. The solvent must be identified in either the nutrition label or ingredient list. (21 CFR 101.36(b)(3)(ii)(B)).

For dietary ingredients that are extracts from which the solvent has been removed, the weights of the dried extracts must also be included (21 CFR 101.36(b)(3)(ii)(C)).

The list of constituents of a dietary ingredient indented under the dietary ingredient and followed by their quantitative amounts by weight per serving are to be listed as well. The constituents in a column or in a linear display may be declared as well (21 CFR 101.36(b)(3)(iii)).

Proprietary blends are to be identified by use of the term "Proprietary Blend" or an appropriately descriptive term or fanciful name. On the same line, the total weight of all Other Dietary Ingredients contained in the blend must be included. Indented underneath the name of the blend, list the Other Dietary Ingredients in the blend must be recorded, in either a column or linear manner, in descending order of predominance by weight. These ingredients should be followed by a symbol referring to the footnote Daily Value Not Established. Dietary ingredients having RDIs or DRVs must be listed separately and the individual weights declared (21 CFR 101.36(b)(2) and (c)).

If the product contains two or more packets of supplements (e.g., a packet of capsules for the morning and a different packet for the evening), the information for each packet may be presented in an individual nutrition label or may include an aggregate nutrition label. For two packets, this would consist of five columns. All of the dietary ingredients should be listed in the first column. The amounts and percentages of the morning packet should be listed in the second and third columns and similar information for the evening packet in the fourth and fifth columns (see the illustration of aggregate nutrition labeling in 21 CFR 101.36(e)(10)(iii); see also 21 CFR 101.36(e)(8)).

LABELING COMPLIANCE

The FDA will collect a composite of 12 subsamples (consumer packages) or 10% of the number of packages in the same inspection lot, whichever is smaller. The FDA will randomly select these packages (21 CFR 101.36(f)(1)). The FDA may permit use of an alternative means of compliance or additional exemptions in accordance with 21 CFR 101.9(g)(9). If a firm needs such special allowances, the request must be made in writing (to Office of Nutritional Products, Labeling, and Dietary Supplements (HFS-800), The Food and Drug Administration, 5100 Paint Branch Parkway, College Park, Maryland 20740-3835) (21 CFR 101.36(f)(2)). For dietary ingredients that are specifically added, the product must contain 100% of the volume or weight that has been declared on the label, with the exception of a deviation that is attributable to the analytical method. Products that contain less than this amount of such a dietary ingredient would be misbranded and in violation of the law. Dietary ingredients that are naturally occurring must be present at 80% of the declared value. For example, if vitamin C is added that was isolated from a natural source or made synthetically to the dietary supplement product, it would be subject to the 100% rule. However, if rose hips were added to a product, the vitamin C in the rose hips is naturally occurring and must be present at least 80% of the declared value (21 CFR 101.9(g)(3) and (g)(4)).

The dietary supplement product is not required to have a Supplement Facts panel if any of the following apply:

1. The company is a small business that has not more than $50,000 gross sales made or business done in sales of food to consumers or not more than $500,000 per year from total sales in accordance with 21 CFR 101.36(h)(1).
2. The company sells less than 100,000 units of the product annually, the firm has fewer than 100 full-time equivalent employees in accordance with 21 CFR 101.36(h)

(2) and the company files an annual notification with the FDA as specified in 21 CFR 101.9(j)(18)(iv).

3. The company ships the product in bulk form, does not distribute it to consumers in such form, and supplies it for use in the manufacture of other dietary supplements in accordance with 21 CFR 101.36(h)(3).

The two exemptions for small businesses and low-volume products (items 1 and 2 above) apply only if the products' labels bear no claims or other nutrition information (21 CFR 101.36(h)(1)-(3)).

Special Labeling Provisions

On products for children less than two years of age, other than infant formula, the following are not be included on the label: calories from fat, calories from saturated fat, saturated fat, polyunsaturated fat, monounsaturated fat, and cholesterol. Also, on products for children less than four years of age, % DVs for total fat, saturated fat, cholesterol, total carbohydrate, dietary fiber, vitamin K, selenium, manganese, chromium, molybdenum, chloride, sodium, or potassium may not be included (21 CFR 101.36(b)(2)(iii) and (i)(1)).

Required in the footnote is the statement "Percent Daily Values Are Based on a 2,000 Calorie Diet" when total fat, saturated fat, total carbohydrate, dietary fiber, or protein are declared (21 CFR 101.36(b)(2)(iii)(D)).

If there is insufficient space for the Supplement Facts panel on the information panel or the principal display panel, it may be located on other panels that can readily be seen by consumers in accordance with 21 CFR 101.9(j)(17) (see 21 CFR 101.36(i)(2)(iii) and (i)(5) and 21 CFR 101.9(j)(17)).

The Supplement Facts panel may be omitted on individual units if nutrition information is fully provided on the outer package of the multiunit pack and the unit containers are securely enclosed and are not intended to be separated for retail sale. Each individual unit must be labeled with the statement "This Unit Not Labeled For Retail Sale" in accordance with 21 CFR 101.9(j)(15) (see 21 CFR 101.36(i)(3) and 21 CFR 101.9(j)(15)).

If dietary supplements are sold from bulk containers, the retailer must display a Supplement Facts panel clearly at the point of purchase (e.g., on a counter card, sign, tag affixed to the product, or some other appropriate device). Alternatively, the required information may be placed in a booklet, loose-leaf binder, or some other appropriate format that is available at the point of purchase (21 CFR 101.36(i)(4), 21 CFR 101.9(a)(2) and (j)(16)).

SAMPLE LABELS

See Label 10.1 for dietary supplement containing multiple vitamins (see 21 CFR 101.36(e)(10)(i)). See Label 10.2 for dietary supplement containing multiple vitamins for children and adults (see 21 CFR 101.36(e)(10)(ii)). See Label 10.3 for multiple vitamins in packets (see 21 CFR 101.36(e)(10)(iii)). See Label 10.4 for dietary supplement containing dietary ingredients with and without RDIs and DRVs (see 21 CFR 101.36(e)(10)(iv)).

INGREDIENT LABELING

The Dietary Supplement Health and Education Act uses the term "ingredient" to refer to the compounds used in the manufacture of a dietary supplement. For instance, when calcium carbonate is used to provide calcium, calcium carbonate is an "ingredient" and calcium is a "dietary ingredient." The term ingredient also refers to substances such as binders, colors, excipients, fillers, flavors, and sweeteners (Public Law 103-417, 60 Federal Register 67194 at 67199 (December 28, 1995)).

Ingredients that are sources of dietary ingredients may be listed within the Supplement Facts panel, for example, "Calcium (as calcium carbonate)." When ingredients are listed in this way, they do not have to be listed again in the ingredient statement (also called an ingredient list) (21 CFR 101.36(d)). If all source ingredients are placed in the Supplement Facts panel and there are no other ingredients, such as excipients or fillers, an ingredient statement is not necessary (21 CFR 101.4(a)(1)).

How to Identify the Ingredient List

To identify the ingredient list, the ingredient list must be preceded by the word "Ingredients," except that words "Other Ingredients" must be used when some ingredients have been identified (i.e., as sources) within the nutrition label (21 CFR 101.4(g)).

When present, the ingredient list must be placed on dietary supplements immediately below the nutrition label or, if there is insufficient space below the nutrition label, immediately contiguous and to the right of the nutrition label (21 CFR 101.4(g)).

The ingredients are to be listed in descending order of predominance by weight. This means that the ingredient that weighs the most is first and the ingredient that weighs the least is last (21 CFR 101.4(a)). Also, spices, natural flavors, or artificial flavors must be declared in the ingredient lists by using either specific common or usual names or by using the declarations "spice," "natural flavor," or "artificial flavor," or any combination thereof (21 CFR 101.22(h)(1) and 21 CFR 101.4(a)(1)).

Paprika, turmeric, saffron, and other spices that are also colorings may be declared by either name or the term "spice and coloring." For example, paprika may be listed as "paprika" or as "spice and coloring" (21 CFR 101.22(a)(2)).

Declaration of an artificial color depends on whether or not the artificial color is certified. List a certified color by its specific or abbreviated name, e.g., "FD&C Red No. 40" or "Red 40." A color that is not certified may be listed as an "Artificial Color," "Artificial Color Added," "Color Added," or by its specific common or usual name (21 CFR 101.22(k)(1) and (k)(2)).

When a blend of fats and/or oils is not the predominant ingredient of the product and the makeup of the blend varies, the following "and/or" labeling or language must be used, such as the following: *"INGREDIENTS: ... vegetable oil shortening (contains one or more of the following: cottonseed oil, palm oil, soybean oil)"* (21 CFR 101.4(b)(14)).

Any added water must be identified in the list of ingredients in descending order of predominance by weight, for example: *"Ingredients: Cod liver oil, gelatin, water, and glycerin"* (21 CFR 101.4(a) and (c) and 21 CFR 101.36(e)(10)(iv)).

When using a chemical preservative, the common or usual name of the preservative must be listed, followed by a description that explains its function e.g., "preservative," "to retard spoilage," "a mold inhibitor," "to help protect flavor," or "to promote color retention" (21 CFR 101.22(j)).

DIETARY SUPPLEMENT LABELING CLAIMS

A nutrient content claim is a claim that expressly or by implication characterizes the level of a nutrient in a dietary supplement (21 CFR 101.13(b)).

The nutrient levels needed to use nutrient content claims are shown in Appendix D of the FDA's proposed labeling guide. Only those nutrient content claims, or their synonyms, that are specifically defined in regulations may be used (21 CFR 101.13(b)).

The regulations for specific nutrient content claims may be found in 21 CFR 101, Subpart D (Specific Requirements of Nutrient Content Claims) as follows:

- § 101.54(b), "high" claims
- § 101.54(c), "good source" claims
- § 101.54(e), "more" claims
- § 101.54(f), "high potency" claims
- § 101.54(g), "antioxidant" claims
- § 101.56, "light" or "lite" claims
- § 101.60, "calorie" or "sugar" claims
- § 101.61, "sodium" or "salt" claims
- § 101.62, "fat, fatty acids, and cholesterol" claims
- § 101.65, implied nutrient content claims
- § 101.65(d), "healthy" claims
- § 101.67, use of nutrient content claims for butter

A nutrient content claim may be no larger than twice the type size of the statement of identity (the name of the food) and may not be unduly prominent in style compared with the statement of identity (21 CFR 101.13(f)).

A Supplement Facts panel is required if a nutrient content claim is made (21 CFR 101.13(n)).

A disclosure statement is a statement that calls the consumer's attention to one or more nutrients (other than the nutrient that is the subject of the claim) in a dietary supplement (e.g., "See nutrition information for fat content") (21 CFR 101.13(h)(1)).

A disclosure statement must be used when making a nutrient content claim and the food (including dietary supplements) contains one or more of the nutrients listed in Table 10.1 in excess of the levels listed below per reference amount customarily

Table 10.1 Minimum Values for Disclosure Statement

Fat	13.0 g
Saturated fat	4.0 g
Cholesterol	60 mg
Sodium	480 mg

consumed, per labeled serving, or, for a product with a reference amount of 30 g or less or 2 tablespoons or less, per 50 g (21 CFR 101.13(h)(1)).

The disclosure statement must be presented in easily legible boldface print or type, in distinct contrast to other printed or graphic matter (21 CFR 101.13(h)(4)(i)).

The disclosure statement is to be placed immediately adjacent to (i.e., right next to) the claim with no intervening material (such as vignettes or other art work) other than information in the statement of identity or any other information that is required to be presented with the claim (21 CFR 101.13(h)(4)(ii)).

Omission of the disclosure statement is permitted from the panel bearing the nutrition information when the nutrient content claim appears on more than one panel of a label (21 CFR 101.13(h)(4)(ii)).

Only one disclosure statement is required per panel when making multiple claims on a panel. The statement is required to be adjacent to the claim printed in the largest type on that panel (21 CFR 101.13(h)(4)(iii)).

A "high" claim may be made when the dietary supplement contains at least 20% of the DV (i.e., the RDIs or DRVs) of the nutrient that is the subject of the claim per reference amount customarily consumed. A "good source" claim may be made when the dietary supplement contains 10–19% of the Daily Value (21 CFR 101.54(b)(1) and (c)(1)).

A statement a nutrient for which there is no established DV is also allowed as long as the claim specifies only the amount of the nutrient per serving and does not imply that there is a lot or a little of that nutrient in the product (e.g., "x grams of phosphatidylserine"). The dietary ingredient for which there is no DV and the quantitative amount of that dietary ingredient in the Supplement Facts panel must be listed in the section below the nutrients with DVs. These dietary ingredients must be identified as having no DVs by the use of the footnote Daily Value Not Established (21 CFR 101.13(i)(3) and 21 CFR 101.36(b)(3)).

Statements using the words "contain" and "provides" may be used for nutrients without DVs if, and only if, the specific amount of the nutrient is included (e.g., "Contains X grams of phosphatidylserine per serving" or "Provides X g of phosphatidylserine") (21 CFR 101.13(i)(3) and 101.54(c)(1)).

A statement outside of the Supplements Facts panel that describes the percentage of the RDI of a vitamin or mineral in a dietary supplement product are considered nutrient content claims and are not exempt from bearing a disclosure statement when required (21 CFR 101.13(b)(1), (c) and (i)).

If a similar dietary supplement is normally expected to contain a nutrient and the dietary supplement is specially processed, altered, formulated, or reformulated as to lower the amount of the nutrient in the food, remove the nutrient in the food, or not include the nutrient, then it is permitted to make a "low" or "free" claim as applicable (21 CFR 101.13(e)(1)).

However, a "low" or "free" claim may not be allowed for the dietary supplement product if it is normally low in or free of a nutrient. However, a claim may be used if it is indicated to refer to all products of that type and not merely to that particular brand (21 CFR 101.13(e)(2)).

Are Claims Such as "100 Percent Milk Free" and "Contains No Preservatives" Subject to the Nutrient Content Claim Requirements?

Claims such as "100 percent milk free" and "contains no preservatives" are not nutrient content claims as long as they are not used in a nutrient context that would make them an implied claim under 21 CFR 101.13(b)(2). The statement "100 percent milk free" is generally a claim to facilitate avoidance of milk products. "Contains no preservatives" is a claim about a substance that does not have a nutritive function (21 CFR 101.65(b)(1) and (b)(2)).

A "no sugar" content claim is subject to the nutrient content claim requirements (21 CFR 101.60(c)(1)).

To avoid misleading consumers, the term "no added sugar" should be limited to dietary supplements containing no added sugars that are normally expected to contain them (21 CFR 101.60(c)(2)(iv)).

A "sugar free" dietary supplement may not claim "low calorie," except when an equivalent amount of a dietary supplement that the labeled dietary supplement resembles and for which it substitutes (e.g., another protein supplement) normally exceeds the definition for "low calorie" (21 CFR 101.60(c)(1)(iii)(A)).

Antioxidant Claims

An antioxidant claim is a nutrient content claim that characterizes the level of one or more antioxidant nutrients present in a dietary supplement (21 CFR 101.54(g)).

When making an antioxidant nutrient claim, the names of the nutrients that are the subject of the claim must be included as part of the claim (e.g., "high in antioxidant vitamins C and E"). Alternatively, the term "antioxidant" or "antioxidants" may be linked in a nutrient content claim (as in "high in antioxidants") by a symbol (e.g., an asterisk) that refers to the same symbol that appears elsewhere on the same panel, followed by the name or names of the nutrients with recognized antioxidant activity. This list should be in letters at least 1⁄16 of an inch in height or no smaller than half the type size of the largest nutrient content claim, whichever is larger (21 CFR 101.54(g)(4)).

To qualify as an antioxidant, the nutrient or dietary ingredient must have an RDI, except as noted above (21 CFR 101.54(g)(1)).

Nutrients that are the subject of the claim must have recognized antioxidant activity. In addition, the level of each nutrient that is the subject of the claim must be sufficient to qualify for either "high" claims in 21 CFR 101.54(b), "good source" claims in 21 CFR 101.54(c), or "more" claims in 21 CFR 101.54(e). For example, for a product to qualify for a "high in antioxidant vitamin C" claim, it must contain 20% or more of the RDI for vitamin C. That is, it must meet the level for "high" defined in § 101.54(b). For a product to qualify for a "good source of antioxidant vitamin C" claim, it must contain 10–19% of the RDI for vitamin C (21 CFR 101.54(g)(2) and (g)(3)).

Recognized antioxidant activity means that there is scientific evidence that, after absorption from the GI tract, the substance participates in physiological, biochemical, or cellular processes that inactivate free radicals or prevent free-radical-initiated chemical reactions (21 CFR 101.54(g)(2)).

A claim may be made for beta-carotene, which does not have an RDI, when the level of vitamin A present as beta-carotene is sufficient to qualify for the claim. For example, a company may make the claim "good source of antioxidant beta-carotene" when 10% or more of the RDI for vitamin A is present as beta-carotene (21 CFR 101.54(g)(3)).

When making additional claims that describe the antioxidant properties of the product, a statement, subject to § 403(a) of The Act (the false and misleading provisions), that describes how a dietary ingredient that does not have an RDI participates in antioxidant processes may be made. Likewise, structure/function claims may be made about antioxidants as long as such claims are not false or misleading and, if appropriate, are made in accordance with § 403(r)(6) of The Act (the provisions for statements of nutritional support). For example, a claim that reads "________, involved in antioxidant processes" would be acceptable as long as it is (1) truthful and not misleading and (2) meets the requirements of § 403(r)(6) of The Act (62 FR 49868 at 49873 (September 23, 1997)).

High Potency Claims

The term "high potency" may be used on the dietary supplement labels to describe individual vitamins or minerals that are present at 100% or more of the RDI per reference amount customarily consumed (21 CFR 101.54(f)(1)(i)).

The term high potency can be used for combination products, such as botanicals with vitamins. However, when using the term high potency to describe individual vitamins or minerals in the product that contains other nutrients or dietary ingredients, the vitamin or mineral that is being described by the term high potency must be clearly identified (e.g., "Botanical X with high potency vitamin E") (21 CFR 101.54(f)(1)(ii)).

The high potency may be used on the multinutrient product to describe the product if it contains 100% or more of the RDI for at least two-thirds of the vitamins and minerals that are listed in 21 CFR 101.9(c)(8)(iv) and that are present in the product at 2% or more of the RDI (e.g., "High potency multivitamin, multimineral dietary supplement tablets") (21 CFR 101.54(f)(2)).

Percentage Claims

A percentage claim is a statement that characterizes the percentage level of a dietary ingredient for which an RDI or DRV has not been established. A percentage claim may be made on products without a regulation that specifically defines such a statement. These statements must be accompanied by any disclosure statement required under 21 CFR 101.13(h). There are simple percentage claims and comparative percentage claims (21 CFR 101.13(q)(3)(ii)).

A simple percentage claim is a statement that characterizes the percentage level of a dietary ingredient for which there is no RDI or DRV (e.g., omega-3 fatty acids, amino acids, and phytochemicals). The statement of the actual amount of the dietary ingredient per serving must be declared next to the percentage statement (e.g., "40 percent omega-3 fatty acids, 10 mg per capsule") (21 CFR 101.13(q)(3)(ii)(A)).

A comparative percentage claim is a statement that compares the percentage level of a dietary ingredient for which there is no RDI or DRV in a product with the amount of the dietary ingredient in a reference food. The reference food must be clearly identified, the amount of that food must be identified, and the information on the actual amount of dietary ingredient in both the dietary supplement and reference food must be declared (e.g., "twice the omega-3 fatty acids per capsule (80 mg) as in 100 mg of menhaden oil (40 mg)") (21 CFR 101.13(q)(3)(ii)(B)).

Health Claims

A health claim is an explicit or implied characterization of a relationship between a substance and a disease or a health-related condition. This type of claim requires significant scientific agreement and must be authorized by the FDA. The claim can be a written statement, a "third party" reference, a symbol, or a vignette (21 CFR 101.14(a)(1) and (c)).

A health claim is different from a structure/function claim in that a health claim describes the effect a substance has on reducing the risk of or preventing a disease, for example, "calcium may reduce the risk of osteoporosis." A health claim requires the FDA evaluation and authorization before its use. A structure/function claim describes the role of a substance intended to maintain the structure or function of the body. Structure/function claims do not require preapproval by the FDA (21 CFR 101.14(a)(1) and (c) and 21 CFR 101.93(f)).

An updated list of FDA authorized health claims is maintained on the FDA website at http://www.cfsan.fda.gov/~dms/flg-6c.html#upd. In addition to these authorized health claims, there are certain "qualified" health claims permitted by the FDA. Qualified health claims are listed on the FDA website at http://www.fda.gov/oc/nutritioninitiative/list.html.

A qualified health claim is a claim supported by less scientific evidence than an authorized health claim. The FDA requires that qualified claims be accompanied by a disclaimer that explains the level of the scientific evidence supporting the relationship.

Unlike authorized health claims, the FDA does not issue regulations for qualified health claims (see the FDA's *Guidance for Industry, Interim Procedures for Qualified Health Claims in the Labeling of Conventional Human Food and Human Dietary Supplements*, July 2003).

The FDA will permit the use of a qualified health claim provided that the following apply:

1. The FDA has issued a letter stating the conditions under which it will consider exercising enforcement discretion for the specific health claim.
2. The qualified claim is accompanied by an agency-approved disclaimer.
3. The claim meets all the general requirements for health claims in 21 CFR 101.14, except for the requirement that the evidence for the claim meet the validity standard for authorizing a claim and the requirement that the claim be made in accordance with an authorizing regulation (see the FDA's *Guidance for Industry, Interim Procedures for Qualified Health Claims in the Labeling of Conventional Human Food and Human Dietary Supplements*, July 2003).

An agency-approved disclaimer is a statement that discloses the level of scientific evidence used to substantiate the health claim (see *FDA Task Force Final Report: Consumer Health Information for Better Nutrition Initiative, Attachment E—Interim Procedures for Qualified Health Claims in the Labeling of Conventional Human Food and Human Dietary Supplements*, July 2003).

To use additional health claims, an individual must submit a health claim petition in accordance with 21 CFR 101.70. A new health claim may be used only after the FDA issues either an authorizing regulation or a letter stating enforcement discretion conditions for a qualified health claim (21 CFR 101.14 and 21 CFR 101.70).

Structure/Function Claims

A company may make the following types of structure/function claims under § 403(r)(6) of The Act:

1. A statement that claims a benefit related to a classical nutrient deficiency disease and that discloses the prevalence of such disease in the United States;
2. A statement that describes the role of a nutrient or dietary ingredient intended to affect the structure or function in humans or characterizes the documented mechanism by which a nutrient or dietary ingredient acts to maintain such structure or function; or
3. A statement that describes the general well-being from consumption of a nutrient or dietary ingredient (21 U.S.C. 343(r)(6)).

When making structure/function claims, the claim must (1) have substantiation that such statement is truthful and not misleading, (2) include the disclaimer, and (3) notify the FDA no later than 30 days after the first marketing of the product that the company is making the statement in accordance with 21 CFR 101.93.

The following text must be used for the disclaimer, as appropriate:

1. Singular: "This statement has not been evaluated by the Food and Drug Administration. This product is not intended to diagnose, treat, cure, or prevent any disease;" or
2. Plural: "These statements have not been evaluated by the Food and Drug Administration. This product is not intended to diagnose, treat, cure, or prevent any disease."

The wording of these disclaimer may not be modified (21 CFR 101.93(c)). The disclaimer must be placed immediately adjacent to the claim with no intervening material or elsewhere on the same panel or page that bears the statement. In the latter case, the disclaimer must be placed in a box and linked to the statement by a symbol (e.g., an asterisk) placed at the end of each statement that refers to an identical symbol placed adjacent to the disclaimer (21 CFR 101.93(d)).

The notification procedures require that a manufacturer, packer, or distributor making such a statement must do the following:

1. Notify the FDA within 30 days of first marketing a product whose label or labeling bears a statement made under § 403(r)(6) of The Act.
2. Submit an original and two copies of the notification to the Office of Nutritional Products, Labeling, and Dietary Supplements (HFS-800), Center for Food Safety and Applied Nutrition, Food and Drug Administration, 5100 Paint Branch Parkway, College Park, Maryland 20740-3835.
3. The notification must be signed by a person who can certify that the information in the notification is complete and accurate and that the notifying firm has substantiation that the § 403(r)(6) statement is truthful and not misleading (21 CFR 101.93(a)(1) and (a)(3)).

When reporting to the FDA, there is no official form to use. The notification may be made by a letter containing the required information in any format that is convenient.

The following information must be included in the notification:

1. The name and address of the manufacturer, packer, or distributor of the dietary supplement that bears the statement.
2. The text of the statement that the company is making.
3. The name of the dietary ingredient or supplement that is the subject of the statement.
4. The name of the dietary supplement (including its brand name) on whose label, or in whose labeling, the statement appears (21 CFR 101.93(a)(2)).

REGULATION AROUND THE WORLD

The FDA is not the only governing body trying to improve the quality and reliability of dietary supplements. The governing bodies of Canada, Europe, India, Japan, and numerous others are all trying to acknowledge the health benefits of supplements, while protecting the general public. Each governing body has a slightly different approach to controlling manufacturing and labeling, but, ultimately, these governing bodies are moving in the same direction. As the world economy expands, the manufacturers and governing bodies will continue to approach one unified set of standards that will reduce manufacturer confusion and increase consumer confidence.

REFERENCE

Food and Drug Administration. http://www.cfsan.fda.gov/~dms/dslg-toc.html. Last accessed August 24, 2009.

CHAPTER 11

Nutraceuticals for the Cardiovascular System

Hieu T. Tran and Kimberly K. Daugherty

CONTENTS

The number one cause of death in the United States is cardiovascular disease. One of every 2.8 deaths in the United States in 2005 was attributable to cardiovascular disease or a cardiovascular event [Lloyd-Jones et al. 2008]. Patients are turning more and more to herbal medications to treat their cardiovascular diseases. The market for herbal medications in the United States was around $590.9 million in 2001 [Izzo et al. 2005]. Various studies report 3–93% usage of herbal medications [Vora and Mansoor 2005]. A 1990 survey showed that 34% of U.S. adults used at least one alternative therapy, and, by 1997, the CAM therapy usage had increased by 25%. Less than 30% of patients tell their physicians, however, that they are taking alternative medications [Miller, Liebowitz, and Newby 2004]. These statistics just underlie the importance of understanding what alternative medications are available to treat cardiovascular diseases, what are the data regarding these treatments, what cardiovascular adverse effects are associated with these treatments, and what, if any, drug interactions that patients and healthcare professionals should be aware of.

GARLIC

Allium sativum (garlic) has been studied for a variety of cardiovascular benefits and is the most widely used supplement in the world [DeBusk 2000; Knox and Gaster 2007].

Content and Effects

Garlic has been shown in numerous studies to lower lipids and blood pressure, reduce atherosclerosis, decrease coagulation and platelet aggregation, and increase fibrinolysis of blood clots [DeBusk 2000; Hermann 2002].

Garlic is made up of two different active ingredients: allicin and alliin [Hermann 2002]. Garlic bulbs contain an odorless, sulfur-containing amino acid known as alliin. After garlic is crushed, alliin is converted by alliinase to allicin, which is highly odoriferous. Allicin seems to be the component that causes the cardiovascular benefits seen with garlic [Frishman, Grattan, and Mamtani 2005]. Garlic is thought to lower blood pressure by opening calcium ion channels in the vascular smooth muscle, resulting in vasodilatation [Khosh and Khosh 2001]. The hyperlipidemic properties of garlic are thought to be attributable to garlic's HMG-CoA reductase activity, increased catabolism of fatty acids such as triglycerides, and retardation of the absorption of cholesterol from the intestine [Caron and White 2001; Mamtani and Mamtani 2005].

Usage

There have been a variety of garlic doses and products studied. The recommended dose seems to depend on the indication the drug is being used for and the product being used (see Table 11.1). The following doses and product(s) are the most commonly effective in clinical trials [Bordia 1981; Lau, Lam, and Wang-Cheng 1987; Vorberg and Schneider 1990; Jain et al. 1993; Adler and Holub 1997; Morcos 1997; Kannar et al. 2001; Durak et al. 2004; Jeyaraj et al. 2005]:

1. Hyperlipidemia: garlic powder (Kwai®), 900 mg/day (1.3% alliin = 0.6% allicin release)
2. Hypertension: garlic powder (Kwai®), 900 mg/day

Adverse Drug Effects

Garlic's number one side effect is its odor and taste, which makes it hard to do blind, controlled studies. Other side effects include flatulence, halitosis, and GUI upset. It has also been shown to cause more serious adverse effects, such as anaphylaxis, spontaneous bleeding, asthma, myocardial infarction, and small intestinal obstructions [Bordia 1981; Lau, Lam, and Wang-Cheng 1987; Vorberg and Schneider 1990; Jain et al. 1993; Adler and Holub 1997; Morcos 1997; Caron and White 2001; Kannar et al. 2001; Durak et al. 2004; Frishman, Grattan, and Mamtani 2005; Jeyaraj et al. 2005].

Table 11.1 Garlic Studies

Reference	Study Design	Key Findings	Limitations
Garlic-positive studies			
Kannar et al. 2001	Double-blind, randomized, placebo-controlled 46 patients 12 weeks Enteric coated Australian garlic powder tablets (2.4 mg allicin-releasing per tablet); 22 patients, 2 tablets twice daily Placebo: 24 patients	Garlic: significant reduction in total cholesterol (−4.2%) and LDL (−6.6%) Placebo: nonsignificant increase in total cholesterol (+2.0%) and LDL (+3.7%); HDL was significantly increased in placebo (+9.1%) compared with garlic group (decrease of 0.9%) No significant change in triglycerides in either group	Diet counseling Small sample Short time period
Jeyaraj et al. 2005	Randomized, placebo controlled, unblinded 32 patients (16 each group) 30–60 years old 60 days 600 mg of fish oil + 500 mg of garlic pearls (garlic oil) per day Placebo	Significant reductions were seen in all parameters except for HDL, which was nonsignificantly increased compared with placebo Total cholesterol: combination, −20% compared with placebo, +1.8% LDL: combination, −21% compared with placebo, +3.4% Triglyceride: combination, −37% compared with placebo, −0.9% HDL: combination, +5.1% compared with placebo, −1.6%	Diet counseling Small sample size Short time period
Bordia 1981	Group A: 20 healthy volunteers garlic for 6 months and then 2 months without garlic Group B (62 patients): divided randomly and blinded into two groups for 10 months; one group received garlic and other placebo Garlic dose: 0.25 mg/kg oil per day divided into two doses in gelatin capsules	Group A: serum cholesterol decreased 17%, triglycerides decreased 20%, HDL increased by 41%; changes were statistically significant. Group B: total cholesterol significantly decreased by 18% by end of study in garlic arm; triglycerides also were significantly decreased in garlic arm; little change in the control group.	No diet mentioned Small sample size
Morcos 1997	Single-blind, placebo-controlled crossover study 40 subjects 4 week washout between arms Each arm was 4 weeks Placebo for 1 month and then fish oil (1,800 mg EPA/1,200 mg DHA) + garlic powder 1,200 mg capsules daily for 1 month	Supplementation for 1 month resulted in 11% decrease in cholesterol, 34% decrease in triglycerides, and 10% decrease in LDL. Also trend toward increase in HDL (32%). No significant effect while on placebo: cholesterol, −1%; triglycerides, −2%; LDL, −4%; HDL, −2%	No true diet control Small sample size Short study period Crossover study

Adler and Holub 1997	50 male subjects 12 weeks 900 mg of garlic placebo per day + 12 g of fish oil placebo per day 900 mg of garlic per day + 12 g of fish oil placebo per day 900 mg of garlic placebo per day + 12 g fish oil per day 900 mg garlic per day + 12 g fish oil per day	Placebo group showed no significant change in total cholesterol, LDL, or triglycerides compared with baseline. Total cholesterol was significantly lower with garlic + fish oil (–12.2%) and garlic (–11.5%) but not with fish oil alone LDL concentrations were also significantly reduced with garlic + fish oil (–9.5%) and with garlic (–14.2%) but were raised with fish oil (+8.5%). Triglycerides were significantly reduced with garlic + fish oil (–34.3%) and fish oil alone (–37.3%); garlic group was not significantly changed	Diet controlled Small sample size Short study period
Durak et al. 2004	23 subjects Hypertensive group (13) and normotensive group (10) 4 months Garlic extract 1 ml/kg/day (about 10 g garlic/day)	Serum total, LDL, triglycerides were significantly lower after extract use and HDL was elevated; no changes in liver function, bilirubin, urea, creatinine, protein, electrolytes, or calcium levels	Small sample size Short time period
Jain et al. 1993	Double-blind, placebo-controlled 42 health adults 12 weeks Standardized garlic powder tablets 300 mg three times a day Placebo	Garlic treatment significantly lowered total cholesterol levels (262 ± 34 to 247 ± 40 mg/dl) Garlic reduced LDL levels by 11% and placebo lowered them by 3% No significant changes in HDL or triglycerides	No diet control Healthy patients Short time periods
Lau, Lam, and Wang-Cheng 1987	Randomized, placebo-controlled Liquid garlic extract at 250 mg dry weight/ml of active garlic component; 4 capsules (1 ml/capsule) per day 6 months 56 subjects (32 male and 24 female) Three studies: hyperlipidemic group (32 patients), normolipidemic group (14 patients), additional hyperlipidemic group (10 patients with no placebo group)	Hyperlipidemic arm: serum cholesterol increased at 1 month in 10 subjects, and at 2 months 13 of 15 subjects showed increased in cholesterol in the garlic arm; by 6 months, 11 subjects had more than a 10% lowering of cholesterol (12–31%); only 2 of 12 had a drop in cholesterol over 10% Normolipidemic patients: there were no significant changes noted for either the garlic or placebo groups Additional hyperlipidemic group: 6 of the 10 subjects had a greater than 10% drop from baseline in their cholesterol levels	Normal diet habits Small sample sizes No placebo group in last arm

(*Continued*)

Table 11.1 *(Continued)*

Reference	Study Design	Key Findings	Limitations
Garlic-positive studies			
Vorberg and Schneider 1990	Double-blind, placebo controlled 40 patients 4 months Garlic powder 900 mg/day (equivalent to 2700 mg fresh garlic/day)	Garlic showed significantly lower total cholesterol (21% versus 3%), triglycerides (25% versus 5%), and blood pressure than placebo	No mention of diet Small sample size
Garlic-negative studies			
Peleg et al. 2003	Randomized, prospective, double-blind, placebo-controlled 16 weeks 33 patients Garlic powder tablets (5,600 grams alliin/tablet) 2 tablets twice daily (13 patients) Placebo (20 patients)	No significant changes were seen in total cholesterol, LDL, HDL, or triglycerides Garlic versus placebo Total cholesterol, –0.7 ± 9.9 versus –2.8 ± 11.2 %; LDL, –0.1 ± 14.7 versus –2.7 ± 15.1%; HDL, –7.7 ± 10.7 versus –0.2 ± 9.8%; triglycerides, +30.1 ± 53.3 versus –6.9 ± 29.0%	Diet counseling Small sample size Short time period High drop out rate in the garlic group
Gardner et al. 2007	Parallel, randomized, placebo-controlled 192 patients Raw garlic, 4.0 g blended Powdered garlic, 4 Garlicin tablets Aged garlic extract, 6 tablets Placebo All products were taken 6 days/week for 6 months Garlic component in each arm was equivalent to an average-sized garlic clove	No statistically significant effect from the 3 forms of garlic on LDL: raw garlic, +0.4 mg/dl; powdered garlic, +3.2 mg/dl; aged garlic extract, +0.2 mg/dl; placebo, –3.9 mg/dl No statistically significant effects on HDL, triglycerides, or total cholesterol Raw versus powdered versus aged versus placebo HDL, +2.3 versus +1.0 versus –0.3 versus –0.8 mg/dl Triglycerides, –5.2 versus –6.6 versus –2.0 versus +6.4 mg/dl Total cholesterol, –0.11 versus –0.02 versus 0.0 versus –0.04 mg/dl	Diet controlled
Tanamai, Veeramanomai, and Indrakosas 2004	Randomized, double-blind crossover 100 subjects (45 in trial group and 55 in control) Garlic tablets each contained 1.5% allicin Trial group took garlic tablets in first 3 months then placebo for 3 months then nothing for 3 months Control group took 3 months of placebo then 3 months of garlic then 3 months of placebo	No significant differences were found in total cholesterol between the two groups at the end of three or six months Side effects: headache, itching, garlic smell; no effects related to liver, kidney function, or hematologic effects	No specific dose given for the garlic Crossover study No washout period mentioned

Berthold, Sudhop, and Bergmann 1998	Randomized, double-blind, placebo-controlled, crossover 25 patients 12 weeks with washout periods of 4 weeks Garlic oil preparation (5 mg twice daily) = equivalent of about 4–5 g of fresh garlic cloves daily (4,000 U of allicin equivalents per day) Placebo	No difference was seen between the garlic group and placebo in terms of cholesterol, LDL, HDL, or triglyceride levels Difference between placebo and active drug: cholesterol, 3.3 mg/dl; LDL, 0.04 mg/dl; HDL, 1.9 mg/dl; triglycerides: 4.2 mg/dl	No specific diet Crossover study Small sample size Short treatment period
Isaacsohn et al. 1998	Randomized, double-blind, placebo-controlled, parallel 12 weeks 300 mg garlic powder tablets three times per day (equivalent to 2.7 g of fresh garlic per day), 28 patients Placebo, 22 patients	No significant changes in lipid parameters in either arm of the study or between arms of the study Garlic versus placebo: cholesterol, +1.64 versus +0.07%; LDL, +1.82 versus –0.35%; HDL, +2.90 versus –1.73%; triglycerides, +0.38 versus +7.66%	Diet controlled Small sample size Short treatment period
Superko and Krauss 2000	Double blind, randomized, placebo-controlled 50 patients 3 months Standardized garlic tablet 300 mg three times daily Placebo	No significant changes in total cholesterol, LDL, HDL, or triglycerides Garlic versus placebo: cholesterol, –2.5 ± 26.4 versus +1.2 ± 19.0 mg/dl; LDL, –1.7 ± 25.5 versus –3.2 ± 14.9 mg/dl; HDL, –0.1 ± 6.6 versus +0.2 ± 8.9 mg/dl; triglycerides, –4.3 ± 43.8 versus +21.5 ± 96.8 mg/dl	Diet controlled Small sample size
Dhawn and Jain 2004	Double-blind, randomized Two groups: essential hypertension (EH) and normotensive (control) 40 patients 8 weeks Garlic pearls (garlic oil 2.5% weight to volume) 2 per day	EH group showed a significant reduction in both systolic blood pressure (148 ± 12/140 ± 16 mmHg) and diastolic blood pressure (94 ± 15/85 ± 23 mmHg) from baseline No significant change in either systolic blood pressure (130 ± 22/127 ± 17 mmHg) or diastolic blood pressure (76 ± 12/74 ± 20 mmHg) in the control group No significant change in either group in total cholesterol, LDL, HDL, or triglycerides	Small sample size Short time period No placebo group

(*Continued*)

Table 11.1 (*Continued*)

Reference	Study Design	Key Findings	Limitations
Garlic-negative studies			
Simons et al. 1995	Double-blind, placebo-controlled, randomized, crossover 30 patients 12 week cycles 28 day washout period Garlic powder tablets 300 mg three times daily Placebo	No significant differences in total cholesterol, LDL, HDL, or triglycerides	Diet controlled Crossover Short time period Small sample size
Neil et al. 1996	Double-blind, randomized, parallel 6 month 115 subjects Dried garlic tablets (1.3% allicin) 300 mg three times daily Placebo	No significant difference was seen in any lipid parameter between the garlic and placebo group	Diet controlled
Plengvidhya et al. 1988	Double-blind, randomized 30 patients Group 1 (16 patients): Placebo capsule twice daily for 2 months then 1 capsule twice daily for 2 months Group 2 (14 patients): Garlic capsules given first then placebo Sprayed garlic preparation	Group 1: no statistically significant changes were seen in total cholesterol, triglycerides, or HDL Group 2: same results	Diet controlled No specifics given on the type of garlic Small sample size Short time period
Luley et al. 1986	Randomized, double-blind 6 weeks Study 1: 34 patients, 198 mg three times daily Study 2: 51 patients, 450 mg three times daily Dried garlic	No change was seen in any lipid parameter studied	Small sample size Short time period

Drug Interactions

It may inhibit multiple isoforms of the P450 enzyme system, which means it may have multiple drug interactions, including warfarin [Knox and Gaster 2007].

Overall Recommendation

Other than some potential drug interactions, overall garlic appears to be safe and somewhat effective. Patients should be encouraged to only take products that contain an appropriate amount of allicin and should alert their physician they are taking garlic, particularly if they are receiving other medications.

GUGGULU

Commiphora mukul (guggulu) belongs to the *Burseraceae* family of plants, which are native to India. This plant resembles a gum-like resin when cut. This plant was used in ancient times as a treatment for obesity and skin disease. It is now thought to contain the presence of antihyperlipoproteinemic compounds called guggulsterones [Singh, Niaz, and Ghosh 1994; Caron and White 2001; Szapary et al. 2003; Mamtani and Mamtani 2005; Knox and Gaster 2007].

Content and Effect

Guggulu has been studied to treat hyperlipidemia mainly. Guggulsterones are thought to exert their antihyperlipoprotenemic properties by antagonizing the farnesoid X receptor, which is activated by bile acids. The farnesoid X receptor controls the levels of bile acids, thus regulating cholesterol levels. Guggulsterones are also thought to increase LDL receptor reuptake and increase receptor binding sites. Guggulsterones are also thought to inhibit platelets and have anti-inflammatory properties [Singh, Niaz, and Ghosh 1994; Caron and White 2001; Szapary et al. 2003; Mamtani and Mamtani 2005; Knox and Gaster 2007].

Usage

There is no consistent dose or product that has been studied (Table 11.2).

Adverse Drug Effects

Guggulu has not been shown to have many side effects. The most common side effects seen in the clinical trials performed to date include diarrhea, loose stools, hiccups, and rashes. These side effects are usually mild and infrequent. One major problem with most guggulu supplements is the potentially dangerous levels of heavy metals that they may contain [Knox and Gaster 2007].

Table 11.2 Guggulu Studies

Reference	Study Design	Key Findings	Limitations
Kuppurajana et al. 1978	Randomized, placebo controlled Three treatment groups (total of 120 patients): obese (>20% of ideal weight), hypercholesterolemic (total cholesterol >300 mg/100 ml), and hyperlipidemic (total lipids >750 mg/100 ml) Given for 21 days 1. Purified guggulu 2 g three times daily 2. Fraction "A" petroleum ether guggulu 0.5 g twice daily 3. Placebo 2 g three times daily 4. Clofibrate 500 mg three times daily	Obese: Cholesterol: purified guggulu statistically lower on day 10 but not significantly lower by day 21; no significant change at any point for the petroleum guggulu or clofibrate Lipids: no significant change in any group Hypercholesterolemic: Cholesterol: purifed guggulu was statistically lower at day 10; petroleum guggulu was statistically significant at both days 10 and 21; clofibrate was statistically significant at both days 10 and 21 Petroleum guggulu showed a significant decrease in both cholesterol and total lipids compared with placebo at both day 10 and day 21 Lipids: Decrease in total lipids by both purified guggulu and petroleum guggulu was statistically significant. No significant difference between groups Final conclusion: Petroleum guggulu significantly lowers serum cholesterol and serum lipids significantly. Results are similar to those seen by clofibrate.	Not applicable
Agarwal et al. 1986.	Phase I study: 21 patients (14 males and 7 females) Mean age 44 years 400 mg three times daily for 4 weeks Phase II study: 19 patients (13 male and 6 female) 30–65 years Primary hyperlipidemia (total cholesterol >250 mg/dl and triglycerides >200 mg/dl 500 mg three times daily for 12 weeks followed by 6 weeks of dietary control)	Phase I: 1 patient complained of epigastric fullness three days after starting therapy; no abnormalities were seen in the hematological parameters, liver function tests, blood urea, or blood sugar levels Phase II: 78.9% of patients were responders to drug therapy by the end of the 12 week treatment (fall in cholesterol and triglycerides at the end of 12 weeks was greater than the day-to-day variation without treatment) Cholesterol level started falling within 2–4 weeks of starting treatment; levels tended to rise after withdrawal of the drug, although they were still statistically lower Average total cholesterol decrease was 17.5 ± 9.9%; average decrease in triglycerides was 30.3 ± 18.4% Phase I: drug is safe and does not produce changes in hepatic or renal function, blood sugar levels, hematological parameters, or electrocardiograms. Phase II: gugulipid significantly lowered serum cholesterol and triglycerides	Small sample size Phase I and II studies only No good explanation of how the gugulipd was prepared

Szapary et al. 2003	Double-blind, randomized, placebo-controlled, parallel design 1. 103 patients 2. 8 weeks 3. Guggul extract (2.5% guggulsterones) standard dose (1000 mg) (33 patients) versus high dose (2000 mg) (34 patients) versus placebo (36 patients); all three times daily	Placebo: LDL decreased by 5% Standard dose and high dose guggulipid raised LDL levels by 4%; the results were significantly different compared with placebo No significant changes in total cholesterol, HDL, triglycerides, or very LDL 6 guggulipid patients developed a rash Guggulipid does not appear to improve serum cholesterol over the short term and might even raise LDL; it may also cause a hypersensitivity rash	No diet control Short treatment plan (8 weeks) Product used may not have contained enough guggulsterones
Singh, Niaz, and Ghosh 1994	Double-blind, randomized, placebo-controlled 1. 12-week diet stabilization period, 24-week treatment period, 12-week washout period 2. 61 patients 3. Guggulipid (25 mg of guggulsterones) 50 mg twice daily	Guggulipid decreased total cholesterol by 11.7%, LDL decreased by 12.5%, triglycerides decreased by 12.0% Placebo levels were unchanged; results were significantly different compared with the treatment group Levels increased substantially in the guggulipid during the washout period (cholesterol increased 6.5%, LDL increased 6.6%, and triglycerides increased 7.7%) compared with insignificant changes in the control group Side effects of guggulipid: headache, mild nausea, eructations, and hiccups	Diet controlled
Verma and Bordia 1988	Randomized, placebo-controlled 1. 40 patients 2. Age 40–60 years 3. 16 weeks Purified gum guggulu 4.5 grams in 2 divided doses	Gum guggulu results: Cholesterol: decreased by 7.8% end of 4th week, 15.78% end of 8th week, and 21.75% end of 16th week (significant decrease by 16th week) Triglycerides: decreased by 6.7% end of 4th week; 17.1% end of 8th week, and 27.1% end of 16th week (significant decrease by 16th week) HDL gradually increased by 35.8% by end of 16th week (significant change)	Small sample size Purified sample

(*Continued*)

Table 11.2 (*Continued*)

Reference	Study Design	Key Findings	Limitations
Malhotra, Ahuja, and Sundaram 1977	Randomized, active comparator 1. 75 weeks 2. 51 subjects (45 male and 6 females) 3. 75 weeks Fraction "A" guggulu 1.5 g/day (41 subjects) Clofibrate 2 g/day (10 subjects)	Both agents decreased cholesterol and triglycerides significantly. Guggulu at 71–80 weeks: cholesterol, 36.8% decrease; triglycerides, 50.4% decrease Clofibriate at 71–80 weeks: Cholesterol, 43.5% decrease; triglycerides, 50.2% decrease No significant difference was seen between the two agents Side effects: diarrhea in 5 of the Guggulu patients Study concluded that both agents were equally effective in treating hyperlipidemia	No specific diet requirements Small number of subjects

Drug Interactions

This agent has also been show to induce CYP3A4 genes, leading to many drug interactions [Knox and Gaster 2007].

Overall Recommendation

Although this product does show some promising results in clinical trials, there is no specific dose or product that can be recommended. This product also has numerous side effects and potential drug interactions.

HAWTHORN

Hawthorn is derived from the berries of the *Crataegus oxyacantha* L. plant, which has different species such as *laevigata* and *monogina*. The plant consists of leaves with flowers or fruit and is in the family Rosacae [Fishman, Grattan, and Mamtani 2005].

Potential Indications

Hawthorn has been used in Europe for years to treat hypertension, myocardial dysfunction, and tachycardia [Walker et al. 2006]. Studies have demonstrated significant improvement in patients with congestive heart failure, New York Heart Association classification II and III [Tauchert 2002; Degenring et al. 2003].

Pharmacology

The therapeutic effects of Hawthorn are thought to be attributable to oligomeric procyanidins, in addition to flavonoids, catechin, and epicatechin, which cause direct vasodilation of the coronary vascular smooth muscles [Walker et al. 2002, 2006]. These effects are slow to develop. This herbal also has some inotropic and chronotropic effects with the potential to irritate the myocytes [Tauchert 2002; Degenring et al. 2003].

Dosing Recommendations/Products Studied

The dosing recommendations and products studied are as follows [Tauchert 2002; Degenring et al. 2003]: flavones, 5 mg; total flavonoids, 10 mg; oligomeric procyanidin as epicatechin, 5 mg. For congestive heart failure, the ethanolic extract ratio is used.

Adverse Drug Effects

The treatment was well tolerated by patients; however, because of potential hypotensive effects, close monitoring is recommended [Tauchert 2002; Walker et al. 2002, 2006; Degenring et al. 2003].

Overall Recommendation

Although this herbal has shown some promising results, the use of this agent is debatable and should only be used under close supervision of a healthcare professional.

DIGITALIS

Digitalis, also known as *Digitalis purpurea* L. and *lanata* belongs to the Scrophulariaceae family. Digitoxin is from both *D. purpurea* and *D. lanata*. Digoxin is only found from *D. lanata*. The dry leaves of the plant are thought to contain the pharmacologic action [Frishman, Grattan, and Mamtani 2005].

Potential Indications

This agent is used mainly for treatment of heart failure [Fishman, Grattan, and Mamtani 2005].

Pharmacology

This herbal is thought to work through positive ionotropic effects on the heart [Frishman, Rattan, and Mamtani 2005].

Dosage Recommendation/Products Studied

This agent requires professional titration and kinetic dosing for use [Frishman, Grattan, and Mamtani 2005]

Overall Recommendation

This agent has numerous studies showing its positive effect, but, because of its need for close kinetic monitoring, it should only be used under the supervision of a healthcare provider trained in its use.

Other Plants with the Same Effects

Other plants with the same side effects include the following: *Adonis vernalis*, *Apocynum cannabinum*, *Helleborus niger* L., *Selenicereus grandiflorus* L., *Convallaria majalis* L., *Nerium oleander* L., *Urginea maritima* L., and *Strophantus kombe*.

GINKGO

Ginkgo biloba belongs to the family Ginkgoaceae. Clinical trials have been conducted using a standardized, concentrated acetone-water extract of dried leaves, prepared to a potency of 24% flavone glycosides and 6% terpenes [DeBusk 2000].

Potential Indications

This agent has been officially declared by the German E Commission to be used in cerebral and peripheral arterial circulatory disturbances [DeBusk 2000].

Pharmacology

This agent is thought to work because of its mixture of flavonol and flavone glycosides (of quercetin and kaempferol, also rutin). The effects from the mixture of these ingredients include reduction of capillary fragility and free radical scavengers. This agent is also thought to work through inhibition of the platelet-activating factor from ginkgolides [DeBusk 2000].

Dosage Recommendation/Products Studied

Dosage recommendation/products studied include the following: extract as tablet, liquid, and intravenous forms; one tablet (40 mg/tab) is to be taken three times a day with meals.

Adverse Drug Effects

Adverse drug effects are thought to not be significant but may include GI disturbances, headache, and allergic skin reactions [DeBusk 2000].

Overall Recommendation

This agent is approved for use as food supplement in the United States. It can be recommended for consumers.

HORSE CHESTNUT SEED

Horse chestnut seed (*Aesculus hippocastanum* L.) belongs to the family Hippocastanaceae. This herbal is a large, globular, brown seed [DeBusk 2000].

Potential Indications

This herbal has been studied and approved by the German Commission E for improvement of venous tone and to reduce the risk of varicose vein formation [DeBusk 2000].

Pharmacology

Horse chestnut seed is a complex mixture of triterpenoid saponin glycosides (aescin) with the flavonoids quercetin and kaempferol. Aescin can reduce lysosomal activity by 30% (stabilize the cholesterol-containing membranes of the lysosomes) and restrict edema (reduce water and protein leakage with light diuretic effect) [DeBusk 2000].

Dosing Recommendations/Products Studied

Dosing recommendations/products studied [DeBusk 2000] include the following: for varicose veins, as an aqueous-alcoholic extract of aescin; initial dosage, 90–150 mg of aescin orally; with improvement, reduce to 53–70 mg daily, oral administration; questionable use of ointment, liniment, and of hydroalcoholic extract.

Adverse Drug Effects

Adverse drug events effects include GI disturbances, such as constipation with oral consumption (rare), isolated reports of renal and hepatic toxicity, and anaphylaxis reaction after intravenous administration [DeBusk 2000].

Drug Interactions

This agent has been shown to have a coumarin content; therefore, it should not be used with warfarin or other blood thinners attributable to the risk of increased bleeding [DeBusk 2000].

Overall Recommendation

Despite this agent's drug interaction with warfarin and other blood thinners, it is considered safe for use by the German Commission E. Healthcare professionals should counsel patients about its potential drug interaction.

BUTCHER'S BROOM

Butcher's broom (*Ruscus aculeatus* L.) affects are thought to be from the rhizome and root of the plant [DeBusk 2000].

Potential Indications

The main indication for this herbal is treatment of venous insufficiency [Frishman, Grattan, and Mamtani 2005].

Pharmacology

This agent is thought to work because of its mixture of steroidal saponins, which have anti-inflammatory and vasoconstricting effects on the venous vasculature [Frishman, Grattan, and Mamtani 2005].

Dosage Recommendation/Products Studied

Dosage recommendation/products studied include the following [Frishman, Grattan, and Mamtani 2005]: capsule or tablet of approximately 300 mg of dried extract; ointment and suppositories for hemorrhoids.

Overall Recommendation

Clinical safety and efficacy remain to be established.

GINSENGS

Ginsengs (*Panax ginseng, P. quinquefolius* L. [American ginseng], *Ginseng Radix Rubra* [Korean ginseng, are made by drying several species from China, Korea, Russia, and Japan and are of the family Araliaceae [Hammond and Whitworth 1981; Han et al. 1998; Sung et al. 2000; Stavro et al. 2005].

Potential Indications

This agent is being studied for use in hypertension [Hammond and Whitworth 1981; Han et al. 1998; Sung et al. 2000; Stavro et al. 2005]

Pharmacology

Triterpenoid saponin glycosides consist of ginsenosides or panaxosides [Hammond and Whitworth 1981; Han et al. 1998; Sung et al. 2000; Stavro et al. 2005]. The content of ginsenosides varies with the age of the root, the habitat, the harvesting season, and the method of curing or drying. In Chinese usage, the whole root is better than any of its parts.

In general, ginseng properties include a tonic or adaptogenic effects. Animal studies have shown these effects: increase endurance, prevent stress-induced ulcer, stimulate hepatic ribosome production, boost the immune system, and stimulate protein synthesis [Sung et al. 2000].

Red ginseng might be used to improve vascular endothelium because of the release of NO [Sung et al. 2000].

Dosage Recommendation/Products Studied

Dosage recommendation/products studied include the following [Hammond and Whitworth 1981; Han et al. 1998; Sung et al. 2000; Stavro et al. 2005]: tea, capsules, extracts, tablets, roots, chewing gum, cigarettes, and candies; for red ginseng, 1.5 g three times per day (each capsule contains 300 mg of red ginseng) orally.

Adverse Drug Effects

Ginseng abuse syndrome, which consists of potential hypertension, has been associated with Chinese ginseng [Hammond and Whitworth 1981]. Other potential adverse effects include nervousness and irritability. Usually, ginseng is not associated with serious adverse reactions [Han et al. 1998; Sung et al. 2000; Stavro et al. 2005].

Overall Recommendation

Ginseng can be used moderately. However, it should be used with caution in patients prone to hypertension or nervousness.

TREE BARK

Tree bark (*Terminalia arjuna*) is derived from *Combretaceae* family. Tree bark used has been used in Indian Pharmacopoeia [Gupta et al. 2001; Bharani et al. 2002].

Potential Indications

The bark extract is supposed to have cardiotonic, anti-ischemic and anti-heart failure properties [Gupta et al. 2001; Bharani et al. 2002].

Pharmacology

Tree bark is a combination of arjunic acid and terminic acid. Bark extract also contains strong antioxidants (flavones, tannins, and oligomeric proanthocyanidins), glycosides (arjunetin arjunosides I–IV), and minerals [Gupta et al. 2001; Bharani et al. 2002].

Dosing Recommendations/Products Studied

Clinical studies have shown that Arjuna potentially has displayed anti-ischemic effects similar to isosorbide mononitrate at 40 mg/day [Bharani et al. 2002] and also possessed potent antioxidant action compared with vitamin E as well as significant hypocholesterolemic effects [Gupta et al. 2001]: for anti-ischemic effect, bark extract at 500 mg every 8 h by mouth; for antioxidant and hypocholesterolemic effect, capsule of 500 mg pulverized powder of Arjuna bark daily by mouth.

Adverse Drug Effects

The bark extract or powder were well tolerated, and no side effects were reported [Gupta et al. 2001; Bharani et al. 2002].

Overall Recommendation

This agent can be used with monitoring for angina and high cholesterol levels (used independently or in association with vitamin E).

REFERENCES

Adler, A. J. and B. J. Holub. 1997. Effect of garlic and fish-oil supplementation on serum lipid and lipoprotein concentrations in hypercholesterolemic men. *Am. J. Clin. Nutr.* 65:445–450.

Agarwal, R. C., S. P. Singh, S. K. Saran, S. K. Das, N. Sinha, O. P. Asthana, P. P. Gupta, S. Nityanand, B. N. Dhawan, and S. S. Agarwal. 1986. Clinical trial of gugulipid: A new hypolipidemic agent of plant origin in primary hyperlipidemia. *Indian J. Med. Res.* 84:626–634.

Berthold, H. K., T. Sudhop, and K. Bergmann. 1998. Effect of a garlic oil preparation on serum lipoproteins and cholesterol metabolisms: A randomized controlled trial. *JAMA* 279:1900–1902.

Bharani, A., A. Ganguli, L. K. Mathur, Y. Jamra, and P. G. Raman. 2002. Efficacy of *Terminalia arjuna* in chronic stable angina: A double-blind, placebo-controlled, crossover study comparing *Terminalia arjuna* with isosorbide mononitrate. *Indian Heart J.* 54:170–175.

Bordia, A. 1981. Effect of garlic on blood lipids in patients with coronary heart disease. *Am. J. Clin. Nutr.* 34:2100–2103.

Caron, M. F. and C. M. White. 2001. Evaluation of the anihyperlipidemic properties of dietary supplements. *Pharmacotherapy* 21:481–487.

DeBusk, R. M. 2000. Dietary supplements and cardiovascular disease. *Curr. Atherosclerosis Rep.* 2:508–514.

Degenring, F. H., A. Suter, M. Weber, and R. Saller. 2003. A randomized double blind placebo controlled clinical trial of a standardized extract of fresh Crataegus berries (Crataegisan) in the treatment of patients with congestive heart failure NYHA II. *Phytomedicine* 10:363–369.

Dhawan, V. and S. Jain. 2004. Effect of garlic supplementation on oxidized low density lipoproteins and lipid peroxidation in patients of essential hypertension. *Mol. Cell. Biochem.* 266:109–115.

Durak, I., M. Kavuteu, B. Aytac, A. Avei, E. Devrim, H. Ozbek, and H. S. Oztürk. 2004. Effects of garlic extract consumption on blood lipid and oxidant/antioxidant parameters in humans with high blood cholesterol. *J. Nutr. Biochem.* 15:373–377.

Frishman, W. H., J. G. Grattan, and R. Mamtani. 2005. Alternative and complementary medical approaches in the prevention and treatment of cardiovascular disease. *Curr. Probl. Cardiol.* 30:383–459.

Gardner, C. D., L. D. Lawson, E. Block, L. M. Chatterjee, A. Kiazand, R. R. Balise, and H. C. 2007. Effect of raw garlic versus commercial garlic supplements on plasma lipid concentrations in adults with moderate hypercholesterolemia: A randomized clinical trial. *Arch. Intern. Med.* 167:346–353.

Gupta, R., S. Singhal, A. Goyle, and V. N. Sharma. 2001. Antioxidant and hypocholesterolaemic effects of *Terminalia arjuna* tree-bark powder: A randomised placebo-controlled trial. *JAPI* 49:231–235.

Hammond, T. G. and J. A. Whitworth. 1981. Adverse reactions to ginseng. *Med. J. Australia* 1:492.

Han, K. H., S. C. Choe, H. S. Kim, D. W. Sohn, K. Y. Nam, B. H. Oh, M. M. Lee, Y. B. Park, Y. S. Choi, J. D. Seo, and Y. W. Lee. 1998. Effect of red ginseng on blood pressure in patients with essential hypertension and white coat hypertension. *Am. J. Chin. Med.* 26:199–209.

Hermann, D. D. 2002. Naturoceutical agents in the management of cardiovascular disease. *Am. J. Cardiovasc. Drugs* 2:173–196.

Isaacsohn, J. L., M. Moser, E. A. Stein, K. Dudley, J. A. Davey, E. Liskov, and H. R. Black. 1998. Garlic powder and plasma lipids and lipoproteins: A multicenter, randomized, placebo-controlled trial. *Arch. Intern. Med.* 158:1189–1194.

Izzo, A. A., G. Di Carlo, F. Borrelli, and E. Ernst. 2005. Cardiovascular pharmacotherapy and herbal medicines: The risk of drug interaction. *Int. J. Cardiol.* 98:1–14.

Jain, A. K., R. Vargas, S. Gotzkowsky, and F. G. McMahon. 1993. Can garlic reduce levels of serum lipids? A controlled clinical study. *Am. J. Med.* 94:632–635.

Jeyaraj, S., G. Shivaji, S. D. Jeyaraj, and A. Vengatesan. 2005. Effect-of-combined supplementation of fish oil with garlic pearls on the serum lipid profile in hypercholesterolemic subjects. *Indian Heart J.* 57:372–331.

Kannar, D., N. Wattanapenpaiboon, G. S. Savige, and M. L. Wahlqvist. 2001. Hypocholesterolemic effect of an enteric-coated garlic supplement. *J. Am. Coll. Nutr.* 3:225–231.

Khosh, F. and M. Khosh. 2001. Natural approach to hypertension. *Altern. Med. Rev.* 6:590–600.

Knox, J. and B. Gaster. 2007. Dietary supplements for the prevention and treatment of coronary artery disease. *J. Alt. Compl. Med.* 13:93–95.

Kuppurajan, K., S. S. Rajagopalan, T. K. Rao, and R. Sitaraman. 1978. Effect of guggulu (*Commiphora mukul* Engl.) on serum lipids in obese, hypercholesterolemic and hyperlipemic cases. *Jr. Assoc. Phys. Ind.* 26:367–373.

Lau, B. H. S., F. Lam, and R. Wang-Cheng. 1987. Effect of an odor modified garlic preparation on blood lipids. *Nutr. Res.* 7:139–149.

Lloyd-Jones, D., R. Adams, M. Camethon, G. De Simone, T. B. Ferguson, K. Flegal, E. Ford, et al. 2008. Heart disease and stroke statistics 2009 update: A report from the American Heart Association Statistics Committee and Stroke Statistics Subcommittee. *Circulation* 119: 1–162. http://circ.ahajournal.org.

Luley, B. C., W. Lehmann-Leo, B. Moller, T. Martin, and W. Schwartzkopff. 1986. Lack of efficacy of dried garlic in patient with hyperlipoproteinemia. *Arzneimittel-Forschung.* 36:766–768.

Malhotra, S. C., M. M. Ahuja, and K. R. Sundaram. 1977. Long term clinical studies on the hypolipidemic effect of *Commiphora mukul* (Guggulu) and Clofibrate. *Indian J. Med. Res.* 65:390–395.

Mamtani, R. and R. Mamtani. 2005. Ayurveda and yoga in cardiovascular diseases. *Cardiol. Rev.* 13:155–162.

Miller, K. L., R. S. Liebowitz, and L. K. Newby. 2004. Complementary and alternative medicine in cardiovascular disease: A review of biologically based approaches. *Am. Heart J.* 147:401–411.

Morcos, N. C. 1997. Modulation of lipid profile by fish oil and garlic combination. *J. Natl. Med. Assoc.* 89:673–678.

Neil, H. A., C. A. Silagy, T. Lancaster, J. Hodgeman, K. Vos, J. W. Moore, L. Jones, J. Cahill, and G. H. Fowler. 1996. Garlic powder in the treatment of moderate hyperlipidaemia: A controlled trial and meta-analysis. *J. R. Coll. Physician Lond.* 30:329–334.

Peleg, A., T. Hershcovici, R. Lipa, R. Anbar, M. Redler, and Y. Beigel. 2003. Effect of garlic on lipid profile and psychopathologic parameters in people with mild to moderate hypercholesterolemia. *Isr. Med. Assoc. J.* 5:637–640.

Plengvidhya, C., S. Sitprija, S. Chinayon, S. Pasatrat, and M. Tankeyoon. 1988. Effects of spray dried garlic preparation on primary hyperlipoproteinemia. *J. Med. Assoc. Thailand* 71:248–252.

Simons, L. A., S. Balasubramaniam, M. von Konigsmark, A. Parfitt, J. Simons, and W. Peters. 1995. On the effect of garlic on plasma lipids and lipoproteins in mild hypercholesterolaemia. *Atherosclerosis* 113:219–225.

Singh, R. B., M. A. Niaz, and S. Ghosh. 1994. Hypolipidemic and antioxidant effects of *Commiphora mukul* as an adjunct to dietary therapy in patients with hypercholesterolemia. *Cardiovasc. Drugs Ther.* 8:659–664.

Stavro, P. M., M. Woo, T. F. Heim, L. A. Leiter, and V. Vuksan. 2005. North American ginseng exerts a neutral effect on blood pressure in individuals with hypertension. *Hypertension* 46:406–411.

Sung, J., K. H. Han, J. H. Zo, H. J. Park, C. H. Kim, and B. H. Oh. 2000. Effects of red ginseng upon vascular endothelial function in patients with essential hypertension. *Am. J. Chin. Med.* 28:205–216.

Superko, H. R. and R. M. Krauss. 2000. Garlic powder, effect on plasma lipids, postprandial lipemia, low-density lipoprotein particle size, high-density lipoprotein subclass distribution and lipoprotein (a). *J. Am. Coll. Cardiol.* 35:321–326.

Szapary, P. O., M. L. Wolfe, L. T. Bloedon, A. J. Cucchiara, A. H. DerMarderosian, M. D. Cirigliano, and D. J. Rader. 2003. Guggulipid for the treatment of hypercholesterolemia: A randomized controlled trial. *JAMA* 290:765–772.

Tanamai, J., S. Veeramanomai, and N. Indrakosas. 2004. The efficacy of cholesterol-lowering action and side effects of garlic enteric coated tablets in man. *J. Med. Assoc. Thailand* 87:1156–1161.

Tauchert, M. 2002. Efficacy and safety of crataegus extract WS 1442 in comparison with placebo in patients with chronic stable New York Heart Association class-III heart failure. *Am. Heart J.* 143:910–915.

Verma, S. K. and A. Bordia. 1988. Effect of *Commiphora mukul* (gum guggulu) in patients of hyperlipidemia with special reference to HDL-cholesterol. *Indian J. Med. Res.* 87:356–360.

Vora, C. K. and G. A. Mansoor. 2005. Herbs and alternative therapies: Relevance to hypertension and cardiovascular diseases. *Curr. Hypertens. Rep.* 7:275–280.

Vorberg, G. and B. Schneider. 1990. Therapy with garlic: Results of a placebo-controlled double-blind study. *Br. J. Clin. Pract. Suppl.* 69:7–11.

Walker, A. F., G. Marakis, A. P. Morris, and P. A. Robinson. 2002. Promising hypotensive effect of hawthorn extract: A randomized double-blind pilot study of mild, essential hypertension. *Phytother. Res.* 16:48–54.

Walker, A. F., G. Marakis, E. Simpson, J. L. Hope, P. A. Robinson, M. Hassanein, and H. C. Simpson. 2006. Hypotensive effects of hawthorn for patients with diabetes taking prescription drugs: A randomized controlled trial. *Br. J. Gen. Pract.* 56:437–443.

CHAPTER 12

Nutraceuticals in Diabetes Management

Maria Lourdes Ceballos-Coronel

CONTENTS

THE DISEASE: DIABETES MELLITUS

Overview: Incidence, Prevalence, Demographics, and Statistical Data

Diabetes mellitus (DM), the all-American disease, is currently the seventh major cause of death in the United States, with a staggering healthcare cost of $174 billion dollars annually [Beecher 1999; Houston and Egan 2005; Lovelady 2005; American Diabetes 2008]. Press releases issued by the Centers for Disease Control and Prevention in June 2007 and 2008 stated that the number of people with DM increased to 24 million. This is an increase of more than 3 million in a two-year time interval. According to these data, 8% of our population is afflicted with this disease, and 25% are 60 years of age and older. Concurrently, 57 million are estimated to have pre-diabetes, placing them at a higher risk for DM.

The statistical impact of DM in the U.S. population is as follows:

- Native Americans and Alaskan Natives, 16.5%
- Blacks, 11.8%
- Hispanics, 11.4%
- Puerto Ricans, 12.6%
- Mexican Americans, 11.9%
- Cubans, 8.2%
- Asian Americans, 7.5%
- Whites, 6.6%

Treatment of diabetes is challenging not because of lack of an available, safe, and effective method; rather, unawareness and denial of this disease add to the treatment challenge. The good news is the incidence of people who were previously unaware or denying they have diabetes has decreased from 30 to 25% over a two-year period. This suggests that efforts to increase awareness are working and people are better prepared to manage the disease and its complications.

DM is a chronic condition that results from an inadequate ability or failure to metabolize carbohydrates, fats, and proteins. The pancreas produces insulin, an important peptide hormone released from the beta cells of the islets of langerhans in response to high glucose levels. The mechanism by which insulin facilitates availability and utilization of glucose may be pictured as "lock and key" interplay. Insulin serves as the key that opens the doors of the cell and allows entry of available glucose from the bloodstream for cellular fuel (ATP) or energy usage. Insufficient amounts or the absence of insulin in the body, therefore, causes a failure to unlock the cells, resulting in increased glucose in the bloodstream, resulting in hyperglycemia. The inability of the cells to receive and utilize any glucose sources from the bloodstream causes the cells to be exceedingly deprived of energy or glucose sources.

TYPES OF DIABETES MELLITUS

Type I: Insulin-Dependent Diabetes Mellitus

Type I, or insulin-dependent DM (IDDM), accounts for 10% of our population, and individuals with IDDM are incapable of producing any insulin before the age of 40. It is treated with insulin injections, diet change, and regular exercise.

There have been studies that suggest individuals with this type, not genetically predisposed to this condition, develop the disease as a result of previous exposure to a potent viral infection, causing the total destruction of the pancreas.

Type II: Non-Insulin-Dependent Diabetes Mellitus

Type II, or non-insulin-independent DM, accounts for 93% of all cases of diabetes. This group can usually produce some insulin; however, it could be inadequate or ineffective because of insulin resistance. This type is associated with being overweight, often obese, and usually occurs after one reaches 40 years of age. Obesity decreases insulin sensitivity; conversely, exercise increases it [Langin 2001; Catena et al. 2003; Pittas 2003; Houston and Egan 2005].

MANAGEMENT OF DIABETES MELLITUS: DIET, EXERCISE, AND MEDICATIONS

Treatment for this group is composed of diet modification, exercise, and oral hypoglycemic agents. There have been instances in which type II could be insulin requiring. Recent findings at the 68th Scientific Session by the American Diabetes Association held in San Francisco in June 2008 disclosed that the lack of expected response from oral hypoglycemic agents or the inadvertent failure of oral medications despite the intake of maintenance-prescribed regimens leads to continued chronic loss of beta cells, ultimately resulting in pancreatic failure. Unfortunately, early detection of beta cell dysfunction cannot be foreseen before signs and symptoms appear; therefore, 80% of these cells have already been destroyed.

NUTRACEUTICALS

The current approach to diabetic management is through a combination of dietary choices and supplements. A new trend in the treatment of diabetes is marketed in the form of nutraceuticals. Nutraceuticals are defined as food or food products compounded or manufactured in the form of capsules, tablets, tinctures, etc. This product is often confused with "functional foods." The latter are consumed foods and not in dosage form. Although nutraceuticals show great promise in the management of diabetes, scientific data and studies are not conclusive enough to establish the optimal effects. Interestingly enough, a report in 2002 observed that diabetics were open to either complementary or alternative medicine, which uses nutraceutical and functional foods [Ames et al. 1993; Egede et al. 2002; Yeh et al. 2002].

World-wide consumption of nutraceuticals is higher in Europe and Japan. To date, more Americans use nutraceuticals (herbals/botanicals) than in the past. The American Diabetes Association (ADA) recommends the use of nutraceuticals and food additives to control blood glucose levels, hoping to deter the onset of the disease and its complications among pre-diabetics and diabetics. The U.S market for diabetogenic natural products is estimated at $50 million, with a yearly increase in demand of 20–30% [Hasler 1998]. In light of the baby-boomer generation entering the age of 50 and above, it is foreseeable that nutraceutical supplements could be in higher demand. This warrants a closer look by the FDA to further their quest into the safety of these products, improving regulations, and lifting current restrictions.

Nutraceutical Vitamins, Minerals, and Enzymes

- ALA and its reduced derivative, dihydrolipoic acid, improve insulin sensitivity, glucose tolerance in type II DM, and diabetic neuropathy. However, in clinical trials, it did not show any significant alteration in the fasting glucose and insulin concentration. This poses a question as to its reliability as a therapeutic agent that would provide significant improvement of glycemic control in DM [Joseph et al. 1999].

- Biotin increases glucokinase activity, thereby improving glucose tolerance and insulin sensitivity. The recommended dose is 16 mg/day in type II DM.
- Carnitine improves glucose metabolism and disposal. The recommended dose is 1–2 g twice daily.
- Chromium is an essential micronutrient acting as a cofactor in numerous insulin regulatory steps. It reduces fasting glucose, postprandial glucose, hemoglobin A1C, C-peptides, fasting insulin, and insulin resistance. Conversely, it increases cellular insulin binding, the number and activation of insulin receptors, and insulin growth factor-I receptor. The recommended dose is 8 mcg/kg/day.
- CoQ_{10} reduces fasting glucose, postprandial glucose, and hemoglobin A1C. The recommended dose is 100 mg twice daily.
- Copper increases insulin sensitivity and improves glucose levels. However, excessive intake of copper may induce or produce insulin resistance.
- Flavonoids enhance insulin secretion, improve insulin sensitivity, reduce serum glucose, and inhibit sorbitol accumulation in the lens of the eye and nerves.
- Folate and vitamin B_{12} (cyanacobalamin) have no significant effects on glucose metabolism; however, both are noted to improve symptoms of diabetic peripheral neuropathy.
- Gamma linoleic acid improves glucose tolerance, improves insulin resistance, and protects, as well as improves, diabetic neuropathy. The recommended dose is 500–1,000 mg/day.
- Glutathione is the most potent intracellular antioxidant. Decreasing the levels of glutathione results in insulin resistance, glucose intolerance, and increased oxidative stress.
- Inositol or myoinositol is required for normal nerve function. Consumption improves diabetic neuropathy.
- Magnesium improves insulin sensitivity and secretion. The recommended dose for those with normal kidney function is 500 mg twice a day with 50–100 mg of vitamin B_6.
- Manganese is an important cofactor in many glycolytic enzymes; it improves insulin synthesis and insulin sensitivity and serves as "insulin." An intact pancreatic beta cell is required for manganese to be effective. The optimal dose is 5–10 mg/day.
- Monounsaturated fats improve glycemic control. The recommended dose is extra virgin olive oil, four tablespoons per day, or whole olives, 12–16 per day.
- *N*-acetyl cysteine improves insulin secretion, reduces insulin resistance, lowers serum glucose, and prevents diabetic cataracts. The recommended dose is 2 g daily.
- Niacinamide improves insulin action and sulfonylurea action. Long-term evidence indicates that niacinamide alone improves glucose tolerance in safe doses at less than 3 g daily.
- Omega 3 fatty acids improve insulin sensitivity and insulin secretion and reduce serum glucose. The recommended dose is 900 mg of EPA and 600 mg of DHA with a total daily dose of EPA plus DHA lower than 3 g.
- Potassium improves insulin secretion, insulin sensitivity, and glucose tolerance when administered orally or intravenously.

- Pycnogenol has been found to lower plasma glucose and hemoglobin A1C, improve glutathione levels, and reduce oxidative stress. The recommended dose is 100 mg/day.
- Selenium is an important antioxidant; it serves as an "insulin-mimetic," reduces fasting glucose, and protects against diabetic retinopathy. The recommended dose is 200 mcg/day.
- Taurine improves glucose tolerance and insulin sensitivity, reduces glycosylation of proteins and hemoglobin, and improves symptoms of diabetic neuropathy. The recommended dose is 1.5–3 g twice daily.
- Thiamine and B_6 administration improves symptomatic peripheral neuropathy within four weeks, reduces pain in 88.9%, reduces numbness in 82.5%, and reduces paresthesias. The recommended dose is 50–100 mg twice a day [Abbas and Swai 1997; Tamai 1999].
- Vanadate is a protein-tyrosine phosphatase inhibitor that reduces glucose, increases glucose transport and uptake, improves insulin sensitivity, prolongs insulin action, and increases intracellular magnesium. The recommended dose is 40–80 mcg/L.
- Vitamin B_6 (pyridoxine) serves as a coenzyme in carbohydrate metabolism. It prevents diabetic neuropathy, improves symptoms, and inhibits glycosylation.
- Vitamin C (ascorbic acid) reduces glycosylation of proteins and reduces sorbitol accumulation but was found not to have any direct effect on glucose.
- A vitamin E derivative has shown to improve insulin action, reduce resistance, improve glucose control, and reduce glycosylation of proteins. Optimal doses are unclear, but 200–400 IU of a mixture of tocopherols and tocotrienols are recommended.
- Zinc improves insulin binding and insulin sensitivity; increases insulin synthesis, secretion, and utilization; protects beta cells; reduces glucose; and improves diabetic retinopathy. The recommended dose is 30–50 mg daily.

Nutraceutical Herbs and Botanicals

Indian gooseberry, Jambal fruits, Bengal gram, black gram, mango leaves, parsiane, string beans, cucumbers, celery, and onions are vegetables and fruits found useful in the treatment of diabetes [Diet Health Club 2008]. Others worth mentioning include the following:

- Bittermelon contains an extract called "plant insulin." It is also known as bitter gourd or gourdin. Ayurvedic medicine uses the extract, which was found to activate in inactive insulin present in the blood as well as rejuvenate the pancreas, thus being beneficial to patients with diabetes. When it is administered, it is documented to lower sugar levels 15 min after intake. Bitter gourd "Karela" juice is a popular remedy for diabetics in the tropics. Consumption of 50 ml of raw Karela juice daily improves blood glucose tolerance in type II DM [Indiadiets 2008].
- Cinnamon (*Cinnamomum aromaticum*) cassia is the most common related species sold in most supermarkets in the United States and has been reported to have remarkable pharmacological effects in the treatment of type II DM. The mechanism of action includes the activation of glycogen synthase in glucose uptake, the inhibition of glycogen synthase, the activation of insulin receptor kinase, the inhibition of dephosphorylation of insulin receptor, and its antioxidant effects. A study

of 60 patients with type II DM showed a reduction of fasting glucose by 18–29% at a cinnamon dose of 1, 3, and 6 g/day for 40 days ["Honey and Cinnamon" 2008]. Adversely, cinnamon contains a toxic component, courmarin, that causes hepatic and renal damage as well as blood-thinning effects in high concentrations [Wong 2007; Wikipedia 2008]. A cinnamon and honey mixture has beneficial health effects. Ayurvedic as well as Yunani medicine have been using honey as a vital medicine for centuries. Currently, scientific studies suggest that, although honey is sweet, if taken in the right dosage as a medicine, it does not harm diabetic patients. The mixture causes a cleansing effect in the digestive system, which eliminates parasites, fungi, and bacteria. These microorganisms slow down digestive processes, which leads to toxic buildup. Cleansing will cause weight loss, ultimately removing a DM risk factor. The mixture is known to lower gastric emptying time and rate, which significantly lessens postprandial blood sugar levels [*Science Daily* 2008]. The recommended use specifically for weight loss is one part cinnamon to two parts raw honey. Use half a teaspoon of cinnamon to one teaspoon of honey at a ratio of 1:2 ["Honey and Cinnamon" 2008].

- Fenugreek seeds contain an alkaloid called trigonelline that has been shown to lower blood glucose and prevent cataracts attributable to diabetes [Indiadiets 2008].
- Garlic (*Allium sativum*) is best noted to minimally decrease systolic blood pressure [*Allium ursinum* 2008]. It has beneficial vascular effects and has antibiotic potential [Bergner 1995]. Studies done and reported at the 68th ADA June 2008 convention found a correlation between diabetes and periodontal "gum disease." Uncontrolled diabetes produces high glucose levels in the saliva, which helps bacteria to thrive. Diabetes reduces the body's resistance to infection, and the gums are among the tissues most likely affected. Periodontal disease, and loss of more teeth, is often linked to the control of diabetes. In a full-study finding published in the July 2008 issue of *Diabetes Care*, periodontal disease was found to be an independent predictor in the incidence of type II DM [Indiadiets 2008]. Generally, teeth are covered with plaque and a sticky film of bacteria. After the consumption of a meal, snack, or beverage that contains sugars or starches, bacterial reactions to these substances cause the release of acids that attack tooth enamel. Repeated attacks can cause the enamel to break down, resulting in cavities [About.com 2006; American Diabetes 2008]. Recurrent untreated dental cavities leads to recurrent abscess formation. Subsequently, this leads to chronic periodontitis and ultimately the loss of teeth. Additional studies are required to establish the long-term beneficial effects of garlic application to gingival tissues and oral intake in the systemic control of periodontal disease in diabetes.
- Grapes have a naturally present compound known as resveratrol, which can protect against cellular damage to blood vessels caused by the high production of glucose in diabetes. This is found naturally in grape skins as well as in grape seeds, peanuts, and red wine. Research was performed by scientists at the Peninsula Medical School in the southwest part of England. The findings were published in the science journal *Diabetes, Obesity and Metabolism*. The elevated levels of glucose that circulate in the blood of diabetic patients cause microvascular and macrovascular complications resulting from damaged mitochondria, an organelle responsible for generating energy. The mitochondrial damage causes leakage of electrons and the formation of highly damaging free radicals. Resveratrol protects against damage by the production of a protective enzyme that prevents leakage of electrons and toxic free radicals [*Science Daily* 2008]. The custom of drinking wine for centuries enhanced the

scientific research into the health benefits derived from red grapes, which began in Europe in the mid- to late 20th century. Supplemental oligomeric proanthocyanidins (OPCs), "pycnogenols," have been used in Europe since 1950 and have been found to reduce diabetic retinopathy. Generally, the more intense the color of the grape skin, the more OPCs it contains, which explains why red wine has greater health benefits than white wine. Grape juice also contains OPCs; however, researchers have found that the beneficial effects are not as comparable [Slomski 1994].

- Green tea and epigallocatechin-3-gallate (EGCG) are substances that reduce fasting and postprandial glucose, fructosamine, and hemoglobin A1C and improve insulin resistance. They increase protection from beta cell destruction by inhibiting inducible NO gene expression and nuclear factor inhibition. The recommended dose is 500 mg twice daily.
- Malunggay or kamunggay (Philippines), Moringa or horse radish tree (English), or Sajina (India) is a wonderful world-known herb that recently entered the consumer and nutritional markets in the United States. The herb grows wildly in abundance in warm areas or tropical climates. Mounting scientific evidence collected by modern scientists on the herb's beneficial effects just proves what millions of people thousands of years ago have always known. Its benefits include being an excellent source of nutrition, a natural non-sugar-based energy booster that lowers blood pressure and promotes relaxation and calmness. It also possesses detoxifying effects that lower blood sugar levels. The multitude of medicinal benefits it offers earned it the name "nature's medicine cabinet" [Kumar 2008; Malunggay 2008]. West African physicians have used this herb for the treatment of diabetes for centuries. This "miracle vegetable" received numerous accolades from various scientific journals and the WHO. According to Market Manila, "Malunggay is consumed in huge quantities every day across the archipelago" [Market Manila 2005]. Malunggay can possibly be the future centerfold of the nutraceuticals. Earnest work is contemplated to provide this herb plant in a pill, tablet, or any other dosage form.

CONCLUSIONS

A documented new syndrome, "syndrome X," also known as metabolic syndrome, represents a constellation of problems that are both cardiovascular in nature and have a predominant degree of insulin resistance. The United States is currently faced with a pandemic of obesity and an epidemic of metabolic syndrome. Increasing rates of obesity amongst children and adults inadvertently lead to glucose intolerance, insulin resistance, DM, and other complications.

Although drug therapy may be required for the treatment of diabetes and metabolic syndrome, appropriate lifestyle changes such as weight loss, exercise, healthy diet, and nutraceutical supplements are the cornerstone for the clinical management and prevention of both conditions.

These advocated modalities are supported by many scientifically proven studies that relate to improvement in the quality of life and an increased lifespan. Treatment and management of diabetes in the past 50 years have changed it from a disease known to produce early disability and demise to one that is complicated to manage but with an excellent prognosis. The transformation is a result of a better understanding of the pathophysiology, an awareness of the disease, and the early detection of diabetes before onset of complications [Beecher 1999; Buse 2008; *Diabetes Dispatch* 2008].

The country's imminent dilemma of increasing healthcare costs is a threat to our continued efforts to progress and improve outcomes in the management of these conditions. Dissatisfied consumers met with exorbitant drug costs, lack of improvement in conventional therapies, poor therapeutic alternatives for chronic diseases such as diabetes, inadequate rapport between medical providers and patients in managed care, desired personalized medicines, an enlarging population trying to prevent the effects of aging, and new insights into the concept of preventive medicine have initiated the emergence of the use of nutraceuticals as an alternative venue to promote wellness and the prevention of ailments. Data show that more than 40% of Americans are using alternative medical therapies, nutraceuticals, herbals, and botanicals [DeFelice 2008]. The surmounting concern in reference to the historical use of nutraceuticals by itself is not enough to ensure safety even if historical use is characterized by consumption or by use in folk medicine all over the world [Degan et al 2005]. Despite these stumbling blocks, the future and continued use of nutraceuticals look bright as a supplement in addition to conventional medically proven drugs in the control and management of this chronic disease, DM.

FOR MORE INFORMATION AND RESEARCH

Bergner, P. 2000. New textbooks in medical herbism. *Medical Herbalism: A Journal for the Clinical Practitioner* 11:17–18.

Hareyan. 2007. New nutraceutical stabilizes blood sugar naturally. *Diabetes Care*. http://www.emaxhealth.com/23/16453.html. Accessed August 8, 2008.

REFERENCES

Abbas, Z. G. and A. B. Swai. 1997. Evaluation of the efficacy of thiamine and pyridoxine in the treatment of symptomatic diabetic peripheral neuropathy. *East Afr. Med. J.* 74:803–808.

About.com Senior Health. 2006. Diabetes and Periodontal Disease. http://seniorhealth.about.com/cs/oralconditions/a/diabetic_gum.htm?p=1. Accessed August 9, 2008.

Allium ursinum: Plants for a Future Database Report 1996–2008. http://www.pfaf.org/database/plants.php. Accessed August 8, 2008.

American Diabetes Association. 2008. 68th Scientific Session, San Francisco, CA. June 6–10, 2008.

American Diabetes Association. http://www.ada.org/public/topics/diabetes_faq.asp. Accessed August 9, 2008.

Ames, B. N., M. K. Shigenaga, and T. M. Hagan. 1993. Oxidants, antioxidants and the degenerative disease of aging. *Proc. Natl Acad Sci USA* 90:7915–7922.

Beecher, G. R. 1999. "Phytonutrients" role in metabolism: Effect on resistance to degenerative processes. *Nutr. Rev.* 57:S3–S6.

Bergner, P. 1995. *Allium sativum*: Antibitotic and immune properties,1995. http://medherb.com/Materia_Medica/Allium_sativum__Antibiotic_and_ImmuneProperties.html. Accessed August 2008.

Buse, J. 2008. 68th Scientific Sessions ADA. *Diabetes Dispatch.*

Catena, C., G. Giacchetti, M. Novello, G. Colussi, A. Cavarape, and L. A. Sechi. 2003. Cellular mechanisms of insulin resistance in rats with fructose induced hypertension. *Am. J. Hypertens.* 16:973–978.

DeFelice, S. 2008. Foundation for innovation medicine. http://www.fimdefelice.org/bio.html. Accessed August 9, 2008.

Degan, F. 2005. DSHEA, The IOM, and empowering data-based decision making, editorial commentary. *JANA* 8:2. http://www.ana-jana.org.

Diabetes Dispatch. June 6–10, 2008. http://www.diabetes dispatch.org/pro.

Diet Health Club. http://www.diethealthclub.com. Accessed August 8, 2008.

Egede, L. E., X. Ye, D. Zheng, and M. D. Silverstein. 2002. The prevalence and pattern of complementary and alternative medicine use in individuals with diabetes. *Diabetes Care* 25:325–329.

Hasler, C. M. 1998. Functional foods: Their role in disease prevention and health promotion. *Food Technol.* 52:63–70.

Honey and cinnamon: Natural cures and nature's remedy. http://www.angelfire.com/az/sthurston/honeyandcinnamon.html. Accessed August 7, 2008.

Houston M. C. and B. M. Egan. 2005. Metabolic syndrome: Contributing factors. *JANA* 8:2. http://www.ana-jana.org.

Indiadiets Diets. Eat to beat illness. http://www.indiadiets.com. Accessed August 9, 2008.

Joseph, J. A., A. B. Shukitt-Hale, N. A. Denisova, D. Bielinski, A. Martin, J. J. McEwen, and P. C. Bickford. 1999. Reversals in age-related declines in neuronal signal transduction cognitive, and motor behavioral deficits with blueberry, spinach or strawberry dietary supplementation. *J. Neurosci.* 19:8114–8121.

Langin D. 2001. Diabetes, insulin secretion, and the pancreatic beta cell mitochondrion. *N. Engl. J. Med.* 345:1772–1774.

Lovelady, S. 2004. Nutraceuticals world. October 1, 2004. http://www.desertharvest.com/physicians/documents/HB-32.pdf. Accessed August 9, 2008.

Malunggay. http://www.malunggay.com. Accessed August 9, 2008.

Market Manila, November 5, 2005. http://www.marketmanila.com. Accessed August 9, 2008.

Pati, K. http://www.bestnutrition.com. Accessed August 8, 2008.

Pittas, A. G. 2003. Nutrition interventions for prevention of type 2 diabetes and the metabolic syndrome. *Nutr. Clin. Care* 6:79–88.

Science Daily, March 20, 2008. http://www.sciencedaily.com. Accessed July 23, 2008.

Science Daily, August 8, 2008. http://www.sciencedaily.com/releases/2008/08/080806184905.htm2008.

Slomski, G. 1994. Prevalence of beta allele of the insulin gene in type II diabetes mellitus. *Human Genetics* 93:25–28. http://www.molecularstation.com/research/prevalence-of-beta-allele-of-the-insulin-genein-type-ii-diabetes-mellitus-8125485.html. Accessed July 23, 2008.

Tamai, H. 1999. Diabetes and vitamin levels. *Nippon Rinsho* 57:2362–2365.

Wikipedia, The Free Dictionary. *Cinnamon.* http://en.wikipedia.org/wiki/Cinnamon. Accessed August 15, 2008.

Wong, C. 2007. Health benefits of cinnamon, About.com, October 27, 2007. http://altmedicine.about.com/od/cinnamon/a/cinnamon.htm. Accessed August 12, 2008.

Yeh, G. Y., D. M. Eisenberg, R. B. Davis, and R. S. Phillips. 2002. Uses of complementary and alternative medicine among persons with diabetes mellitus: Results of a national survey. *Am. J. Public Health* 92:1648–1652.

CHAPTER 13

Curcumin: A Versatile Nutraceutical and an Inhibitor of Complement

Girish J. Kotwal

CONTENTS

INTRODUCTION

Turmeric, a golden spice referred to as the Indian Gold, is prepared from underground rhizomes, of which the turmeric root (*Curcuma longa*) and the underground shoots form an interconnected underground, complex, finger-like structure. Turmeric belongs to the ginger family (*Zingiberaceae*) and for centuries has been considered in the Indian systems of medicine, such as Ayurveda and traditional Chinese medicine, to have therapeutic benefit. It is also a key ingredient of Indian, Malaysian, Chinese, Polynesian, and Thai curries, as well as western mustard preparations. In recent decades, the compound curcumin, a polyhydroxy phenolic compound that forms 4% of turmeric, has been found to have broad-spectrum antitumor activity [Sisodia et al. 2005], anti-neuroinflammatory activity [Kulkarni et al. 2005], and anti-Abeta plaque activity [Cole et al. 2004]. It is currently being marketed by several companies as a nutraceutical in treatment with possible benefits in Alzheimer's disease, arthritis, aging, a number of cancers, cardiovascular health, diabetes, obesity, and more. Some of the companies that made unsubstantiated claims about the commercial curcumin products have come under close scrutiny by the FDA. Therefore, curcumin products often indicate that they are natural health supplements that are not intended to diagnose, treat, cure, or prevent any disease. Dr. Bharat Aggarwal and his group at the M. D. Anderson Cancer Center in Houston, Texas, has worked and published extensively on the laboratory studies on the mechanism of action of curcumin [Aggarwal and Harikumar 2009]. They have demonstrated that curcumin mediates its effects by modulation of several important molecular targets, including transcription factors (e.g., NF-κB and epidermal growth factor-1), enzymes (e.g., COX2, 5-lipoxygenase, and iNOS), and cytokines (e.g., TNF, IL-1, IL-6, and chemokines) [Aggarwal and Harikumar 2009]. Recent studies have shown that curcumin is able to inhibit both pathways of complement activation [Kulkarni et al. 2005], and, because unregulated complement activity can cause serious autoimmune tissue damage in diseases such as rheumatoid arthritis, Alzheimer's disease, macular degeneration, and atherosclerosis, curcumin has a potential for benefit in such conditions. Most of the polydroxy phenolic compounds of herbal origin are small-sized complement regulatory molecules that offer a great potential in the development of small-sized neuroprotective agents with an ability to cross the blood-brain barrier.

DISTRIBUTION OF TURMERIC: THE CURRENT SOURCE OF ALL CURCUMIN

Documented use of curcumin dates back to some several thousand years. Turmeric is grown in hot, moist climates and is cultivated and consumed in the Indian subcontinent; it is grown and prepared commercially in the southern Indian state of Kerala, as well as several other states in India. Turmeric is also grown in Africa, Australia, China, Indonesia, Peru, and the West Indies.

PREPARATION OF CURCUMIN

Curcumin can be extracted from the powdered turmeric by multiple ways such as a soap extraction process for producing water and oil-soluble curcumin [Stransky 1979], microwave-assisted extraction [Mandal et al. 2007], and others. The preferred extraction method would depend on the available facilities and the yield and purity required. Curcumin (from Sigma) used in our experiments was dissolved using minimum amount of ice-cold 0.1 M sodium hydroxide solution, and the final concentration was made to 2 mg/ml using ice-cold phosphate-buffered saline (0.154 M NaCl, pH 7.2). The solution was prepared in the dark, and the tubes used for holding the solution were covered with a metal foil to avoid photo degradation of curcumin.

ACTIVITIES OF CURCUMIN

Curcumin has several activities that influence, either alone or in concert, the health benefits of curcumin. It inhibits inflammation in several ways by inhibiting NF-κB-dependent gene transcription, by inhibiting COX2 and iNOS induction [Aggarwal and Harikumar 2009], and by inhibiting complement [Kulkarni et al. 2005]. The interaction of curcumin with the complement system has been studied extensively by our group and should serve as a model for other nutraceuticals interactions with human proteins and molecules. The complement system is made of 30 proteins, which when activated can form a cascade of events that can result in the destruction of bacteria, the neutralization of viruses, the recruitment of inflammatory cells, or the targeting of microorganisms for phagocytosis. The complement system is a powerful first line of defense against microorganisms, but, when unregulated or poorly regulated, it can cause destruction of the tissues and result in autoimmune damage. Our studies have revealed information regarding nature of binding of curcumin to the complement components as well as its mode of action on the complement pathway. To determine the interaction of curcumin with complement, Q-sense (D-300), a device based on quartz crystal microbalance with dissipation monitoring technology (QCMD), was used. In Q-sense, the change in frequency of a quartz-based sensor crystal resonating at its resonant frequency during adsorption of an adsorbing moiety is correlated with the change in mass of the adsorbed moiety. The energy dissipated during this adsorption is also recorded. The changes in frequency (f) and dissipation (D) are specific for a particular system comprising an adsorbing moiety and the ratio of change in dissipation to the change in frequency (dD/dF) gives information regarding the viscoelastic properties and adsorption kinetics of the system. QCMD is a well-established technique that offers advantages of real-time monitoring, speed, simplicity, sensitivity, and economy compared with the other routinely used techniques. It does not involve labeling of protein molecules, and therefore the chances of changes in conformation of the protein under optimal experimental conditions are less. Thus, it was decided to use Q-sense to study the interaction of curcumin with the complement components and compare with the vaccinia virus complement control protein, a well-characterized complement regulatory protein. Both curcumin and viral complement inhibitor from vaccinia virus (VCP) are known to inhibit the alternative and classical pathways of complement activation; hence, it was important to investigate their effect on the complement components C3 and C3b, which are central to both the pathways of complement activation. Our studies indicated that curcumin and VCP bound to both C3 and C3b with a greater affinity to C3 [A. P. Kulkarni and G. J. Kotwal, unpublished observation]. Some other activities of curcumin are as an antioxidant and inhibitor of apoptosis, tumor invasion, apoptosis, and an inducer of cell cycle arrest [Aggarwal and Harikumar 2009].

PROBLEMS WITH CURCUMIN AND HUMAN TRIALS

Several problems with curcumin usage need to be resolved before it becomes widely acceptable as an approved therapy with FDA-approved clinical trial-substantiated claims. Curcumin has also quite often worked in laboratory studies but failed to show adequate therapeutic benefit in human trials [AIDS.org 2007]. Also, it is advisable that curcumin not be used in conjunction with chemotherapy, but very often that means that curcumin is recommended after chemotherapy has failed and precious time has been lost. The other problem is the high doses required to have a therapeutic effect as well as the problem with solubility and bioavailability [Higdon 2005]. It is 2,000-fold less active than VCP [Kulkarni et al. 2005b]. Because of the curcumin being metabolized in the body after being conjugated in the liver and intestine to form metabolites, such metabolites are possibly less active than the curcumin itself. Also, the absorption of curcumin happens mainly in the GI tract, making it less suitable for influencing organs outside the GI tract; therefore, there is a need to use some new delivery techniques, such as nanotechnology, to achieve better distribution to affected sites. Several investigators are attempting to better deliver curcumin by generating complexes of curcumin with various substances. One report uses the casein-micelle complexation to deliver curcumin as a drug nanocarrier to cancer cells [Sahu et al. 2008].

FUTURE FOR CURCUMIN

Curcumin is a nutraceutical that is extracted from turmeric. Because of considerable research in laboratories around the world, it has been postulated to have wide-ranging health benefits based on its activities studied. The characterization of its multiple interactions requires use of emerging technologies. An example of one such technology, QCMD, to characterize interactions of curcumin with the third component of the complement system has been mentioned above. The future of curcumin for the time being may be limited to its consumption as a part and parcel of turmeric added on a regular basis in generous amounts to food preparations, as has been done for centuries. Turmeric has stood the test of time and has a number of different additional applications in lotions and soaps. As more preclinical and clinical studies of pure curcumin are performed and the therapeutic benefit is evaluated in rigorous double-blind randomized trials, curcumin could find its potential promise in the laboratory getting realized. Its bioavailability and solubility issues will have to be resolved before a formulation of curcumin can replace the toxic chemotherapy that many cancer patients find difficult to tolerate.

ACKNOWLEDGEMENTS

The author thanks Dr. Amod Kulkarni for a useful discussion on the use of QCMD in studying the interactions of curcumin with complement components and for sharing unpublished data. Details of the work by Kulkarni et al. will be published in the *Open Biochemistry Journal* in 2010.

REFERENCES

Aggarwal, B. B. and K. B. Harikumar. 2009. Potential therapeutic effects of curcumin, the anti-inflammatory agent against neurodegenerative, cardiovascular, pulmonary, metabolic, autoimmune and neoplastic diseases. *Int. J. Biochem. Cell Biol.* 41:40–59.

AIDS.org. 2007. Curcumin: Clinical trial finds no antiviral effect. 2007. http://www.aids.org/atn/a-242-01.html. Accessed July 9, 2009.

Cole, G. M., T. Morihara, G. P. Lim, F. Yang, A. Begum, and S. A. Frautschy. 2004. NSAID and antioxidant prevention of Alzheimer's Disease: Lessons from in vitro and animal models. *Ann. N.Y. Acad. Sci.* 1035:68–84.

Higdon, J. 2005. *Curcumin.* Corvallis, OR: Linus Pauling Institute at Oregon State University. http://lpi.oregonstate.edu/infocenter/phytochemicals/curcumin.

Kulkarni, A. P., L. A. Kellaway, and G. J. Kotwal. 2005a. Herbal complement inhibitors in the treatment of neuroinflammation: Future strategy for neuroprotection. *Ann. N.Y. Acad. Sci.* 1056:413–429.

Kulkarni, A. P., L. A. Kellaway, and G. J. Kotwal. 2005b. Curcumin inhibits the classical and the alternate pathways of complement activation. *Ann. N.Y. Acad. Sci.* 1056:100–112.

Mandal, V., Y. Mohan, and S. Hemalatha. 2007. Optimization of curcumin extraction by microwave designed extraction process and HPTLC analysis. *Pharmacognosy Mag.* 3:132–138.

Sahu, A., N. Kasoju, and U. Bora. 2008. Fluorescene study of the curcumin-casein micelle complexation and the application as a drug nanocarrier to cancer cells. *Biomacromolecules* 10:2905–2912.

Shisodia, S., G. Sethi, and B. B. Aggarwal. 2005. Curcumin: Getting back to the roots. *Ann. N.Y. Acad. Sci.* 1056:206–217.

Stransky, C. E. 1979. Process for producing water and oil soluble curcumin coloring agents. *Free Patents Online.* http://www.freepatentsonline.com/4138212.html. Accessed November 2008.

CHAPTER 14

Probiotics and Prebiotics as Nutraceuticals

Seema Y. Pathak, Cathy Leet, Alan Simon, and Yashwant Pathak

CONTENTS

GENESIS OF HUMAN MICROBIOTA

It is estimated that 500–1,000 species of bacteria live in the human body [Sears 2005]. Bacterial cells are much smaller than human cells, and there are at least 10 times as many bacteria as human cells in the body (approximately 10^{14} versus 10^{13} [10^{13} = 1 trillion]) [Savage 1977; Berg 1996]. Although normal flora are found on all surfaces exposed to the environment (on the skin and eyes, in the mouth, nose, small intestine, and colon), the vast majority of bacteria live in the large intestine. The terms intestinal "microflora" or "microbiota" refer to the microbial ecosystem colonizing the GI tract. Ninety-nine percent of the bacteria isolated from human fecal specimens will not grow in the presence of atmospheric oxygen [Savage 1977]. Bacteria make up most of the flora in the colon [University of Glasgow 2005] and 60% of the dry mass of feces [Guarner and Malagelada 2003]. This fact makes feces an ideal source to test for gut flora.

The stomach and proximal small intestine contain relatively low numbers of microbes (10^3–10^5 bacteria/g or ml content) because of a low pH and rapid flow in this region. Acid-tolerant lactobacilli and streptococci predominate in the upper small intestine. The distal small intestine (ileum) maintains a more diverse microbiota and higher bacterial numbers (10^8/g or ml content) than the upper bowel and is considered a transition zone preceding the large intestine. The large intestine (colon) is the primary site of microbial colonization because of slow turnover and is characterized by large numbers of bacteria (10^{10}–10^{11}/g or ml content), low redox potential, and relatively high short-chain fatty acid (SCFA) concentrations. In addition to an increasing gradient of indigenous microbes from the stomach to the colon, there are also characteristic spatial distributions of organisms within each gut compartment. At least four microhabitats have been described: the intestinal lumen, the unstirred mucus layer or gel that covers the epithelium of the entire tract, the deep mucus layer found in intestinal crypts, and the surface of mucosal epithelial cells [Lee 1984; Berg 1996]. The processes involved in the establishment of microbial populations are complex, involving microbial succession as well as microbial and host interactions and eventually resulting in dense, stable populations inhabiting specific regions of the gut.

Babies are born with no bacterial presence and are practically sterile. Microbes that originate from the surroundings and from the mother establish themselves with the course of time. *Escherichia coli* from the mother's feces start contaminating the infant in vaginal delivery. The length of delivery process is an important contributing factor [Bettelheim et al. 1974; Brook et al. 1979]. Bacteria from the mother's cervix also colonize the alimentary canal of the baby. The naso-pharynx of the baby receives bacteria from the mother's vagina [MacGregor and Tunnesseen 1973]. In cesarean delivery, the baby's first exposure to microbes comes from air; nursing staff and other surgical equipment act as a vector in this situation [Bettelheim et al. 1974; Lennox-King et al. 1976a,b]. The transmission of microbes from generation to generation is ensured by human expressions of neonatal care, such as kissing, touching, and sucking [Tannock 1994].

There is an obvious difference in exposure and acquisition of bacteria in babies born in developing countries. Enterobacteria and streptococci can be the first group established in most cases. *E. coli* is established within 48 h after contamination [Mata and Urrutia 1971]. The infants in developing countries get exposed to environmental bacteria, regardless of the mode of delivery, compared with developed countries. This explains the absence of certain groups of intestinal bacteria in the babies born in western countries. After the exposure to the breast milk, the gut now is in a continuous process of acquiring new microbes [Moughan et al. 1992].

The staphylococci, streptococci, corynebacteria, lactobacilli, micrococci, propionibacteria, and bifidobacteria all originate from the mothers nipple, surrounding skin, and milk ducts [Asquith and Harrod 1979; West, Hewitt, and Murphy 1979]. In formula-fed infants, the exposure to bacteria comes from dried powder, water, and equipment used in the manufacturing process.

Cooperstock and Zedd [1983] divide the development of intestinal bacteria in infants into four different stages. The phase one is the initial acquisition phase lasting over the first two weeks. Breast feeding period is phase two. Weaning and introduction of supplements is phase three. Phase four starts after the weaning is complete. The initial colonization of *E. coli* in large numbers is later responsible for the establishment of anaerobic genera, *Bacteriodes*, *Bifidobacterium*, and *Clostridium*. This happens during the time period of four to seven days. *Bifidobacteria* have dominance in breast-fed infants. This changes once the dietary supplementation begins in the breast-fed infants and *Bifidobacteria* are no longer the prominent genera.

By the second year of life, with the introduction of solid food almost complete, the fecal microbiota of the baby resembles the adult fecal microbiota [Stark and Lee 1982; Copperstock and Zedd 1983].

In phases three and four, other bacterial groups, including eubacteria, veionella, staphylococci, propionibacteria, bacilli, fusobacteria, and yeast, establish themselves along with the *Bacteriods* and anaerobic gram-positive cocci [Conway 1997].

Antimicrobial and antibiotic agents have a significant influence in the microflora of infants [Bennet et al. 1982, 1986; Bennet and Nord 1989]. A specific component of the microbiota becomes more vulnerable than others because of this exposure. Mutations in the microbiota can be seen after finishing the drug treatment. The effects of the drug regimen can be persistent [Finegold, Mathisen, and George 1983].

Diet is the most powerful tool to influence intestinal microbiota. Targeted ingredients in the formula, such as oligosaccharides, affect colon fermentation [Knol et al. 2005]. Breast milk itself contains antimicrobial activity, which helps to stimulate development and maturation of intestinal mucosa. This phenomenon now further promotes stability and decreased intestinal disturbances [Palmer et al. 2007].

PROBIOTICS

The word probiotic comes from pro, meaning "for," and bios, meaning "life." So probiotic literally means "for life." The term probiotic is used to describe nutritional supplements and other products that contain live bacteria. There are literally hundreds of probiotic supplements available to buy. Although they all promise to help restore and replenish our gut microflora, many are so deficient in living or viable bacteria that the package they come in has more value than what is inside [Dekker et al. 2007].

In 1905, Dr. Elie Metchnikoff, a Russian scientist working at the famous Institut Pasteur in Paris, was the first to write about the health benefits of probiotics. Dr. Metchnikoff, who later won a Nobel Prize for his research on the immune system, wrote that Bulgarian peasants who consumed large amounts of yogurt lived long, healthy lives. Examination of the yogurt by Dr. Metchnikoff led to his discovery of a unique lactic-acid-producing bacteria that helped digestion and improved the immune system [Dekker et al. 2007]. The historical association of probiotics with fermented dairy products led to extensive research validating Dr. Metchnikoff's early observations. Investigations during the past several decades have demonstrated numerous health-supportive properties of probiotics on human health [Isolauri 2001; Goossens et al. 2003; Porth 2004; The United States Probiotics Organization 2007].

Structurally, the GI or digestive tract is a hollow tube that runs from the mouth to the anus. Mastication (the chewing of food), peristalsis (the movement of food), enzymes, and stomach fluids break down the food into small, absorbable molecules. In the lining of the small intestines, specialized cells act as a barrier, separating needed nutrients from the molecules. By the time food leaves the small intestine and enters the colon, all of the nutrients in food will have entered the bloodstream [Berg 1996]. In addition to digestion, the GI tract also provides several important immune response activities. Large numbers of white blood cells (lymphocytes) reside under the tonsils and in the appendix. Clumps of lymphocytes and lymphatic tissues make up Payer's patches, immune system structures located within the small intestine. Within the walls of the large intestine, huge numbers of immune system modulators and regulators reside [Goossens et al. 2003; Guyton and Hall 2005]. Evidence has demonstrated that the health status of the tonsils, appendix, and the small and large intestine have an impact on the health of the immune system. Numerous strains of gut microflora reside in significant numbers in the small intestine (10^6–10^8/g of small intestinal contents) and even greater numbers in the colon (10^{11}–10^{12}/g of colon contents) or large intestine. Microflora of the large intestine perform several activities beneficial to human health, including supporting healthy digestion through fermentation, promoting healthy bacterial and yeast balance, and stimulating certain immune system components [Goossens et al. 2003].

Probiotics, as defined by the United States Probiotics Organization, are "live microorganisms administered in adequate amounts which confer a beneficial health effect on the host" [Guyton and Hall 2005]. Probiotics bacteria are frequently, but not always, chosen from bacteria that normally inhabit the GI system of humans. The genera *Lactobacillus acidophilus* (LA) and *Bifidobacterium longum* (normal inhabitants of the healthy intestine) are the most clinically validated of all probiotics strains [Kailasapathy and Chin 2000; Goossens et al. 2003; Porth 2004; The United States Probiotics Organization 2007]. Scientific study has shown repeatedly that two-strain probiotics supplements containing *L. acidophilus* and *B. longum* are highly effective in the following:

- Supporting overall human health [Bai and Ouyang 2006; Quigley and Flourie 2007]
- Responding to small daily challenges [Gopal et al. 2001; Reid et al. 2001; Marelli, Papaleo, and Ferrari 2004]

When humans are under increased physical, emotional, or intellectual stress, changes often occur within the GI environment [Kailasapathy and Chin 2000]. Examples of these changes include slowed secretary responses, increased formation of reactive oxygen species, increased transit times of fecal material, disruption of mucosal cells, and altered epithelium tissues. These changes often result in occasional gas, bloating, and constipation and may interfere with probiotics functionality [Dong and Kaunitz 2006; Miyake, Tanaka, and McNeil 2006; Davidson, Kritas, and Butler 2007].

The effectiveness of all probiotics is dependent on the ability of the organisms to reach the large intestine in a viable state and adhere to the intestinal wall. Only then can colonization of the microflora succeed. Researchers have discovered recently that certain broad-spectrum probiotic combinations are able to function well within altered GI environments [Miyake, Tanaka, and McNeil 2006]. To date, the probiotic combination of *L. acidophilus*, *L. rhamnosus*, *B. bifidum*, *B. breve*, *B. longum*, and *B. lactis* shows great promise in the following:

- Supporting long-term colon care
- Providing deep intestinal support
- Helping the body respond during times of increased physical, emotional, or mental stress [Karimi and Pena 2003; Collado, Meriluoto, and Salminen 2007a]

The probiotic supplement should be proven to function *in vivo*, tolerate harsh intestinal environments, and successfully adhere to the intestinal wall. Table 14.1 gives the list of the strains useful in probiotics formulations, and Table 14.2 describes their applications in various disease conditions.

PREBIOTICS

Gibson and Roberfroid [1995] define prebiotics as "non-digestible food ingredients that beneficially affect the host by selectively stimulating the growth and/or activity of one or a limited number of bacteria in the colon, and thus improve host health." Prebiotics are simply the "food" for beneficial bacteria.

Prebiotics modify the balance of the intestinal microbiota by stimulating the activity of beneficial bacteria, such as *Lactobacilli* and *Bifidobacteria* [Gibson and Roberfroid 1995; Collins and Gibson 1999]. There is now considerable evidence that manipulation of the gut microbiota by prebiotics can beneficially influence the health of the host [Ginson and Roberfroid 1995; Roberfroid 1999; Delzenne and Kok 2001; Sartor 2004; Rastall et al. 2006; Parracho, McCartney, and Gibson 2007]. In particular, many attempts have been made to control serum triacylglycerol concentrations through modification of dietary habits with regard to consumption of prebiotics and probiotics [Delzenne and Kok 2001; Parracho, McCartney, and Gibson 2007]. Furthermore, unlike probiotics, prebiotics are not subject to biological viability problems and thus can be incorporated into a wide range of alimentary products (such as milk, yogurt, and infant formula), and they target organisms that are natural residents of the gut microbiota [Gibson and Roberfrodi 1995]. For example, oligosaccharides

Table 14.1 Commonly Used Strains in Probiotic Products

Probiotic	Structure/Function Claim	Reference
Lactobacillus acidophilus	Helps alleviate occasional gas, constipation, and lactose intolerance symptoms in children	Salazar-Lindo et al. 2007
	Supports healthy bowel movements while traveling	McFarland 2007
	Works with the body's own ability to modulate occasional intestinal discomfort	Rousseaux et al. 2007
Lactobacillus plantrum	Used postoperative immune stimulation	Sanders 2007
Lactobacillus reuteri	Immune stimulation against diarrhea	Sanders 2007
Lactobacillus rhamnosus	Immune stimulation, alleviates atopic eczema	Sanders 2007
Lactobacillus salivarius	Positive effects with intestinal ulcers and inflammation	Sanders 2007
Lactobacillus rhamnosus	Supports healthy balance of enterococci	Manley et al. 2007
	May support healthy skin integrity	Sawada et al. 2007
	Relieves occasional abdominal discomfort in school children	Gawrońska et al. 2007
Lactobacillus casei	Reduces symptoms of lactose intolerance, prevents bacterial overgrowth in small intestine	Sanders 2007
Lactobacillus johnsonii	Immune stimulation and active against *Helicobacter pylori*	Sanders 2007
Bifidobacterium bifidum	Supports healthy immune system responses	DeSimone et al. 1992
	Prevents occasional loose stools	Saavedra et al. 1994
Bifidobacterium lactis	Supports healthy intestinal colonization	Sanders 2006
	Restores healthy immune responses in the elderly	Gill et al. 2001
Bifidobacterium breve	Maintains healthy gut microflora colonies	Wang et al 2007; Li et al. 2004
Bifidobacterium longum	Supports healthy liver enzyme activity	Malaguarnera 2007
	Supports healthy bowel movements in adults	Amenta et al. 2006
	Supports the body's natural anti-inflammatory response	Xiao et al. 2007
	In laboratory research, *Bifidobacterium longum* removed lead and cadmium from water	Halttunen et al. 2007
	Supports healthy development of cells	Xu et al. 2007
Bifidobacterium animalis	Stabilizes intestinal passage, immune stimulation, improves phagocytic activity, alleviates atopic eczema, prevents diarrhea in children and traveler's diarrhea	Sanders 2007
Bifidobacterium lactis	Immune stimulation	
Escherichia coli	Immune stimulation	

Table 14.2 Common Applications of Probiotics Products in Different Disease Conditions

Disease	Probiotics	Research Results	Reference
Colon cancer	*Lactobacillus rhamnosus* and *Bifidobacterium longum*	In a small but well-designed Irish study, 80 people who had either colon cancer or benign polyps were randomly assigned to receive either a probiotic or a placebo to determine the effects on their tumors, growths, and intestines. The probiotic contained two types of bacteria, *Lactobacillus rhamnosus* and *Bifidobacterium longum*; the placebo was an inactive pill. After 12 weeks, the patients who received the probiotics showed decreased DNA damage in the lining of the colon and decreased growth and reproduction of colon cells.	Rafter et al. 2007
Antibiotic-induced diarrhea	*Lactobacillus acidophilus*	To see whether probiotics could prevent or reduce the diarrhea that often occurs during treatment with antibiotics, researchers randomized 135 hospitalized patients to receive either a probiotic drink (57 patients) containing *Lactobacillus* or a placebo (56 patients) twice a day while being treated with antibiotics and for one week after the course finished. Only seven patients (12%) in the probiotic group developed diarrhea compared with 19 patients (34%) in the placebo group.	Hickson et al. 2007
Chemotherapy-induced diarrhea	*Lactobacillus rhamnosus*	For most, diarrhea is an uncomfortable and embarrassing problem that almost always resolves after two to three days. For people being treated with chemotherapy, however, diarrhea can be deadly. It can lead to dehydration, hospitalization, and, if severe enough, discontinuation of the chemo drugs. In a Swedish study, colon cancer patients were randomly assigned to receive daily supplements of *Lactobacillus rhamnosus* during chemotherapy or a placebo. Those who received the probiotic *Lactobacillus* supplements had significantly less severe diarrhea (grades 3 and 4) than those who did not: 22 versus 37%. They also had less abdominal pain, were hospitalized less often, and needed fewer chemo dose reductions attributable to bowel problems.	Osterlund et al. 2007

(*Continued*)

Table 14.2 (*Continued*)

Disease	Probiotics	Research Results	Reference
Traveler's diarrhea	*Lactobacillus acidophilus* and *Bifidobacterium bifidum*	Traveler's diarrhea is most often caused by a bacterial infection, such as *Escherichia coli*, *Campylobacter*, *Shigella*, or *Salmonella* and is transmitted in undercooked or raw foods, contaminated food, contaminated water, or contaminated ice cubes. In a review of 12 clinical studies, a mixture of *Lactobacillus acidophilus* and *Bifidobacterium bifidum* worked the best to prevent and reduce the severity of traveler's diarrhea. No serious adverse reactions to the probiotics were reported in the trials.	McFarland 2007
Crohn's disease	*Bifidobacterium* and *Lactobacillus*	Ten active Crohn's outpatients with diarrhea and abdominal pain were enrolled into a probiotic study. All 10 had received prescription drugs to reduce their symptoms, but they remained painful. They took probiotics containing *Bifidobacterium* and *Lactobacillus* for four months. By the end of therapy, seven patients had improved clinical symptoms after combined probiotics and prebiotic therapy. Six patients had a complete response.	Fujimori et al. 2007
Ulcerative colitis	*Lactobacillus acidophilus*	To determine how effective *L. acidophilus* alone or in combination with mesalazine, an ulcerative colitis medication, is in achieving remission. A total of 187 ulcerative colitis patients were randomized to receive *L. acidophilus* (65 patients), mesalazine alone (60 patients), or *L. acidophilus* with mesalazine (62 patients). After 12 months, treatment with *L. acidophilus* was more effective than standard treatment with mesalazine in prolonging the relapse-free time.	Zocco et al. 2006
Stomach pain in children	*Lactobacillus acidophilus*	A total of 104 children who complained of "idiopathic tummy aches," or stomach pain without any identifiable cause, were enrolled in a double-blind, randomized controlled trial in which they received *L. acidophilus* (n = 52) or placebo (n = 52) for four weeks. The results showed that the children in the probiotics group were more likely to have no pain than those in the placebo group (25 versus 9.6%).	Gawrońska et al. 2007

have been suggested to represent the most important prebiotic dietary factor in human milk, promoting the development of a beneficial intestinal microbiota [Bode 2006]. When oligosaccharides are consumed, the undigested portion serves as food for the intestinal microflora. Two common supplemental sources are fructo-oligosaccharides (FOS) and inulin. FOS, which are found in many vegetables, consist of short chains of fructose molecules. Inulin has a much higher degree of polymerization than FOS and is a polysaccharide. FOS and inulin are found naturally in Jerusalem artichoke, burdock, chicory, leeks, onions, and asparagus. FOS products are derived from chicory root that contain significant quantities of inulin. Inulin is considered a soluble fiber. As a soluble dietary fiber, inulin also shortens fecal transit time, slightly increases fecal bulk, reduces constipation, has been shown to reduce both serum and hepatic cholesterol and triglycerides, and may provide improved absorption of minerals such as calcium, magnesium, iron, and phosphate. Furthermore, unlike FOS, inulin's longer chain length makes it more easily tolerated by the human intestinal system [Tokunaga, Oku, and Hosoya 1986]. Other benefits noted with FOS or inulin supplementation include increased production of beneficial SCFAs.

Regarding SCFAs, about 60 g of carbohydrate is fermented by the bacteria each day to SCFAs, which are rapidly absorbed. The SCFAs produced include acetic acid, propionic acid, and butyric acid. These acids have important actions in the colon and in the body as a whole. Acetic acid is an energy source for the body and is a substrate for fat synthesis in the liver. Propionic acid is also an energy source for the liver, is gluconeogenic (i.e., can be used to make glucose), and may reduce cholesterol synthesis. Butyric acid is the major fuel for colonic cells and has been shown to stimulate differentiation and programmed cell death of cancer cells. SCFA enemas have been used effectively in the treatment of ulcerative colitis. SCFAs produced in the colon increase cell proliferation throughout the whole gut. SCFAs are also very important because they promote water absorption and prevent osmotic diarrhea. SCFAs inhibit the growth of pathogenic bacteria [University of Glasgow 2005].

Another supportive factor in modulation of healthy microbiota is lactoferrin. Research shows that supplemental lactoferrin modulates the release of messenger proteins known as essential nutrients, trypsin, and protease inhibitors that protect it from destruction in the GI tract [Orsi 2004]. It is also rich in antioxidants, and its receptors have been found on most immune cells, including lymphocytes, monocytes, macrophages, and platelets [Orsi 2004]. Its presence in neutrophils suggests that lactoferrin is also involved in phagocytic immune responses [Lonnerdal and Iyer 1995]. Because of its iron-binding properties, lactoferrin has been proposed to play a role in iron uptake by the intestinal mucosa [Legrand et al. 2005].

Because it strongly binds iron, lactoferrin supports healthy modulation of gut microflora and assists the attachment of beneficial bacteria to the intestinal wall [Legrand et al. 2005]. Lactoferrin may support healthy development and differentiation of T lymphocytes. Preliminary research suggests that it supports the healthy production of cytokines and lymphokine, such as TNF-α and IL-6 [Ward, Paz, and Conneely 2005]. More recently, lactoferrin receptors have been found in both a variety of immune system cells, including natural killer cells, and intestinal tissue [Legrand et al. 2005; Ward, Paz, and Conneely 2005]. This discovery demonstrates that supplemental lactoferrin might have a profound impact on immune health.

To be effective, however, supplemental lactoferrin must be digested in the small intestine. Breast milk has a multitude of biological activities benefiting the newborn infant. To reach the small intestine (in which nutrients are digested and released into the infant's bloodstream), the nutrients must be able to withstand exposure to stomach fluids. Although the pH of an infant's stomach fluids rarely dips below 4–5 for the first six months of life, the pH of an adult's gastric fluids ranges between 1 and 2. This high acidity is lethal to lactoferrin. Therefore, just as probiotics need protection to survive the normal GI fluids of the stomach, so do lactoferrins [Orsi 2004].

SOURCES OF PROBIOTICS

There are many sources available from food to dietary supplement liquids and pills. Bacteria must be viable in storage and viable in delivery to the intestines. To date, scientists have shown that, as a result of manufacturing processes, storage environment, and inhibition through the digestive tract, many products are not delivering label claims.

Probiotic supplements need to be protected from the environment. They cannot be exposed to air, sunshine, artificial light, or moisture [The Centers for Disease Control 2008]. In addition, probiotics bacteria need to be protected from the digestive juices and enzymes in the stomach [Collado, Meriluoto, and Salminen 2007b]. Research has shown as much as 90% of a supplemented probiotic is destroyed in stomach gastric secretions and/or 50% loss in exposure to environment in storage.

ENTERIC COATING OF PROBIOTICS

Enteric coating of probiotics is intended to allow the passage of a tablet or capsule through the gastric fluids of the stomach to prevent the release of product contents before it reaches the intestines. However, because of the complexities involved with applying an enteric coating on a tablet or capsule, some enteric coatings do not entirely inhibit stomach acid from entering the encapsulation. As a result, stomach acid can interact with the sensitive bacteria, leading to a significant decrease in viability. In addition, the enteric coating manufacturing process frequently uses solvents such as methacrylate copolymers. The tablets and capsules are sprayed with these solvents at high temperatures to create the enteric coating. This type of application further exposes microbes to conditions that can dramatically reduce the product shelf-life. There is a new technology used in protecting viable ingredients that is a patented, encapsulation technique, known as True Delivery™ Technology, which results in a product that is stable at room temperature for up to 18 months. Additionally, the unique coating protects the bacteria from harsh stomach acid so they can be released live and intact in the intestines, in which they need to arrive in live form to perform their beneficial function. Research may show that other methods support viability; for example, combining strains with specific fibers to buffer the probiotic from stomach acidity has shown some results. However, to date, the patented encapsulation has demonstrated the best effects on stability and protection. The important factors related to probiotics stability are as follows:

- Stable: guaranteed to deliver live, intact probiotics throughout product shelf-life, not just at time of manufacture
- Protected: protects the probiotics from harsh stomach acid
- Effective: contains clinically studied probiotics bacteria and has been shown to colonize in the intestines

OTHER TYPES OF PROBIOTICS

Probiotics are defined as live microorganisms that confer a health benefit on the host. Although most probiotics are bacteria, one strain of yeast, *Saccharomyces boulardii*, has been found to be an effective probiotics in double-blind clinical studies. Studies in areas of antibiotic include diarrhea, traveler's diarrhea, acute diarrhea in children, recurrent clostridium difficile-associated diseases, and IBDs. *S. boulardii* does not colonize in the intestines but will be eliminated in stool within five to seven days after usage is discontinued. Because of this being a noncolonizing strain that is not naturally present in human gut flora, questions are highlighted regarding its long-term safety of this or any other non human strain of bacteria being introduced to the human gut flora. *S. boulardii* is administrated to patients in a lyophilized form, and the treatment is well tolerated. However, some rare cases of *S. boulardii* fungemias have been reported in patients with an indwelling central venous catheter [Kotowska, Albrecht, and Szajewska 2005; Llanos et al. 2006]. The origin of the fungemia is thought to be either a digestive tract translocation or a contamination of the central venous line by the colonized hands of health workers [Kotowska, Albrecht, and Szajewska 2005]. This raises the question of the risk-benefit ratio of *S. boulardii* in critically ill or immunocompromised patients. Thus, administration of *S. boulardii* should be contraindicated for patients of fragile health, as well as for patients with central venous catheter [Herbrecht and Nivoix 2006].

SOME INTERESTING CLINICAL APPLICATIONS OF PROBIOTICS

Probiotics, Infection, and Immunity

The review by Macfarlane et al. [2002] summarizes the most recent contributions to this rapidly developing area. Probiotic bacteria, mainly *Bifidobacteria* and *Lactobacilli* for historical reasons, can prevent or ameliorate some diseases. Many empirical studies have been done, but work to develop the ideal characteristics of probiotics lags behind. Current literature covers survival of probiotics in the gut, mucosal adherence, antibacterial/pathogen mechanisms, effects on immune function, and clinical studies. Probiotics bacteria are effective in preventing and reducing the severity of acute diarrhea in children. They are also useful in antibiotic-associated diarrhea, but not for elimination of *Helicobacter pylori*. In IBD, especially ulcerative colitis, probiotics offer a safe alternative to current therapy. Probiotics have been used to prevent urogenital tract infection with benefit and, perhaps more intriguingly, to reduce atopy in children. Probiotics do not invariably work, and study of mechanisms is urgently needed.

The Potential Role of Probiotics in Pediatric Urology

The research paper by Reid [2002] studied the potential role that probiotics therapy may have in pediatric urology. Many children around the world die of diseases, such as GI infection and HIV, whereas many have urinary tract infections that subsequently recur frequently in adulthood. Until recently, the role of intestinal and urogenital (vaginal, urethral, and perineal) microflora in health and disease has received scant attention. The data available in the literature on this topic were examined, and a personal viewpoint is presented on how they may relate to urology. There is mounting evidence that certain strains of *Lactobacilli* and *Bifidobacteria* have a major part in the maintenance and restoration of health in children and adults. Implications for pediatric urology include a decreased risk of infection and stone disease as well as possible positive effects on preventing and managing inflammatory and some carcinogenic diseases.

High-Dose Oral Bacteria Therapy for Chronic Nonspecific Diarrhea of Infancy

Balli et al. [1992] evaluate the effectiveness of oral bacteriotherapy using a combination of anaerobe fecal *Lactobacilli* for chronic, nonspecific diarrhea of infancy. A double-blind study was performed in a total of 40 children treated with low and high doses of bacteria. The results confirm the importance of fecal flora in this disease and support the hypothesis that oral bacteriotherapy can improve clinical and laboratory presentation, especially when given at high doses.

Bifidobacteria and *Lactobacilli* in Human Health

The gastrointestinal microflora is a complex ecological system, normally characterized by a flexible equilibrium [Orrhage and Nord 2000]. The most important role of the microflora, from the point of view of the host, is probably to act in colonization resistance against exogenous, potentially pathogenic, microorganisms. *Bifidobacteria* and *Lactobacilli* are gram-positive, lactic-acid-producing bacteria constituting a major part of the intestinal microflora in humans and other mammals. Administration of antimicrobial agents may cause disturbances in the ecological balance of the GI microflora, with several unwanted effects such as colonization by potential pathogens. To maintain or reestablish the balance in the flora, supplements of intestinal microorganisms, mainly *Bifidobacteria* and *Lactobacilli*, sometimes called probiotics, have been successfully used. This article reviews the role of *Bifidobacteria* and *Lactobacilli* in human health.

Live Probiotics Protect Intestinal Epithelial Cells from the Effects of Infection with Enteroinvasive *Escherichia coli*

The colonic epithelium maintains a lifelong, reciprocally beneficial interaction with the colonic microbiota. Disruption is associated with mucosal injury. Resta and Barret [2003] proposed that probiotics may limit epithelial damage induced by enteroinvasive pathogens and promote restitution. Human intestinal epithelial cell lines (HT29/cl.19A and Caco-2) were exposed to enteroinvasive *E. coli* (EIEC 029:NM) and/or probiotics (*Streptococcus thermophilus* [ST], ATCC19258; and LA, ATCC4356). Infected cells and controls were assessed for transepithelial resistance, chloride secretory responses, alterations in cytoskeletal and tight junctional proteins, and responses to epidermal growth factor stimulation. Exposure of cell monolayers to live ST/LA, but not to heat inactivated ST/LA, significantly limited adhesion, invasion, and physiological dysfunction induced by EIEC. Antibiotic killed ST/LA reduced adhesion somewhat but were less effective in limiting the consequences of EIEC invasion of cell monolayers. Furthermore, live ST/LA alone increased transepithelial resistance, contrasting markedly with the fall in resistance evoked by EIEC infection, which could also be blocked by live ST/LA. The effect of ST/LA on resistance was accompanied by maintenance (actin, zona occludens-1) or enhancement (actinin, occludin) of cytoskeletal and tight junctional protein phosphorylation. ST/LA had no effect on chloride secretion by themselves but reversed the increase in basal secretion evoked by EIEC. EIEC also reduced the ability of epidermal growth factor to activate its receptor, which was reversed by ST/LA. Live ST/LA interact with intestinal epithelial cells to protect them from the deleterious effect of EIEC via mechanisms that include, but are not limited to, interference with pathogen adhesion and invasion. Probiotics likely also enhance the barrier function of naive epithelial cells not exposed to any pathogen.

Breakdown of Lactose

Lactose is an important sugar that is converted to lactic acid by lactic-acid-producing bacteria, such as *Lactobacillus acidophilus* and *Bifidobacterium longum* [Marteau, Vesa, and Rambaud 1997]. Impaired conversion of lactose to lactic acid can result in symptoms such as occasional gas, bloating, and indigestion, attributable to accumulated non-absorbed lactose in the GI tract [DeSimone et al. 1992; Garman, Coolbear, and Smart 1996 et al]. Lactic acid bacteria can help metabolize the non-absorbed lactose in the GI tract and therefore reduce symptoms of lactose intolerance. In a randomized, controlled clinical trial, *Bifidobacterium longum* was shown to support the breakdown of lactose and reduce flatulence [Lactose 2008].

Immune System Support

Although a normal microflora is associated with good health, changes in intestinal health are associated with altered immune function. A well-functioning GI immune system mediates immune responsiveness at mucosal sites and throughout the entire body via the control of the quality and quantity of foreign substances gaining access to the immune system [Schriffrin et al. 1997]. *Lactobacillus acidophilus* and *Bifidobacterium longum* have been shown to possess immunoprotective and immunomodulatory properties. These benefits include modulation of cytokine and various IL production, autoimmunity, natural killer cell cytotoxicity, lymphocyte proliferation, and antibody production. In an open, randomized, controlled trial, *Lactobacillus acidophilus* and *Bifidobacterium bifidum* were supportive of colon health in older adults. In addition, B cell (important antibody producing immune cells) levels increased compared with the untreated group. The probiotics were very well tolerated, with no significant side effects or variations in clinical chemistry or hematologic parameters [Gibson et al. 1997].

Decrease Occasional Constipation

Constipation is defined as infrequent or difficult defecation that can result from decreased motility of the intestines. It is a common problem, particularly in older adults. When the feces remain in the large intestine for prolonged periods, there is excessive water absorption, making the feces dry and hard [Garman, Coolbear, and Smart 1996]. *Lactobacillus acidophilus* and *Bifidobacterium longum* promote regular bowel movements by contributing to the reestablishment of healthy intestinal flora and stimulation of intestinal peristalsis via lactic acid production [Bennet and Eley 1976].

Support of Putrefactive Processes

When unbalanced conditions are present in the intestines (i.e., unbalanced diet, high acidity, and/or low levels of lactic acid bacteria), organic matter may be putrified (decomposed or rotting) by certain bacteria and produce undesirable compounds [Gibson et al. 1997]. Probiotics promote homeostasis (balance) in both the intestine and the vagina [Hilton et al. 1992; Witsell et al. 1995]. These activities are carried out via support of direct production of antibodies, competition with adhesion to intestinal cells, or indirect modulation of the immune system. Probiotics also support a healthy yeast balance [On-line Medical Dictionary 2000].

Support Digestion

Normal microflora of the large intestine help support and complete digestion via fermentation [Wagner et al. 1997]. Oral ingestion of probiotics produces a stabilizing effect on the gut flora.

Additional Benefits

The benefits of probiotics extend beyond digestion support and immune support. *Lactobacillus acidophilus* and *Bifidobacterium longum* also help support the better utilization and bioavailability of nutrients, including vitamins, minerals, proteins, fats, and carbohydrates [Witsell et al. 1995].

REFERENCES

Amenta, M., M. T. Cascio, P. D. Fiore, and I. Venturini. 2006. Diet and chronic constipation. Benefits of oral supplementation with symbiotic zir fos (*Bifidobacterium longum* W11 + FOS Actilight). *Acta. Biomed.* 77:157–162.

Asquith, M. T. and J. R. Harrod. 1979. Reduction in bacterial contamination in banked human milk. *J. Pediatr.* 95:993–994.

Bai, A. P. and Q. Ouyang. 2006. Probiotics and inflammatory bowel diseases. *Postgrad. Med. J.* 82:376–382.

Balli, F., P. Bertolani, G. Giberti, and S. Amarri. 1992. High-dose oral bacteria-therapy for chronic non-specific diarrhea of infancy. *Pediatr. Med. Chir.* 14:13–15.

Bennet, R., M. Eriksson, C. E. Nord, and R. Zetterström. 1982. Suppression of aerobic and anaerobic faecal flora in newborns receiving parenteral gentamicin and ampicillin. *Acta. Paediatr. Scand.* 71:559–562.

Bennet, R., M. Eriksson, C. E. Nord, and R. Zetterström. 1986. Fecal bacterial microflora of newborn infants during intensive care management and treatment with five antibiotic regimens. *Pediatr. Infect. Dis.* 5:533–539.

Bennet, R. and C. E. Nord. 1989. The intestinal microflora during the first week of life: Normal development and changes induced by caesarean section, preterm birth and antimicrobial treatment. In *The Regulatory and Protective Role of the Normal Microflora.* Edited by Grubb R., T. Midtvedt, and E. Norin. London, UK: Macmillan Press, pp. 19–34.

Bennett, A. and K. G. Eley. 1976. Intestinal pH and propulsion: An explanation of diarrhoea in lactase deficiency and laxation by lactulose. *J. Pharm. Pharmacol.* 28:192–195.

Berg, R. 1996. The indigenous gastrointestinal micro flora. *Trends Microbiol.* 4: 430.

Bettelheim, K. A., A. Breardon, M. C. Faiers, and S. M. O'Farrell. 1974. The origin of O serotypes of *Escherichia coli* in babies after normal delivery. *J. Hyg. (Lond.)* 72:67–70.

Bode, L. 2006. Recent advances on structure, metabolism, and function of human milk oligosaccharide. *Int. J. Probiotics Prebiotics* 1:19–26, 71.

Brook, I., C. Barett, C. Brinkman, W. Martin, and S. Finegold. 1979. Aerobic and anaerobic bacterial flora of the maternal cervix and newborn gastric fluid and conjunctiva: A prospective study. *Pediatrics* 63:451–455.

Bullen, C. L., P. V. Tearle, and M. G. Stewart. 1977. The effect of "humanized" milks and supplemental breast feeding on the faecal flora of infants. *J. Med. Microbiol.* 10:403–413.

Centers for Disease Control. 2008. Get smart: Know when antibiotics work. http://www.cdc.gov/drugresistance/community. Accessed February 15, 2009.

Collado, M. C., J. Meriluoto, and S. Salminen. 2007a. Development of new probiotics by strain combinations: Is it possible to improve the adhesion to intestinal mucus? *J. Dairy Sci.* 90:2710–2716.

Collado, M. C., J. Meriluoto, and S. Salminen. 2007b. Role of commercial probiotic strains against human pathogen adhesion to intestinal mucus. *Lett. Appl. Microbiol.* 45:454–460.

Collins, M. D. and G. R. Gibson. 1999. Probiotics, prebiotics, and synbiotics: Approaches for modulating the microbial ecology of the gut. *Am. J. Clin. Nutr.* 69:1052–1057.

Conway, P. 1997. Development of intestinal microbiota. In *Gastrointestinal Microbiology.* Volume 3. Edited by Mackie, R. I., B. A. White, and R. E. Isaacson. New York, NY: Chapman & Hall, pp. 3–38.

Cooperstock, M. S. and A. J. Zedd. 1983. Intestinal flora of infants. In: *Human Intestinal Microflora in Health and Disease.* Edited by Hentges, D. J. New York, NY: Academic Press, pp. 79–99.

Davidson, G., S. Kritas, and R. Butler. 2007. Stressed mucosa. *Nestle Nutr. Workshop Ser. Pediatr. Program* 59:133–142.

Dekker, J., M. Collett, J. Prasad, and P. Gopal. 2007. Functionality of probiotics: Potential for product development. *Forum Nutr.* 60:196–208.

Delzenne, N. M. and N. Kok. 2001. Effects of fructans-type prebiotics on lipid metabolism. *Am. J. Clin. Nutr.* 73:456–458.

De Simone, C., A. Ciardi, A. Grassi, S. Lambert Gardini, S. Tzantzoglou, V. Trinchieri, S. Moretti, and E. Jirillo. 1992. Effect of *Bifidobacterium bifidum* and *Lactobacillus acidophilus* on gut mucosa and peripheral blood B lymphocytes. *Immunopharmacol. Immunotoxicol.* 14:331–340.

Dong, M. H. and J. D. Kaunitz. 2006. Gastroduodenal mucosal defense. *Curr. Opin. Gastroenterol.* 22:599–606.

Finegold, S. M., G. E. Mathisen, and W. L. George. 1983. Changes in human intestinal flora related to the administration of antimicrobial agents. In *Human Intestinal Microflora in Health and Disease.* Edited by Hentges, D. J. New York, NY: Academic Press, pp. 356–446.

Fujimori, S., A. Tatsuguchi, K. Gudis, T. Kishida, K. Mitsui, A. Ehara, T. Kobayashi, Y. Sekita, T. Seo, and C. Sakamoto. 2007. High dose probiotic and prebiotic co-therapy for remission induction of active Crohn's disease. *J. Gastroenterol. Hepatol.* 22:1199–1204.

Garman, J., T. Coolbear, and J. Smart. 1996. The effect of cations on the hydrolysis of lactose and the transferase reactions catalysed by beta-galactosidase from six strains of lactic acid bacteria. *Appl. Microbiol. Biotechnol.* 46:22–27.

Gawrońska, A., P. Dziechciarz, A. Horvath, and H. A. Szajewska. 2007. Randomized double-blind placebocontrolled trial of *Lactobacillus* GG for abdominal pain disorders in children. *Aliment Pharmacol. Ther.* 25:177–184.

Gibson, G. R. and M. B. Roberfroid. 1995. Dietary modulation of the human colonic microbiodata: Introducing the concept of probiotics. *J. Nutr.* 125:1401–1412.

Gibson, G. R., J. M. Saavedra, S. Macfarlane, and G. T. Macfarlane. 1997. Probiotics and intestinal infections. In *Probiotics 2: Applications and Practical Aspects.* Edited by Fuller, R. London, UK: Chapman & Hall, pp. 10–39.

Gill, H. S., K. J. Rutherford, M. L. Cross, and P. K. Gopal. 2001. Enhancement of immunity in the elderly by dietary supplementation with the probiotic *Bifidobacterium lactis*. *Am. J. Clin. Nutr.* 74:833–839.

Goossens, D., D. Jonkers, E. Stobberingh, A. van den Bogaard, M. Russel, and R. Stockbrugger. 2003. Probiotics in gastroenterology: Indications and future perspectives. *Scand. J. Gastroenterol. Suppl.* 2003:15–23.

Gopal, P. K., J. Prasad, J. Smart, and H. S. Gill. 2001. In vitro adherence properties of *Lactobacillus rhamnosus* DR20 and *Bifidobacterium lactis* DR10 strains and their antagonistic activity against an enterotoxigenic *Escherichia coli*. *Int. J. Food Microbiol.* 67:207–216.

Grutte, F. K., R. Horn, and H. Haenel. 1965. Nutrition and biochemical microecology processes occurring in the colon of infants (in German). *Z. Kinderheilkd.* 93:28–39.

Guarner, F. and J. R. Malagelada. 2003. Gut flora in health and disease. *The Lancet* 361:512–519.

Guyton, A. C. and J. E. Hall. 2005. Secretory functions of the alimentary canal. In *Textbook of Medical Physiology*. 11th edition. Philadelphia, PA: W. B. Saunders Company, p. 738.

Halttunen, T., S. Salminen, and R. Tahvonen. 2007. Rapid removal of lead and cadmium from water by specific lactic acid bacteria. *Int. J. Food Microbiol.* 114:30–35.

Herbrecht, R. and Y. Nivoix. 2006. *Saccharomyces cerevisiae* Fungemia: An adverse effect of *Saccharomyces boulardii* probiotic administration. *Clin. Infect. Dis.* 40:1635–1637.

Hewitt, J. H. and J. Rigby. 1976. Effects of various milk feeds on numbers of *Escherichia coli* and *Bifidobacterium* in the stools of newborn infants. *J. Hyg. (Lond)* 78:85–93.

Hickson, M., A. L. D'Souza, N. Muthu, T. R. Rogers, S. Want, C. Rajkumar, and C. J. Bulpitt. 2007. Use of probiotic *Lactobacillus* preparation to prevent diarrhea associated with antibiotics: Randomized double blind placebo controlled trial. *BMJ* 335:80.

Hilton, E., H. D. Isenberg, P. Alperstein, K. France, and M. T. Borenstein. 1992. Ingestion of yogurt containing *Lactobacillus acidophilus* as prophylaxis for candidal vaginitis. *Ann. Intern. Med* 116:353–357.

Isolauri, E. 2001. Probiotics in human disease. *Am. J. Clin. Nutr.* 73:1142S–1146S.

Kailasapathy, K. and J. Chin. 2000. Survival and therapeutic potential of probiotic organisms with reference to *Lactobacillus acidophilus* and *Bifidobacterium* spp. *Immunol. Cell Biol.* 78:80–88.

Karimi, O. and A. S. Pena. 2003. Probiotics: Isolated bacteria strain or mixtures of different strains? Two different approaches in the use of probiotics as therapeutics. *Drugs Today (Barc).* 39:565–597.

Kotowska, M., P. Albrecht, and H. Szajewska. 2005. *Saccharomyces boulardii* in the prevention of antibiotic-associated diarrhoea in children: A randomized double-blind placebo-controlled trial. *Aliment. Pharmacol. Ther.* 21:583–590.

Lee, A. 1984. Neglected niches: The microbial ecology of the gastrointestinal tract. In *Advances in microbial ecology.* Edited by Marshall, K. New York, NY: Plenum Press, pp. 115–162.

Legrand, D., E. Elass, M. Carpentier, and J. Mazurier. 2005. Lactoferrin: A modulator of immune and inflammatory responses. *Cell. Mol. Life Sci.* 62:2549.

Lennox-King, S. M. J., S. M. O'Farrell, K. A. Bettelheim, and R. A. Shooter. 1976a. Colonization of caesarean section babies by *Escherichia coli. Infection* 4:134–138.

Lennox-King, S. M. J., S. M. O'Farrell, K. A. Bettelheim, and R. A. Shooter. 1976b. *Escherichia coli* isolated from babies delivered by caesarian section and their environment. *Infection* 4:139–145.

Li, Y., T. Shimizu, A. Hosaka, N. Kaneko, Y. Ohtsuka, and Y. Yamashiro. 2004. Effects of *Bifidobacterium breve* supplementation on intestinal flora of low birth weight infants. *Pediatr. Int.* 46:509–515.

Llanos, R. D., A. Querol, J. Pemán, M. Gobernado, and M. T. Fernández-Espinar. 2006. Food and probiotic strains from the *Saccharomyces cerevisiae* species as a possible origin of human systemic infections. *Int. J. Food Microbiol.* 110:286–290.

Lonnerdal, B. and S. Iyer. 1995. Lactoferrin: Molecular structure and biological function. *Annu. Rev. Nutr.* 15:93–110.

Lundequist, B., C. E. Nord, and J. Winberg. 1985. The composition of fecal microflora in breast-fed and bottle fed infants from birth to 8 weeks. *Acta. Paediatr. Scand.* 74:45–51.

MacFarlane, G. T. and J. H. Cummings. 2002. Probiotics, infections and immunity. *Curr. Opin. Infect. Dis.* 15:501–506.

MacGregor, R. R. and W. W. Tunnessen. 1973. The incidence of pathogenic organisms in the normal flora of the neonate's external ear and nasopharynx. *Clin. Pediatr.* 12:697–700.

Malaguarnera, M., F. Greco, G. Barone, M. P. Gargante, M. Malaguarner, and M. A. Toscano. 2007. *Bifidobacterium longum* with fructo-oligosaccharide (FOS) treatment in minimal hepatic encephalopathy: A randomized, double-blind, placebo-controlled study. *Dig. Dis. Sci.* 52:3259–3265.

Manley, K. J., M. B. Fraenkel, B. C. Mayall, and D. A. Power. 2007. Probiotic treatment of vancomycin-resistant enterococci: A randomized controlled trial. *Med. J. Aust.* 186:454–457.

Marelli, G., E. Papaleo, and A. Ferrari. 2004. *Lactobacilli* for prevention of urogenital infections: A review. *Eur. Rev. Med. Pharmacol. Sci.* 8:87–95.

Marteau P., T. Vesa, and J. C. Rambaud. 1997. Lactose maldigestion. In *Probiotics Applications and Practical Aspects.* Edited by Fuller, R. London, UK: Chapman & Hall, pp. 65–88.

Mata, L. J. and J. J. Urrutia. 1971. Intestinal colonization of breast-fed children in a rural area of low socioeconomic level. *Ann. N.Y. Acad. Sci.* 176:93–108.

McFarland, L. V. 2007. Meta-analysis of probiotics for the prevention of traveler's diarrhea. *Travel Med. Infect. Dis.* 5:97–105.

Medical Dictionary, On-line. 2002. *Candida albicans.* http://www.graylab.ac.uk/cgi-bin/omd?query+candida albicans. Accessed February 10, 2000.

Medical Dictionary, On-line. 2008. Lactose. http://www.graylab.ac.uk/cgi-bin/omd. Accessed February 10, 2009.

Mitsuoka, T. and C. Kaneuchi. 1977. Ecology of the *Bifidobacteria. Am. J. Clin. Nutr.* 30:1799–1810.

Miyake, K., T. Tanaka, and P. L. McNeil. 2006. Disruption-induced mucus secretion: Repair and protection. *PLoS Biol.* 4:e276.

Moughan, P. J., M. J. Birtles, P. D. Cranwell, W. C. Smith, and M. Pedraza. 1992. The piglet as a model animal for studying aspects of digestion and absorption in milk-fed human infants. In *Nutritional Triggers for Health and in Disease.* Edited by Simopoulos, A. P. Basel, Switzerland: Karger, pp. 40–113.

Orrhage, K. and C. E. Nord. 2000. A probiotic mixture alleviates symptoms in irritable bowel syndrome patients: A controlled 6-month intervention. *Drugs Exp. Clin. Res.* 26:95–111.

Orsi, N. 2004. The antimicrobial activity of lactoferrin: Current status and perspectives. *Biometals* 17:189–196.

Osterlund, P., T. Ruotsalainen, R. Korpela et al. 2007. *Lactobacillus* supplementation for diarrhea related to chemotherapy of colorectal cancer: A randomized study. *Br. J. Cancer* 97:1028–1034.

Palmer, C., E. M. Bik, D. B. DiGiulio, D. A. Relman, and P. O. Brown. 2007. Development of the human infant intestinal microbiota. *PLoS Biology* 5:e177.

Parracho, H., A. L. McCartney, and G.R. Gibson. 2007. Probiotics and prebiotics in infant nutrition. *Proc. Nutr. Soc.* 66:405–411.

Porth, C. M. 2004. Gastrointestinal tract function. In *Pathophysiology: Concepts of Altered Health States.* 7th edition. Philadelphia, PA: Lippincott, pp. 617–622.

Quigley, E. M. and B. Flourie. 2007. Probiotics and irritable bowel syndrome: A rationale for their use and an assessment of the evidence to date. *Neurogastroenterol. Motil.* 19:166–172.

Rafter, J., M. Bennett, G. Caderni, Y. Clune, R. Hughes, P. C. Karlsson, A. Klinder, M. O'Riordan, G. C. O'Sullivan, B. Pool-Zobel, G. Rechkemmer, M. Roller, I. Rowland, M. Salvadori, H. Thijs, J. Van Loo, B. Watzl, and J. K. Collins. 2007. Dietary synbiotics reduce cancer risk factors in polypectomized and colon cancer patients. *Am. J. Clin. Nutr.* 85:488–496.

Rastall, R. A., G. R. Gibson, H. S. Gill, F. Guarner, T. R. Klaenhammer, B. Pot, G. Reid, I. R. Rowland, and M. E. Sanders. 2006. Modulation of the microbial ecology of the human colon by probiotics, prebiotics and synbiotics to enhance human health: An overview of enabling science and potential applications. *FEMS Microbiol. Ecol.* 52:145–152.

Reid, G., D. Beuerman, C. Heinemann, and A. W. Bruce. 2001. Probiotic *Lactobacillus* dose required to restore and maintain a normal vaginal flora. *FEMS Immunol. Med. Microbiol.* 32:37–41.

Reid, G. 2002. The potential role of probiotics in pediatric urology. *J. Urol.* 168:1512–1517.

Resta-Lenert, S. and K. E. Barrett. 2003. Live probiotics protect intestinal epithelial cells from the effects of infection with enteroinvasive *Escherichia coli* (EIEC). *Gut* 52:988–997.

Roberfroid, M. B. 1999. Concepts in functional foods: The case of inulin and oligofructose. *J. Nutr.* 129:1398–1401.

Rousseaux, C., X. Thuru, Z. Gelot et al. 2007. *Lactobacillus acidophilus* modulates intestinal pain and induces opioid and cannabinoid receptors. *Nat. Med.* 13:35–37.

Saavedra, J. M., N. A. Bauman, I. Oung, J. A. Perman, and R. H. Yolken. 1994. Feeding of *Bifidobacterium bifidum* and *Streptococcus thermophilus* to infants in hospital for prevention of diarrhoea and shedding of rotavirus. *Lancet* 344:1046–1049.

Salazar-Lindo, E., D. Figueroa-Quintanilla, M. I. Caciano et al. 2007. Effectiveness and safety of *Lactobacillus* LB in the treatment of mild acute diarrhea in children. *J. Pediatr. Gastroenterol. Nutr.* 44:571–576.

Sanders, M. E. 2006. Summary of probiotic activities of *Bifidobacterium lactis. J. Clin. Gastroenterol.* 40:776–783.

Sanders, M. E. 2007. Probiotics: Strain matter. *Funct. Foods Nutr. Mag.* June: 36–41.

Sartor, R. 2004. Therapeutic manipulation of the enteric microflora in inflammatory bowel diseases: Antibiotics, probiotics, and prebiotics. *Gastroenterology* 126:1620–1633.

Savage, D. S. 1977. Microbial ecology of the gastrointestinal tract. *Ann. Rev. Microbiol.* 31:107–133.

Sawada, J., H. Morita, A. Tanaka et al. 2007. Ingestion of heat-treated *Lactobacillus rhamnosus* GG prevents development of atopic dermatitis in NC/Nga mice. *Clin. Exp. Allergy* 37:296–303.

Schriffrin, E. J., D. Brassart, A. L. Servin, F. Rochat, A. Donnet-Hughes. 1997. Immune modulation of blood leukocytes in humans by lactic acid bacteria: Criteria for strain selection. *Am. J. Clin. Nutr.* 66 (Suppl.):515S–520S.

Sears, C. L. 2005. A dynamic partnership: Celebrating our gut flora. *Anaerobe* 11:247–251.

Stark, P. L. and A. Lee. 1982. The microbial ecology of the large bowel of breast-fed and formula-fed infants during the first year of life. *J. Med. Microbiol.* 15:189–203.

Tannock, G. W. 1994. The acquisition of the normal microflora of the gastrointestinal tract. In *Human Health: The Contribution of Microorganisms*. Edited by Gibson, S. A. W. London, UK: Springer-Verlag, pp. 1–16.

Tomkins, A. M., A. K. Bradley, S. Oswald, and B. S. Drasar. 1981. Diet and the faecal microflora of infants, children and adults in rural Nigeria and urban UK. *J. Hyg. (Lond)* 86:285–293.

United States Probiotics Organization. 2007. What are probiotics? http://www.usprobiotics.org. Accessed August 1, 2007.

University of Glasgow. 2005. The normal gut flora. http://www.gla.ac.uk/departments/humannutrition/students/resources/meden/Infection. Accessed February 13, 2009.

Wang, C., H. Shoji, H. Sato, S. Nagata, Y. Ohtsuka, T. Shimizu, and Y. Yamashiro. 2007. Effects of oral administration of *Bifidobacterium breve* on fecal lactic acid and short-chain fatty acids in low birth weight infants. *J. Pediatr. Gastroenterol. Nutr.* 44:252–257.

West, P. A., J. H. Hewitt, and O. M. Murphy. 1979. The influence of methods of collection and storage on the bacteriology of human milk. *J. Appl. Bacteriol.* 46:269–277.

Wagner, R. D., C. Peirson, T. Warner et al. 1997. Biotherapeutic effects of probiotic bacteria on candidiasis in immunodeficient mice. *Infect. Immun.* 65:4165–4172.

Ward, P. P., E. Paz, and O. M. Conneely. 2005. Multifunctional roles of lactoferrin: A critical overview. *Cell. Mol. Life Sci.* 62:2540–2548.

Witsell, D. L., C. G. Garrett, W. G. Yarbrough, S. P. Dorrestein, A. F. Drake, and M. C. Weissler. 1995. Effect of *Lactobacillus acidophilus* on antibiotic-associated gastrointestinal morbidity: A prospective randomized trial. *J. Otolaryngol.* 24:230–233.

Xiao, J. Z., S. Kondo, N. Yanagisawa, K. Miyaji, K. Enomoto, T. Sakoda, K. Iwatsuki, and T. Enomoto. 2007. Clinical efficacy of probiotic *Bifidobacterium longum* for the treatment of symptoms of Japanese cedar pollen allergy in subjects evaluated in an environmental exposure unit. *Allergol. Int.* 56:67–75.

Xu, Y. F., L. P. Zhu, B. Hu, G. F. Fu, H. Y. Zhang, J. J. Wang, and G. X. Xu. 2007. A new expression plasmid in *Bifidobacterium longum* as a delivery system of endostatin for cancer gene therapy. *Cancer Gene Ther.* 14:151–157.

Zocco, M. A., L. Z. dal Verme, F. Cremonini, A. C. Piscaglia, E. C. Nista, M. Candelli, M. Novi, D. Rigante, I. A. Cazzato, V. Ojetti, A. Armuzzi, G. Gasbarrini, and A. Gasbarrini. 2006. Efficacy of *Lactobacillus* GG in maintaining remission of ulcerative colitis. *Aliment. Pharmacol. Ther.* 23:1567–1574.

CHAPTER 15

Nutraceuticals and Weight Management

Gwendolyn W. Pla

CONTENTS

INTRODUCTION

Obesity and being overweight are significant risks to health for adults and children. Both overweight and obesity are estimated by calculation of the body mass index (BMI). BMI is the ratio of weight to height and represents the degree of body fatness. Although the BMI does not actually measure body fat, there is a good correlation between body fat and BMI.

HEALTH CONSEQUENCES OF OVERWEIGHT AND OBESITY

The health consequences of overweight and obesity include the following: (1) premature death, because obese individuals have a greater than 50% risk of death than do those of a healthy weight; (2) risks of heart disease, including congestive heart failure, angina, sudden cardiac death, and abnormal heart rhythm are increased for individuals whose BMI is above 25; (3) increased risk of certain cancers, including endometrial, colon, gall bladder, prostate, kidney, and breast; (4) sleep apnea; (5) arthritis; and (6) reproductive complications. A weight gain of 11–18 lb doubles the risk of type 2 diabetes mellitus (DM), and more than 80% of the people with type 2 DM are overweight or obese. Other health consequences include gall bladder disease, incontinence, increased surgical risk, and depression. In children and adolescents, social stigma and depression can also occur [Office of the Surgeon General of the United States 2007].

Overweight children and adolescents are likely to be at risk for hypertension, elevated cholesterol, and type 2 DM. Until recently, type 2 DM was called adult-onset diabetes and was rarely seen in persons under 40 years of age. Type 2 DM is on the rise in children and adolescents and, as in adults, has been linked to overweight and obesity. Additionally, overweight children are likely to become overweight adults.

PREVALENCE OF OBESITY

According to data from the most recent National Health and Nutrition Examination Survey, the National Center for Health Statistics found that, in adults aged 20–74 years, the prevalence of obesity increased from 15% in 1976–1980 to 34% in 2005–2006. Among children and adolescents, ages 2–19 years, the prevalence of obesity in 2005–2006 was 16%. These children were at or above the 95th percentile of the 2000 BMI for age growth charts. Also, in children and adolescents, the prevalence of being overweight increased from 5 to 13.9% in children ages 2–5 years, from 6.5 to 18.8% in children ages 6–11 years, and from 5 to 17.4% in children ages 12–19 years. Furthermore, 31.9% of children ages 2–19 are above the 85th percentile of the 2000 BMI for age growth charts [Centers for Disease Control and Prevention 2008].

As in the United States, the prevalence of obesity is growing worldwide. Pain [2007] notes that obesity is a problem in both developed and developing countries and that starvation, malnutrition, and obesity can all coexist in developing countries as some members of the population become more affluent. Also, the prevalence of obesity in Europe has quadrupled in the past 10 years, from 10 to 40%.

PREVENTION AND TREATMENT

Effective means of prevention and treatment of obesity and overweight remain elusive. Treatment, historically, has focused on reliance on a number of different kinds of diets, including very low-calorie diets, low-calorie diets, diet and exercise programs, pharmaceuticals, surgery, and behavior modification programs [Bray 1998]. Nutraceuticals and functional foods are receiving considerable interest for their role in improving health status, including weight problems and fat distribution. The term nutraceutical comes from a merger of the terms nutrition and pharmaceutical. These are foods or parts of food that have been shown to convey health benefits. If a functional food is used in the prevention or treatment of disease, it is considered a nutraceutical [Kalra 2003]. Nutraceuticals may be naturally occurring or added as supplements. The term functional foods, which originated in Japan in the 1980s, refers to foods that provide nutrients as well as other substances that promote health. They come from both plant and animal sources [Hasler 1998].

A number of reports have been published on the actions of various food components, including nutrients, and herbs on body weight and fat distribution, on fat free mass and metabolic rate. This paper will discuss some of the scientific findings regarding the use of nutraceuticals and functional foods for reduction of body fat and body weight and body fat distribution.

SELECTED NUTRACEUTICALS REPORTED TO INFLUENCE BODY FAT AND BODY WEIGHT

Chromium

Trivalent chromium is an essential nutrient, needed for carbohydrate and fat metabolism, that is found in numerous food sources, including beef, liver, eggs, chicken, brewer's yeast, oysters, wheat germ, green peppers, apples, bananas, and spinach. Despite the widespread availability of chromium in the food supply, Preuss and Anderson [1998] and Anderson [1997] suggest that the majority of people consume less than adequate chromium. Chromium supplementation improves insulin function in diabetics and has been investigated for reduction in body fat and for retention of lean body mass. Novel chromium complexes, chromium picolinate, and niacin bound chromium (NBC), have been investigated for their effects on body weight and body fat distribution. In 2003, Vincent reported the results of a review that found no effect of chromium picolinate on body mass or composition [Vincent 2003]. Other reports have found chromium picolinate to be associated with weight gain [Bagchi 2007]. NBC has been shown to be efficacious in improving body composition and weight status in overweight and obese subjects when combined with a reduced calorie diet and exercise regimen. Reduced food intake has also been observed. NBC was shown to improve body fat loss but retain lean body mass. Additionally, NBC has been found to have the greatest bioavailability compared with other novel chromium compounds. A meta-analysis of 10 studies of chromium picolinate on weight loss in overweight to obese subjects found that, with energy restriction of 3,300 kJ (788 kcal), the amount of weight lost was 1.5–2.5 kg (3.3–5.5 lb)/week, and, at more moderate energy restriction, 5,000 kJ (1194 kcal)/day, the amount of weight reduction was 0.5–0.6 kg (1.1–1.3 lbs/week [Pittler and Ernst 2004]). Regarding safety, chromium picolinate has been found to have mutagenic potential in laboratory animals. No negative safety effects have been associated with NBC [National Institutes of Health/Office of Dietary Supplements].

Conjugated Linoleic Acid

CLA is a family of isomers of linoleic acid, each having different functions. One has anticarcinogenic, antiobesity, and antidiabetic effects; another has anticancer effects. As an antiobesity agent, CLA is believed to act by decreasing food and energy intakes, decreasing lipogenesis, and increasing fat oxidation, lipolysis, and energy expenditure. They are found in dairy products associated with the fat fraction and in ruminant animals, such as beef and lamb. There are also synthetic mixtures [Kong 2007]. Gaullier et al. [2007] reported that CLA was found to be effective in reducing fat mass in the abdomen and legs of overweight and obese women. However, other studies have found no such effects. For example, Lamarche and Desroches [2004] reported that CLA-enriched butter had no effect on body fat distribution in men. Kong [2007] noted that CLA has consistently been shown to "decrease body fat accumulation and increase muscle mass" in several experimental animals, but the results from human trials have not been consistent. Wang and Jones [2004] concluded, after a review of the data of several studies, that the effect on weight and fat accumulation was determined by the particular isomer used and the amount given. In one study, the diet composition (e.g., high-fat versus low-fat diets) was important; greater reductions in weight gain were seen with high-fat diets compared with low-fat diets. However, the effect diminished over time, disappearing by the 12th week. In another study, there was no significant relationship between diet composition and CLA. The results of another study in which a dose of 0.5% decreased fat pad weights in lean rats but the same dose increased fat pad weight in obese rats, led the authors to suspect an animal genotype effect. In most of the studies reviewed, total body fat was reported to be diminished and body protein increased, with the exception of one study in which a decrease of body protein was reported. Despite the favorable effects reported in some studies, there were also some adverse events in some of the studies. The adverse events that were reported were hyperlinsulinemia and fatty livers and spleens. These raise safety concerns.

A meta-analysis of 20 human studies of randomized, longitudinal, double-blind studies with a control group was reported by Whigham et al. [2007]. The studies that were examined included normal weight, overweight, and obese subjects with information on type of isomer used and dosage. Given that animal studies identified the *t*10, *c*12 as the isomer with the largest effect on body fat, they did not include the treatment groups that only received *c*9, *t*11 CLA isomers. Only those studies of effect of CLA (*t*10, *c*12) on body fat with validated body composition techniques were included. On the basis of the meta-analysis, it was determined that there is a small but significant effect of CLA (*t*10, *c*12) on body fat.

Combined Effects of Chromium Picolinate and CLA

The combined effects of chromium picolinate and CLA have been examined in human and animal studies. For example, Diaz et al. [2008] studied the combined effects of chromium picolinate and CLA on changes in the body composition of overweight women who were on energy restricted diets and exercise regimens. They reported no effect of this combination on body composition. In a previous report, Bhattacharya et al. [2006] found that the combination of chromium and CLA did decrease body weight and fat mass in mice fed a high-fat diet.

Calcium

This essential mineral is needed for many functions, including bone formation and integrity, blood clotting, muscle contraction, transmission of nerve impulses, and the action of several enzymes.

There has been a lot of interest in calcium as a nutrient to modulate body weight. Numerous reports from animal and human studies regarding effects of calcium supplementation from dairy products and from calcium supplements have been inconsistent. A number of studies have reported an association of calcium supplementation with reduction in body fat. Others have shown no such effects [Dwyer 2005]. Trowman et al. [2006] conducted a systematic review of the literature and subsequent meta-analysis of trials involving calcium supplementation using calcium supplements or dairy products in persons ages 18 years or older. They concluded that there is no significant benefit of calcium supplementation on body weight. However, they found that some of the studies had flaws in the randomization process, which could affect the results. Gonzalez et al. [2006] reported the results of a study in which retrospective data were used to assess a calcium intake-weight change relationship over an 8- to 10-year period. Women who used supplements gained less weight than those who did not. There were no significant differences in men. They concluded that increased calcium intake may be beneficial to women. Zemel [2007] has reported that isocaloric substitution of three daily servings of dairy products in humans produced an increase in lipolysis. Additionally, it has been demonstrated that a high-calcium diet fed to rats and to mice stimulated a significant increase in fecal fat and energy excretion. A greater antiobesity effect of high-calcium diets was observed in obese mice fed a high-fat diet than in those fed a low-fat diet. In both human and animal studies, a shift in the distribution of body fat was observed. Dairy calcium sources have been shown to be more effective in controlling fat accumulation than supplemental sources of calcium.

Zemel suggested, also, that other bioactive components of dairy products may act independently or synergistically to reduce fat by reducing lipogenesis and increasing lipolysis and lipid oxidation. All of these studies were in animals or adult humans. Moreover, in support of his position of dietary calcium as an antiobesity agent, Zemel proposes a mechanism for this action. The roles of parathyroid hormone and 1,25-$(OH)_2$-D in responding to low-blood calcium are well known. He proposed that dietary calcium suppresses these hormones that favor energy storage to promote energy (fat) loss.

In one study involving children, DeJongh, Binkley, and Specker [2006] examined whether there was an association between change in body fat or in fat mass and total calcium intake in preschool children using data from a previously randomized trial of calcium supplementation. They found no consistent relationship between changes in total percentage body fat or fat mass and total or dietary calcium intake.

Phytoestrogens

Soy protein is a source of isoflavone, a phytoestrogen. Aubertin-Leheudre et al. [2007], in recognition of the many beneficial health effects of the phytoestrogens, conducted a study to investigate whether isoflavone supplementation could affect fat-free mass. They found that supplementation of postmenopausal, obese, sarcopenic women with 70 mg/day isoflavone for six months did significantly increase fat-free mass. This effect is significant because of the relationship of metabolic rate to fat-free mass. Soy protein has been reported to reduce body weight and fat mass in both animals and humans [Valasquez and Bhathena 2007].

Medium-Chain Triglycerides

Medium-chain fatty acids have been used therapeutically since the 1950s for treatment of malabsorption syndromes. Their absorption, solubility, and metabolism differ from long-chain fatty acids. Medium-chain fatty acids, lauric acid, and capric acid are components of coconut oil, palm oil, and milk. Lauric acid is a component of breast milk. The use of coconut oil and palm oil is largely discouraged because of their degree of saturation. However, there are advantages to both. A body of research, recently published, supports the value of coconut oil as an antimicrobial agent [Enig 2007]. Palm oil is a source of antioxidants with some protection against certain cancers [Sundram et al. 2003]. Several studies in the early 1980s found that experimental animals fed medium-chain triglycerides (MCTs) had increased thermogenesis and gained less body weight than those fed long-chain triglycerides (LCTs) [Seaton et al. 1986]. Following these reports, a study was conducted to determine whether metabolic rate could be increased more by a single meal of MCTs than by a single meal of LCTs in adult men. The results indicated that oxygen consumption after the MCT meal increased more than after the LCT meal. Studies comparing effects on body weight and fat distribution in experimental animals have produced different results [Seaton et al. 1986]. St-Onge [2005] reported the results of a study in humans in which energy expenditure from MCTs increased more than it did with LCTs. However, there was a slight variation in the extent to which it occurred. Body weights did not change in women, but there were significant reductions in total adipose tissue, subcutaneous adipose tissue, and upper body adipose tissue in men.

Both animal studies and human studies have shown that MCTs stimulate more diet-induced thermogenesis than LCTs, thus leading to less body fat. This may occur because medium-chain fatty acids are not stored, are oxidized for energy production, raise the body temperature, and use more energy [Enig 2007].

CONCLUSIONS

Overweight and obesity continue to be serious health problems despite the many treatment approaches. As we consider the growing body of literature concerning the effectiveness of nutraceutical and functional foods in meeting this challenge of stemming the prevalence of overweight and obesity, we see that there is not widespread agreement among scientists about the effectiveness of these substances. Some of these studies have produced conflicting results, which may be related to the lack of uniformity in study design, composition of nutraceuticals and mode of delivery, duration of the trial, study variables, and experimental models.

Moreover, there are some adverse events that have been reported with many of the nutraceuticals at high levels. It is time to look very carefully for prevention methods, some of which should be designed to cure the environment and to return to eating practices that were common before the obesity epidemic. There appears to be, based on scientific evidence, a need for more consumption of natural products, such as coconut oil, and increased consumption of milk and other dairy products. For consumers who are lactose intolerant, there is lactose-reduced milk. Also, yogurt has calcium as well as probiotics that may convey additional health benefits [National Institutes of Health/National Center for Complementary and Alternative Medicine]. Hill and Peters [1998] discussed environmental conditions that have fueled the explosion of the obesity epidemic. These include easy availability of low-cost foods that are offered in large portions, excessive consumption of high energy foods, often as fat, and low levels of physical activity. They suggested that we "must cure the environment," by consumer education that would reduce portion sizes, promote fun activities to encourage physical activity in children, increase availability of foods that are more nutrient dense and less energy dense, and provide incentives (e.g., lower-cost health insurance), etc.

The Dietary Guidelines for Americans [U.S. Department of Health and Human Services/U.S. Department of Agriculture 2005] and the diet planning principles [Whitney and Rolfes 2008] give us blueprints for action. It is time for health professionals, families, and communities to improve use of these blueprints.

REFERENCES

Anderson, R. A. 1997. Chromium as an essential nutrient for humans. *Regul. Toxicol. Pharmacol.* 26:S35–S41.

Aubertin-Leheudre, M, C. Lord, A. Khalil, and I. J. Dionne. 2007. Six-month supplementation of isoflavone supplement increases fat-free mass in obese-sarcopenic postmenopausal women: a randomized double-blind controlled trial. *Eur. J. Clin. Nutr.* 61:1442–1444

Bagchi, M., H. G. Preuss, S. Zafra-Stone, and D. Bagchi. 2007. Chromium (III) in promoting weight loss and lean body mass. In *Obesity, Epidemiology, Pathophysiology, and Prevention.* Edited by Bagchi, D. and H. G. Preuss. Boca Raton, FL: Taylor & Francis, pp. 339–347.

Bhattacharya, A., M. M. Rahman, R. McCarter, M. O'Shea, and G. Fernandes. 2006. Conjugated linoleic acid and chromium lower body weight and visceral fat mass in high-fat-diet-fed mice. Lipids 41:437–444.

Bray, G. A. 1998. Historical framework for the development of ideas about obesity. In *Handbook of Obesity.* Edited by Bray, G. A., C. Bouchard, and W. P. T. James. New York, NY: Marcel Dekker, pp. 1–29.

Centers for Disease Control and Prevention. 2005. Overweight and obesity. http://www.cdc.gov/nccdphp/dnpa/obesity. Accessed July 8, 2009.

DeJongh, E. D., T. L. Binkley, and B. L. Specker. 2006. Fat mass gain is lower in calcium-supplemented than in unsupplemented preschool children with low dietary calcium intakes. *Am. J. Clin. Nutr.* 84:1123–1127.

Diaz, M. L., B. A. Watkins, L. Yong, R. A. Anderson, and W. W. Campbell. 2008. Chromium picolinate and conjugated linoleic acid do not synergistically influence diet- and exercise-induced changes in body composition and health indexes in overweight women. *J. Nutr. Biochem.* 19:61–68.

Dwyer, J. T., D. B. Allison, and P. M. Coates. 2005. Dietary supplements in weight reduction. *J. Am. Diet. Assoc.* 105 (Suppl. 1):S80–S86.

Enig, M. G. 2007. Role of medium-chain triglycerides in weight management. In *Obesity, Epidemiology, Pathophysiology, and Prevention.* Edited by Bagchi, D. and H. G. Preuss. Boca Raton, FL: Taylor & Francis, pp. 451–461.

Gaullier, J. M, J. Halse, H. O. Høivik, K. Høye, C. Syvertsen, M. Nurminiemi, C. Hassfeld, A. Einerhand, M. O'Shea, and O. Gudmundsen. 2007. Six months supplementation with conjugated linoleic acid induces regional specific fat-mass decreases in overweight and obese. *Br. J. Nutr.* 97:550–560.

Gonzalez, A. J., E. White, A. Kristal, and L. Littman. 2006. Calcium intake and 10 year weight change in middle-aged adults. *J. Am. Dietet. Assoc.* 106:1066–1073.

Hasler, C. M. 1998. Functional foods: Their role disease prevention and health promotion. *Food Tech.* 52:57–62.

Hill, J. O. and J. C. Peters. 1998. Environmental contributions to the obesity epidemic. *Science* 280:1371–1374.

Kalra, E. K. 2003. Nutraceutical definition and introduction. *AAPS Pharm. Sci.* 1–2.

Kong, Z. L. 2007. Conjugated linoleic acid and weight control: from the biomedical immune viewpoint. In *Obesity, Epidemiology, Pathophysiology, and Prevention.* Edited by Bagchi, D. and H. G. Preuss. Boca Raton, FL: Taylor & Francis.

Lamarche, B. and S. Desroches. 2004. Metabolic syndrome and effects of conjugated linoleic acid in obesity and lipoprotein disorders: The Quebec experience. *Am. J. Clin. Nutr.* 79:1149S–1152S.

National Institutes of Health/National Center for Complementary and Alternative Medicine. Bethesda, MD. http://nccam.nih.gov/health/probiotics/index.htm. Accessed July 8, 2009.

Office of the Surgeon General of the United States. http://www.surgeongeneral.gov/topics/obesity/calltoaction/fact_consequences.htm. Accessed July 8, 2009.

Pain, G. C. 2007. Epidemiology of obesity. In *Obesity, Epidemiology, Pathophysiology, and Prevention.* Edited by Bagchi, D. and H. G. Preuss. Boca Raton, FL: Taylor & Francis, pp. 21–29.

Pittler, M. H. and E. Ernst. 2004. Dietary supplements for body weight reduction. *Am. J. Clin. Nutr.* 79:529–536.

Preuss, H. G. and R. A. Anderson. 1998. Chromium update: Examining recent literature 1997–1998. *Curr. Opin. Clin. Nutr. Metab. Care* 1:509–512.

Seaton, T. B., S. L. Welle, M. K. Warenko, and R. G. Campbell. 1986. Thermic effect of medium chain and long chain triglycerides in man. *Am. J. Clin. Nutr.* 44:630–634.

St-Onge, M. P. 2005. Dietary fats, teas, dairy, and nuts: Potential functional foods for weight control? *Am. J. Clin. Nutr.* 81:7–15.

Sundram, K., R. Sambanthamurthi, and Y. A. Tan. 2003. Palm fruit chemistry and nutrition. *Asia Pac. J. Clin. Nutr.* 12:355–362.

Trowman, R., J. C. Dumville, S. Hahn, and D. J. Torgerson. 2006. A systematic review of the effects of calcium supplementation on body weight. *Br. J. Nutr.* 95:1033–1038.

U.S. Department of Health and Human Services/U.S. Department of Agriculture. 2005. Dietary guidelines for Americans. http://www.nutrition.gov.

Velasquez, M. T. and S. J. Bhathena. 2007. The role of dietary soy protein in obesity. *Int. J. Med. Sci.* 4:72–82.

Vincent, J. B. 2003. The potential value and toxicity of chromium picolinate as a nutritional supplement, weight loss agent, and muscle developing agent. *Sports Med.* 33:213–230.

Wang, Y. and P. J. H. Jones. 2004. Dietary conjugated linoleic acid and body composition. *Am. J. Clin. Nutr.* 79 (Suppl):1153S–1158S.

Whigham, L. D., A. C. Watras, and D. A. Schoeller. 2007. Efficacy of conjugated linoleic acid for reducing fat mass: A meta-analysis in humans. *Am. J. Clin. Nutr.* 85:1203–1211.

Whitney, E. and S. R. Rolfes. 2008. *Understanding Nutrition*. Belmont, CA: Wadsworth.

Zemel, M. B. 2007. Dairy foods, calcium, and weight management. In *Obesity, Epidemiology, Pathophysiology, and Prevention*. Edited by Bagchi, D. and H. G. Preuss. Boca Raton, FL: Taylor & Francis, pp. 477–493.

CHAPTER 16

Nutraceuticals for Bone and Joint Diseases

Meghan Bodenberg and Holly Byrnes

CONTENTS

OSTEOPOROSIS

Bone and joint diseases can lead to disability, immobility, pain, and a reduction in activities of daily living for many patients. Musculoskeletal conditions are estimated to cost over $254 billion annually in the United States and rank number 1 in visits to physicians' offices. Two of the most prevalent bone and joint conditions include osteoporosis and osteoarthritis. As the baby-boomer generation ages, healthcare professionals may see an increasing number of patients seeking treatment for and relief of symptoms from osteoporosis and osteoarthritis with nutraceutical agents. To make informed decisions, it is prudent for healthcare professionals to be knowledgeable regarding the pharmacology, adverse effects, drug interactions, and clinical evidence regarding the most commonly used nutraceuticals for these two conditions.

Osteoporosis, as defined by the National Institutes of Health Consensus Development Program [2000] is "a skeletal disorder characterized by compromised bone strength predisposing to an increased risk for fractures. Bone strength reflects the integration of two main features: bone density and bone quality. ... Bone quality refers to architecture, turnover, damage, accumulation (e.g., microfractures), and mineralization." The hip, spine, and wrist are the most common bones affected by osteoporosis. An estimated 44 million Americans are affected by osteoporosis, and another 34 million Americans have low bone mass and are at an increased risk for developing osteoporosis [Qaseem et al. 2008].

The 2008 American College of Physicians Clinical Practice Guidelines, along with previous guidelines on osteoporosis [United States Department of Health and Human Services 2004], recommend conventional therapies for treatment of osteoporosis, such as bisphosphonates, raloxifene, and hormone replacement therapy treatment. The drawback to these conventional therapies is that they have documented negative side effects. For example, large randomized, placebo-controlled trials revealed an increase in coronary artery disease, stroke, breast cancer, dementia, venous thromboembolic events, and gallbladder disease in patients who used hormone replacement therapy [Hulley et al. 2002; Rossouw et al. 2002]. This evidence has dissuaded many physicians and patients from using hormone replacement therapy for the prevention or treatment of osteoporosis in women [Lawson et al. 2003]. Consumers may look to nutraceutical products as a substitute, perceiving that, because they are natural, they are a safer alternative.

Table 16.1 identifies natural health products that have clinical evidence (randomized clinical trials) to support their claim of use in the management of osteoporosis [Whelan, Jurgens, and Bowles 2006]. Calcium and vitamin D are dietary supplements that are considered standards of therapy for the prevention of osteoporosis and in conjunction with other therapies for the treatment of osteoporosis [United States Department of Health and Human Services 2004; Qaseem et al. 2008].

Table 16.1 Nutraceuticals with Clinical Evidence to Support Claims for Benefits in Osteoporosis

Black tea
Calcium
Copper
Dehydroepiandrosterone
Evening primrose oil
Fish oils
Fluoride
Green tea
Magnesium
Manganese
Oolong tea
Phytoestrogens
Strontium
Vitamin D
Vitamin K

Source: Whelan, A. M., T. M. Jurgens, and S. K. Bowles. 2006. Natural health products in the prevention and treatment of osteoporosis: Systematic review of randomized controlled trials. *Ann. Pharmacother.* 40:836–849.

Reviews of fluoride have shown that, although it may increase bone mass density (BMD) in the spine, it does not reduce the risk of fractures [Haguenauer et al. 2003]. Strontium renalate, the salt form of strontium that is currently being studied for safety and efficacy in osteoporosis, is likely to be a prescription medication instead of a nutraceutical [Meunier et al. 2004]. Evidence that drinking green tea, black tea, and oolong tea for a period of 10 years will increase bone mineral density has mainly developed from population research and will need to be further substantiated by randomized clinical trials [Wu et al. 2002]. Randomized clinical trials for copper, evening primrose oil, fish oils, magnesium, and manganese only studied these agents in combination with other nutraceuticals, making it difficult to determine their individual benefit for treatment of osteoporosis [Whelan, Jurgens, and Bowles 2006] (also see Table 16.2).

Table 16.2 Nutraceticals Claiming Benefit for Osteoporosis but Lacking Clinical Evidence

Alfalfa (*Medicago sativa*)
Avocado
Black cohosh (*Cimicifuga racemosa*)
Black pepper
Boron
Chondroitin
Dandelion
Deer velvet
Dong Quai (*Angelica sinesis*)
Folic acid
Gelatin
Germanium
Ginseng
Horsetail (*Equisetum* spp.)
Iron
Lactase
Licorice
Marshmallow (*Althaea officinalis*)
Melantonin
Methylsulfonylmethane
Oat straw
Papaya
Pigweed
Phosphorus
Silicon
Vitamin B12
Xylitol
Yellow dock (*Rumex crispus*)
Zinc

Source: Whelan, A. M., T. M. Jurgens, and S. K. Bowles. 2006. Natural health products in the prevention and treatment of osteoporosis: Systematic review of randomized controlled trials. *Ann. Pharmacother.* 40:836–849.

Table 16.3 Calcium Adequate Intake

Age	Sex	Calcium
Birth to 6 months	Both	210 mg
6 months to 1 year	Both	270 mg
1–3 years	Both	500 mg
4–8 years	Both	800 mg
9–18 years	Both	1,300 mg
19–50 years	Both	1,000 mg
Pregnant or nursing	Female	1,000 mg
Over 51 years	Both	1,200 mg

Source: Institute of Medicine, Food and Nutrition Board. 1997. Dietary reference intakes: Calcium, phosphorous, magnesium, vitamin D, and fluoride. Washington, DC: National Academy Press.

CALCIUM

Calcium has an important role in formation and maintenance of healthy bones. Table 16.3 includes the recommended daily requirement, or adequate intakes, for calcium based on age and sex. If possible, calcium requirements should be fulfilled by food sources. Table 16.4 contains a list of foods high in calcium. Certain foods, such as fiber, whole-wheat food, fruits, and antacids should not be taken at the same time as calcium-rich foods to prevent a decrease in calcium's bioavailability. Caffeine should also be limited because it increases urinary calcium excretion.

Product Selection

Calcium is available in many salt forms and combination products (see Table 16.5). Calcium carbonate and calcium citrate are the most commonly recommended for

Table 16.4. Calcium Dietary Sources

Food or Beverage Source	Serving Size	Amount of Calcium (mg)
Milk and fortified soymilk	8 oz	291–302
Fortified orange juice	1 cup	350
Plain yogurt	8 oz	345–415
Ice cream	1 cup	200
Cheese	1 oz	205
Cottage cheese	1 cup	211
Sardines with bones	3 oz	372
Salmon with bones	3 oz	167
Broccoli	1 cup	100–136
Spinach	0.5 cup	113
Soybeans	1 cup	131
Collards	1 cup	357
Turnips	1 cup	262
Tofu	4 oz	106

Source: O'Connell, M. B. 1999. Prevention and treatment of osteoporosis in the elderly. *Pharmacotherapy* 19:7S–20S.

Table 16.5 Available Calcium Products

Salt	%	Examples of Products
Calcium carbonate	40	Tums, Caltrate 600, Oscal, Calci-Chew, Chooz 500 mg
Calcium citrate	24	Citracal, Cal-Citrate-225
Tribasic calcium phosphate	39	Posture®
Calcium lactate	18	
Calcium gluconate	9	Cal-G
Combo products with vitamin D Os-Cal 500+D, Caltrate 600+D, Viactiv (some products also contain vitamin K)		

Source: Lexicomp-online. http://www.crlonline.com.

supplementation. Calcium carbonate can be given two or three times daily with food, whereas calcium citrate can be taken with or without food. Calcium citrate requires more tablets, which increases its cost compared with calcium carbonate. Calcium carbonate depends on acid for disintegration and dissolution, whereas calcium citrate is an acid-independent product. For patients on proton pump inhibitors and histamine$_2$ blockers, this is an important consideration. Several products are combination products containing vitamin D, which aids in the absorption of calcium. Other miscellaneous products also contain ingredients such as vitamin K, boron, magnesium, phytoestrogens, and aspirin. These agents are expensive and can have more side effects and drug interactions than agents with calcium alone.

Adverse Effects

Evidence from randomized trials showed no clinically important adverse events associated with the use of calcium [Qaseem et al. 2008]. The most common side effects reported include constipation, bloating, and gas [Lexicomp-online 2009]. Kidney stones rarely occur and are more common in those with a history of calcium stones [O'Connell 1999].

Food and Drug Interactions

Calcium can decrease absorption of iron, floroquinolones, and tetracyclines. Loop diuretics, such as furosemide, increase calcium excretion, and thiazide diuretics, such as hydrochlorothiazide, decrease calcium excretion. Food may increase calcium absorption. Bran, foods high in oxalates, or whole-grain cereals may decrease calcium absorption. Fiber and antacids can also decrease the absorption of calcium.

Calcium: Clinical Evidence

Evidence for the use of calcium alone is difficult to determine, because most studies include the combination of calcium and vitamin D in the treatment regimen.

Results from several randomized trials [Grant et al. 2005; Prince et al. 2006] and a meta-analysis [Shea et al. 2002] containing 15 smaller studies did not reveal a statistically significant difference between calcium and placebo in prevention of vertebral, nonvertebral, and hip fractures in postmenopausal women. However, patient noncompliance to therapy and unknown vitamin D status in some of these studies may have contributed to the lack of significance in the results.

Other randomized clinical trials have shown a reduction in risk of hip fractures and other nonvertebral fractures [Chapuy et al. 1992]. A recent meta-analysis [Tang et al. 2007] has concluded that the relative risk (RR) for fracture with calcium alone was 0.90 (confidence interval [CI], 0.80–1.00). However, this meta-analysis did not include results from the RECORD trial [Grant et al. 2005], a large trial of people aged 70 years or older, which concluded there was no significant difference between calcium and placebo regarding incidence of fractures, death, number of falls, and quality of life.

Recommendation

Although the evidence on use of calcium alone or in combination with vitamin D is mixed, it is currently recommended in the guidelines to add calcium and vitamin D to other osteoporosis treatment regimens [Qaseem et al. 2008]. Also, bisphosphonates, one of the standard treatments for osteoporosis, are contraindicated in patients with low calcium levels.

VITAMIN D

Vitamin D is a fat-soluble vitamin that is naturally present in few foods but is often added to fortified food products, such as milk and cereal [Calvo, Whitting, and Barton 2004]. It is also produced endogenously through exposure of the skin to ultraviolet rays from the sun. Both endogenous and exogenous vitamin D are biologically inert and must undergo two hydroxylations to be converted to active form. In the liver, vitamin D is converted to 25-hydroxyvitamin D, also known as calcidiol. In the kidney, it undergoes hydroxylation to 1,25-dihydroxyvitamin D, also known as calcitriol [Tang et al. 2007].

Vitamin D is needed for calcium absorption in the gut and maintenance of adequate serum calcium and phosphate concentrations to enable normal bone mineralization. It is also necessary for bone growth and remodeling by osteoblasts and osteoclasts. Vitamin D deficiency can cause rickets in children and osteomalacia in adults [Tang et al. 2007].

Recommended Daily Intake

In 2008, the American Academy of Pediatrics (AAP) changed their daily recommended intakes of vitamin D for pediatric and adolescent populations to 400 IU. This exceeded the amount currently recommended by the Food and Nutrition Board

Table 16.6 Dosing Guidelines for Vitamin D

	National Osteoporosis Foundation	Institute of Medicine	AAP
Birth to 50 years	N/A	200 IU	400 IU
51–70 years	800–1000 IU	400 IU	
>70 years	800–1000 IU	600–800 IU	

at the Institute of Medicine of the National Academies for this age group. The AAP's recommendation was based on evidence of recent clinical trials, along with a history of safe use of 400 IU in this age group. This recommendation includes infants who are exclusively and partially breast fed from shortly after birth until the time they are weaned and able to consume over 1000 ml/day fortified vitamin D formula or whole milk [Wagner and Greer 2008]. All formulas sold in the United States contain at least 400 IU/L, ensuring that bottle-fed infants do receive adequate amounts of vitamin D [Gartner and Greer 2003] (see Table 16.6).

Dietary Sources

The main sources of vitamin D in the American diet are fortified foods, such as ready-to-eat cereals, and milk. Although many products, such as yogurt, margarine, juice, and macaroni and cheese, are eligible to be fortified, very few products actually are fortified and available for sale in the U.S. marketplace [Calvo, Whitting, and Barton 2004]. Fish (such as salmon, tuna, and mackerel) and fish oils are the best sources of natural vitamin D [Office of Dietary Supplements 2008]. Other sources include beef liver, cheese, and egg yolks. See Table 16.7 for a list of food sources that contain vitamin D.

Table 16.7 Vitamin D Dietary Sources

Food/Beverage Source	Serving Size	International Units
Cod liver oil	1 tablespoon	1360
Salmon, cooked	3.5 oz	360
Mackerel	3.5 oz	345
Tuna fish, canned in oil	3 oz	200
Sardines, canned in oil, drained	1.75 oz	250
Milk	8 oz	98
Margarine, fortified	1 tablespoon	60
Fortified, ready-to-eat cereal	0.75–1 cup	40
Egg	1 whole	20
Liver, beef, cooked	3.5 oz	15
Cheese, Swiss	1 oz	12

Source: Office of Dietary Supplements. 2008. Dietary supplement fact sheet: Vitamin D. National Institutes of Health, Bethesda, MD. http://ods.od.nih.gov/factsheets/vitamind.asp. Accessed December 28, 2008.

Available Dosage Forms

Vitamin D is available in two forms: D_2 (ergocalciferol) and D_3 (cholecalciferol). Although metabolized differently, these two forms have been considered equivalent based on their ability to cure rickets. However, there is evidence to suggest that vitamin D_3 might be more effective at raising and maintaining 25(OH) D concentrations than vitamin D_2 [Armas, Hollis, and Heaney 2004]. Although many people can obtain adequate amounts of vitamin D through fortified foods and sunlight, some might require dietary supplements to meet their daily requirement of vitamin D.

Cholecalciferol (vitamin D_3) is the form most commonly used in over-the-counter products. It is available in tablets or capsules from 200 to 1,000 IU, in combination with calcium carbonate in doses of 100–400 IU, or in multivitamins, at a dose of 400 IU. Ergocalciferol is available over the counter as liquid drops.

Adverse Effects

Evidence from randomized trials showed no clinically important serious adverse events associated with the use of vitamin D.

Drug Interactions

Vitamin D metabolism can be impaired by corticosteroid medications, such as prednisone and prednisolone, often used to reduce inflammation. When used long term, these effects can further contribute to the loss of bone and the development of osteoporosis. The weight loss drug orlistat and the cholesterol-lowering agent cholestryamine can reduce the absorption of vitamin D and other fat-soluble vitamins. The antiepileptic medications phenytoin and phenobarbital are hepatic enzyme inducers and can increase the hepatic metabolism of vitamin D to inactive compounds and reduce calcium absorption [Gough et al. 1986].

Clinical Evidence

MacLean et al. [2008] included five systematic reviews in their analysis of the efficacy of vitamin D. Four of these meta-analyses did not show any statistically significant reduction in fracture risk with the use of vitamin D_2 or D_3; however, two of the four meta-analyses did find a relative reduction in vitamin D analogs, such as calcidiol and calcitriol. In addition, one meta-analysis showed a statistically significant reduction in nonvertebral and hip fractures for vitamin D_2 or D_3 [MacLean et al. 2008].

Papadimitropoulos et al. [2002] concluded, in a meta-analysis of the efficacy of vitamin D in postmenopausal women, that vitamin D reduced the incidence of vertebral fractures (RR, 0.63; 95% CI, 0.45–0.88) and showed a trend toward reduced incidence of nonvertebral fractures (RR, 0.77; 95% CI, 0.57–1.04). Most patients evaluated in the clinical trials for vertebral fractures were taking hydroxylated vitamin D rather than standard vitamin D_2 or D_3.

A meta-analysis by Bischoff-Ferrari et al. [2005] showed that a vitamin D dose of 700–800 IU/day reduced the RR of hip fracture by 26% (three randomized clinical trials with 5,572 subjects; pooled RR, 0.74; 95% CI, 0.61–0.88) and any nonvertebral fracture by 23% (five randomized clinical trials with 6,098 subjects; pooled RR, 0.77; 95% CI, 0.68–0.87). No significant benefit was seen in randomized clinical trials analyzed in this meta-analysis with lower doses of vitamin D (400 IU/day or less).

Calcium and Vitamin D Combination Therapy: Clinical Evidence

There are several clinical studies and meta-analyses that evaluated the evidence for the combination of calcium and vitamin D rather than the separate effects of these dietary supplements. These studies also revealed mixed results. A large meta-analysis of over 36,000 postmenopausal women taking vitamin D at 400 U and calcium at 1,000 mg showed a small but significant improvement in hip bone density but did not show a significant reduction in hip fracture [Jackson et al. 2006]. An 18-month study of healthy ambulatory women receiving 1,200 mg of elemental calcium and 800 IU of vitamin D showed a reduction in the number of hip fractures and the total number of nonvertebral fractures [Chapuy et al. 1992]. However, the RECORD trial showed no benefit in the combination of 1,000 mg of calcium and 800 IU of vitamin D in previously mobile people over 70 years of age [Grant et al. 2005].

DEHYDROEPIANDROSTERONE

Dehydroepiandrosterone (DHEA) is produced in the adrenal glands and in the liver. It is also secreted by the testes. DHEA levels peak in adults during their early 20s and then begin to decline after age 25. This decline has been associated with several diseases, including osteoporosis. Therefore, administration of DHEA has been proposed to prevent or treat osteoporosis [Whelan, Jurgens, and Bowles 2006; Jellin 2009].

"Dehydroepiandrosterone sulfate (DHEA-S) is the storage form of DHEA. Peripheral tissues and target organs convert DHEA-S to DHEA, which can then be metabolized to androstenedione, the major human precursor to androgens and estrogens. DHEA-S concentrations are 100 to 500 times higher than testosterone and 1000 to 10,000 times higher than estradiol. DHEA is a precursor to other hormones, but it does not have any direct estrogenic or androgenic activity. DHEA can increase estradiol, estrone, osteocalcin, growth hormone, and insulin-like growth factor 1 (IGF-1)" [Jellin 2009].

Dietary Sources

DHEA can be synthesized from natural sources, such as soy and wild yam, by the conversion of constituents such as diosgenin into DHEA [Whelan, Jurgens, and Bowles 2006].

Dosing Recommendations

DHEA orally at 50–100 mg/day has been studied to improve BMD in older women and men with osteoporosis or osteopenia.

Drug Interactions

DHEA can interfere with the antiestrogen effects of anastrozole, exemestane, letrozole, and other aromase inhibitors [Jellin 2009]. Corticosteroid drugs, such as dexamethasone and prednisone, can suppress endogenous DHEA productions.

DHEA may increase levels of drugs metabolized by CYP3A4 as a result of enzyme inhibition; however, the clinical significance of these potential interactions is unknown. Drugs that are substrates of CYP3A4 include alprazolam, amitriptyline, amiodarone, buspirone, citalopram, felodipine, fexofenadine, itraconazole, ketoconazole, lansoprazole, losartan, lovastatin, midazolam, ondansetron, prednisone, sertraline, sibutramine, sildenafil, simvastatin, and verapamil.

DHEA is a potent estrogen agonist and can overcome the estrogen receptor antagonist action of fulvestrant and tamoxifen in estrogen-receptor-positive cancer cells. Patients who have estrogen-receptor-positive cancers, such as breast cancer, uterine cancer, and ovarian cancer, should not take DHEA.

Insulin might potentially decrease the effectiveness of DHEA supplements.

Adverse Effects

Orally, the side effects of DHEA are generally mild at low doses such as 50 mg/day. In one study, two women reported oiliness of facial skin, acne, and hair growth [Morales et al. 1998].

Safety

DHEA is possibly safe in smaller doses and for short term. Studies did not extend beyond 24 months. There is some concern that long-term use or higher doses might increase risk of breast cancer, prostate cancer, or other hormone-sensitive cancers [Jellin 2009]. Data are only available in healthy women over the age of 60.

Clinical Evidence

There is some evidence to support DHEA's ability to improve BMD in older women and men with osteoporosis or osteopenia, but additional clinical trials are needed to determine whether this increase in BMD has an impact on fracture results (see Table 16.8).

Table 16.8 Summary of DHEA Trials

Reference	Design	Subjects	Therapy	Results
Nair et al. 2006	2-year, randomized, double-blind placebo-controlled	87 elderly men with low levels of DHEA sulfate and bioavailable testosterone; 57 women with levels of DHEA sulfate	Men, DHEA at 75 mg, testosterone, or placebo; women, DHEA at 50 mg or placebo	Men, slight but significant ↑ in BMD at femoral neck; women, slight but significant ↑ in BMD at ultradistal femoral; no significant ↑ in BMD at other sites in other groups
Sun et al. 2002	6-month, placebo-controlled	86 men	DHEA at 100 mg, calcium salt at 150 mg three times a per, and vitamin D_3 at 500 once daily or placebo, calcium salt, and vitamin D_3	↑ BMD of L2, L3, L4, femoral neck; total BMD did not ↑ significantly
Baulieu et al. 2000	12-month, double-blind, placebo-controlled	133 healthy women (60–79 years old)	DHEA at 50 mg daily	↑ BMD of femoral neck and Ward's triangle in women <70 years old; ↑ BMD of upper radius and total radius in women >70 years old
Morales et al. 1998	6-month, double-blind, placebo-controlled, crossover	8 health postmenopausal women (50–65 years old)	DHEA at 100 mg daily (7 women used daily hormone replacement therapy)	No changes in BMD of the hip or spine

PHYTOESTROGENS

Also known as daidzein, genistein, isoflavones, kudzu, red clover, and soy, phytoestrogens are plant-derived compounds that can have weak estrogenic effects. There are three major classes of phytoestrogens: isoflavones, lignans, and coumestans. Of these, isoflavones has been most extensively studied [Yao, Dobs, and Brown 2006]. Isoflavones come from metabolized soy products (such as tofu) and soybeans.

Pharmacology

Isoflavones contain plant-derived estrogenic compounds; however, the estrogenic potency has been estimated to be only 1/1000 to 1/100,000 that of estradiol. They have been reported to inhibit bone resorption in postmenopausal women, increase insulin growth factor-1, and promote calcium absorption.

Dosage

Isoflavones, such as Promensil and Novogen, at 40 mg/day have been used to treat osteoporosis.

Adverse Effects

Orally, phytoestrogens are generally well tolerated. In most studies, adverse events did not differ significantly from placebo. In one study, 11 subjects reported mild digestive upsets, and one subject reported having loose stools [Lydeking-Olsen et al. 2004].

Drug Interactions

There is preliminary evidence that phytoestrogens (red clover) might inhibit cytochrome P450 1A2, P450 2C19, P450 2C9, and P450 3A4. However, this interaction has not been reported in humans at this time. Theoretically, concomitant use of large doses of red clover with anticoagulants can increase patient's risk of bleeding attributable to the coumarin content of red clover. There is a theoretical potential for large amounts of red clover to interfere with contraceptive drugs and hormone replacement therapy attributable to competition for estrogen receptors. Patients should avoid using red clover if they are concomitantly taking tamoxifen because there is a concern that red clover has potential estrogenic effects. There is preliminary evidence that genistein, a constituent of red clover, "might antagonize the antitumor effects of tamoxifen" [Jellin 2009].

Safety

Phytoestrogens are likely safe when used orally and at appropriate dosage (amounts commonly used in foods).

Clinical Evidence

Based on the summary of clinical trials evaluating phytoestrogens, the results look promising; however, additional and larger studies need to be conducted (see Table 16.9).

VITAMIN K

Two forms of vitamin K have been studied for the treatment of osteoporosis: vitamin K_1 (phytonadione or phylloquinone) and vitamin K_2 (menaquinone or menatetrenone). These are both naturally occurring forms of vitamin K. Food sources high in vitamin K_1 include green leafy vegetables, such as spinach, cabbage, and mustard and turnip greens. Food sources containing vitamin K_2 are found in meat and fermented food, and vitamin K_2 is synthesized by bacteria in the colon.

Pharmacology

Vitamin K has been shown to promote bone mineralization via osteocalcin [Yao, Dobs, and Brown 2006].

Table 16.9 Clinical Trials Evaluating Phytoestrogens

Reference	Design	Subjects	Therapy	Results
Alekel et al. 2000	6-month, double-blind, placebo-controlled	69 healthy perimenopausal women (49.4–50.9 years old)	SPI+ (isoflavone-rich soy protein delivering aglycone components, 80.4 mg/day); SPI– (isoflavone-poor soy protein delivering aglycone components, 4.4 mg/day); control (whey protein)	No reduction in lumbar spine BMD in SPI+ and SPI– groups; significant losses ($p = 0.0037$) in control group; using regression analysis, SPI+ has significant positive effect (5.6% change, $p = 0.023$) in BMD, whereas SPI– and control had no effect
Atkinson, Compston, and Day 2004	12-month, double-blind, placebo-controlled	177 premenopausal, perimenopausal, and post menopausal women (mean, 55 years old)	Promensil tablets (contain biochanin at 26 mg, formononetin at 16 mg, genistein at 1 mg, daidzein at 0.5 mg), or placebo	Statistically significant ↓ in lumbar spine BMD in the placebo group compared with the isoflavone group (↓ 1.86 ± 0.29% [placebo] versus ↓ 1.08 ± 0.27% [treatment]; $p = 0.05$)
Chen et al. 2003	12-month, double-blind, placebo-controlled	203 healthy, postmenopausal Asian women (48–62 years old)	3 groups: soy at 40 mg, soy at 80 mg, and placebo; all groups: calcium at 500 mg and vitamin D at 125 U/day	No statistically significant differences between groups for total body, lumbar spine, and hip BMD
Kreijkamp-Kaspers et al. 2004	12-month, double-blind, placebo-controlled	175 healthy, postmenopausal women (60–75 years old)	Soy protein powder (containing genistein at 52 mg, daidzein at 41 mg, glycitein at 6 mg, or placebo)	No difference between groups for lumbar spine BMD; 1.31% ↑ in BMD in soy group compared with placebo ($p = 0.02$) in intertrochanter region
Lydeking-Olsen et al. 2004	24-month, double-blind, placebo-controlled	89 post menopausal w/ osteoporosis or ≥3 risk factors	Groups: isoflavone-rich soy milk at 500 ml/day; transdermal progesterone at 25.7 mg; isoflavone-rich soy milk and transdermal progesterone at 25.7 mg; placebo	Mean percentage change in lumbar spine BMD did not decline in soy or progesterone groups, whereas significant losses occurred in the combined and placebo groups; there were no significant differences in hip BMD between groups

"Vitamin K may also decrease bone resorption by decreasing prostaglandin E2 synthesis in osteoclasts, and by effects on calcium balance, and IL-6 production in bone" [Papadimitropoulos et al. 2002].

Dosage

For osteoporosis, vitamin K_1 (phytonadione) at 1 or 10 mg daily or vitamin K_2 (menaquinone) at 45 mg daily has been used in studies.

Adverse Reactions

Adverse effects have not been reported or observed in clinical trials [Whelan, Jurgens, and Bowles 2006].

Drug Interaction

Excessive Vitamin K can reduce the anticoagulant effect of oral anticoagulants, such as warfarin.

Clinical Studies

Research on the effects of vitamin K on BMD and fracture risk in people with osteoporosis have conflicting results. A meta-analysis of 13 randomized, controlled trials revealed that all studies but one showed an advantage of phytonadione and menaquinone in reducing bone loss [Cockayne et al. 2006]. All seven trials that reported fracture effects were Japanese and used menaquinone [Cockayne et al. 2006]. Pooling of the seven trials found an odds ratio (OR) of 0.40 (95% CI, 0.25–0.65) for vertebral fractures, OR of 0.23 (95% CI, 0.12–0.47) for hip fractures, and an OR of 0.19 (CI, 0.11–0.35) for all nonvertebral fractures [Cockayne et al. 2006]. In a recent study in elderly women and men, vitamin K_1 at 500 mcg/day did not affect BMD [Booth et al. 2008] (see Table 16.10).

OSTEOARTHRITIS

Osteoarthritis is the most common joint disease and a leading cause of pain and physical disability. The knee, hand, and hip are the three most common areas affected by osteoarthritis, and conventional treatment includes exercise, weight loss, pharmacologic management with NSAIDs, and, in severe cases, surgery and/or joint replacement [Lohmander and Roos 2007].

GLUCOSAMINE AND CHONDROITIN

Pharmacology

Two of the most widely used and studied supplements for osteoarthritis are glucosamine and chondroitin. Glucosamine sulfate is proposed to stimulate the

Table 16.10 Clinical Studies Evaluating the Effects of Vitamin K on BMD

Reference	Design	Subjects	Therapy	Results
Booth et al. 2008	3-year, double-blind controlled trial	452 men and women (60–80 years old)	Multivitamin with or without 500 mcg of phylloquinone; both groups received 600 mg of elemental calcium and vitamin D at 400 IU daily	No differences in BMD measurements at any of the anatomical sites measured (femoral neck, spine, and total body BMD) between the treatment and control groups
Iwamoto, Takeda, and Ichimura 2001	24-month, randomized	72 postmenopausal women with osteoporosis (53–78 years old)	3 groups: intermittent cyclical etidronate (200 mg/day, 14 days per 3 months); menatetrenone at 45 mg/day; calcium lactate at 2 g/day	Significant ↑ in BMD in menatetrenone group compared with calcium group ($p < 0.0001$); significant increase in etidronate compared with calcium and menatetrenone ($p < 0.0001$ and $p < 0.01$ respectively); rate reduction of occurrence of new vertebral fractures in etidronate and vitamin K groups compared with calcium group were 68.0 and 65.2%, respectively
Shiraki et al. 2000	24-month, randomized, open label	241 osteoporotic patients	45 mg of oral vitamin K_2 (menatetrenone) daily versus placebo	Incidence of clinical fractures during two years of treatment was higher in control group than in treatment group ($p = 0.0273$); change in lumbar BMD at 6, 12, and 24 months were statistically significant in treatment group: at 6 months, $-1.8 \pm 0.6\%$ (control) versus $1.4 \pm 0.7\%$ (treated) ($p = 0.010$); at 12 months, $-2.4 \pm 0.7\%$ (control) versus $-0.1 \pm 0.6\%$ (treated) ($p = 0.0153$); at 24 months, $-3.3 \pm 0.8\%$ (control) versus $-0.5 \pm 1.0\%$ (treated) ($p = 0.0339$)
Iwamoto et al. 1999	1-year, randomized	72 postmenopausal women	4 groups: control, vitamin K_2 at 45 mg/day, vitamin D_3 at 1.0 g/day, and hormone replacement therapy	Vitamin K_2 suppressed the decrease in spinal BMD compared with no treatment group; BMD in women treated with vitamin K2 was inversely correlated with their age ($r = -0.54$; $p < 0.05$)

cartilage by producing large amounts of glycoproteins and glycosaminoglycans and may exhibit mild anti-inflammatory effects at higher doses, whereas chondroitin also claims anti-inflammatory properties and aids in proteoglycan synthesis [Morelli, Naquin, and Weaver 2003].

Glucosamine sulfate has shown greater benefit than glucosamine hydrochloride [Dahmer and Schiller 2008] and is typically dosed at 500 mg by mouth three times daily. The usual dose of chondroitin is 400 mg by mouth three times daily.

Adverse Effects

Glucosamine is generally well tolerated with mild GI adverse effects (epigastric pain, nausea, diarrhea, and constipation) reported. Less than 1% of patients experienced edema, tachycardia, drowsiness, and/or headache. Chondroitin may cause changes in intraocular pressure.

Drug Interactions

When using these agents in the treatment of osteoarthritis, patient allergies and concurrent medications are factors to be considered. Patients who have a shellfish allergy should avoid glucosamine, and both agents are cautioned in those patients with asthma because of a greater risk of exacerbation. Glucosamine may cause a theoretical decrease in the effectiveness of diabetes medications and an increased risk of bleeding in patients on warfarin. No clinically significant drug information is available for chondroitin.

Clinical Evidence

Studies of these agents have provided conflicting results and have often been overshadowed by suboptimal study design and/or industry bias.

The National Institutes of Health-funded Glucosamine/Chondroitin Arthritis Intervention Trial (GAIT) [Sawitzke et al. 2008] originally studied the results of these supplements on joint pain. As published in 2006, the combination of glucosamine and chondroitin did not provide significant relief from osteoarthritis pain, except for a small subset of individuals with moderate to severe pain [Slowing the Progression 2008]. An ancillary GAIT study sought to evaluate whether glucosamine and/or chondroitin would positively impact disease progression as indicated by joint space width loss. The results indicated that, when compared with placebo, neither glucosamine, chondroitin, nor the combination added benefits in slowing the progressive loss of cartilage in osteoarthritis of the knee [Sawitzke et al. 2008].

A few small studies have shown favorable results in topical preparations of glucosamine, chondroitin, shark cartilage, and camphor; however, it is unclear whether the benefit is attributable to the glucosamine/chondroitin or the camphor [Dahmer and Schiller 2008; Sawitzke et al. 2008].

Although glucosamine has demonstrated questionable efficacy, there is not strong evidence against its use. Patients, who respond well to glucosamine sulfate, typically experience a reduction in symptoms within one to three months, and a 60-day trial may be supported by the healthcare provider [Dahmer and Schiller 2008].

S-ADENOSYLMETHIONINE

Pharmacology

S-adenosylmethionine (SAMe) is another agent that has been studied in the treatment of osteoarthritis. SAMe is available as a prescription product in Europe for the treatment of arthritis and depression. The proposed mechanism of action of SAMe in the treatment of osteoarthritis is donation of a methyl group that aids in the reaction that increases chondrocytes and cartilage thickness and may also decrease cytokine-induced chondrocyte damage.

Adverse Effects

The adverse effects of SAMe are rare and include anxiety, headache, urine frequency, pruritis, nausea, and diarrhea.

Drug Interactions

SAMe increases serotonin turnover, which can lead to CNS adverse effects and drug interactions. The concomitant use of SAMe and the tricyclic antidepressants may increase the risk of serotonin syndrome.

Clinical Evidence

Studies have indicated that SAMe is more effective than placebo and may be comparable with NSAIDs in the reduction of pain and stiffness [Morell, Naquin, and Weaver 2003; Gregory, Sperry, and Friedman 2008].

Although this agent appears promising, there are several barriers that may limit the use of SAMe. Several of the studies showing benefit of SAMe were nonrandomized, unbiased, and included flawed statistical analysis [Morelli, Nauqin, and Weaver 2003]. The wide dosage range of 400–1200 mg daily and monthly cost of $60–120, coupled with the product's questionable shelf stability, may lead patients to not pursue SAMe treatment.

DEVIL'S CLAW

Pharmacology

Devil's claw, *harpagophytum procumbens*, is an African plant, and the medicinal properties of the principal compound harpogoside come from the plant tuber. Devil's claw exhibits anti-inflammatory effects via COX2 and lipoxygenase inhibition [Jellin 2009].

Harpadol is a powdered devil's claw root product, which contains 2% harpagoside and is dosed at 2.6 g/day.

Doloteffin, a devil's claw extract, provides 60 mg/day of the harpagoside constituent at 2400 mg daily has also been used.

Adverse Effects

The most frequent adverse effect of devil's claw is diarrhea (8%). Additional adverse effects include abdominal pain and skin reactions. One case report indicated the incidence of purpura in a patient anticoagulated with warfarin [Gregory et al. 2008].

Drug Interactions

Devil's claw may inhibit cytochrome P450 2C19, 2C9, and 3A4. Caution must be taken with concomitant administration of the following medications (see Table 16.11).

Clinical Evidence

Devil's claw alone and/or with NSAIDs decreased the symptoms of osteoarthritic pain [Gregory et al. 2008]. Use of the powdered devil's claw root product decreased the patient's use of NSAIDs for pain relief [Gregory et al. 2008]. More studies and evidence regarding the safety and drug interaction profile of devil's claw are needed before routine use of this nutraceutical can be advocated.

ANTIOXIDANT VITAMINS

Antioxidants have been studied in the treatment of osteoarthritis because of the potential role of reactive oxygen species contributing to pathogenesis. Limited animal and human studies examine the impact of vitamin C, beta-carotene, and vitamin E. The Framingham Osteoarthritis Cohort Study found a threefold lower risk of osteoarthritis progression in humans taking between 120 and 200 mg of vitamin C daily, although there was no significant effect on the incidence of joint pain [Wang et al. 2004; Ameye and Chee 2006]. Adjustment of vitamin C intake showed

Table 16.11 Devil's Claw Cytochrome P450 Drug Interactions

Enzyme	Example Substrate(s)
2C19	omeprazole (Prilosec), lansoprazole (Prevacid), and pantoprazole (Protonix); diazepam (Valium); carisoprodol (Soma); nelfinavir (Viracept)
2C9	NSAIDs, such as diclofenac (Cataflam, Voltaren), ibuprofen (Motrin), meloxicam (Mobic), and piroxicam (Feldene); celecoxib (Celebrex); amitriptyline (Elavil); warfarin (Coumadin); glipizide (Glucotrol); losartan (Cozaar)
3A4	lovastatin (Mevacor), ketoconazole (Nizoral), itraconazole (Sporanox), fexofenadine (Allegra), triazolam (Halcion)

Source: Jellin, J. M., ed. 2009. Natural medicines comprehensive database. http://www.naturaldatabase.com.

a reduced risk of knee osteoarthritis progression with beta-carotene. Any potential effects of lutein effects on cartilage have not been studied [Micromedex].

This study also showed that increased dietary intake of vitamin E decreased the risk of osteoarthritis progression in men. Additional short-term studies of vitamin E at 600 mg daily for 10 days and 400 IU daily for six weeks were more effective than placebo in relieving osteoarthritis-associated pain [Wang et al. 2004].

Overall, additional long-term studies are needed to confirm the benefits of antioxidant vitamins in the treatment of osteoarthritis.

REFERENCES

Alekel, D. L., A. S. Germain, C. T. Peterson, K. B. Hanson, J. W. Stewart, and T. Toda. 2000. Isoflavone-rich soy protein isolate attenuates bone loss in the lumbar spine of perimenopausal women. *Am. J. Clin. Nutr.* 72:844–852.

Ameye, L. G. and W. S. Chee. 2007. Osteoarthritis and nutrition. From nutraceuticals to functional foods: A systematic review of the scientific evidence. *Arthr. Res. Ther.* 8:r127.

Armas, L. A., B. W. Hollis, and R. P. Heaney. 2004. Vitamin D2 is much less effective than vitamin in humans. *J. Clin. Endocrinol. Metab.* 89:5387–5391.

Atkinson, C., J. E. Compston, and N. E. Day. 2004. The effects of phytoestrogen isoflavones on bone density in women: A double-blind, randomized, placebo-controlled trial. *Am. J. Clin. Nutr.* 79:326–333.

Baulieu, E. E., G. Thomas, S. Legrain, N. Lahlou, M. Roger, B. Debuire, V. Faucounau, L. Girard, M. P. Hervy, F. Latour, M. C. Leaud, A. Mokrane, H. Pitti-Ferrandi, C. Trivalle, O. de Lacharrière, S. Nouveau, B. Rakoto-Arison, J. C. Souberbielle, J. Raison, Y. Le Bouc, A. Raynaud, X. Girerd, and F. Forette. 2000. Dehydroepiandrosterone (DHEA), DHEA sulfate, and aging: Contribution of the DHEAge Study to a sociobiomedical issue. *Proc. Natl. Acad. Sci. USA.* 97:4279–4284.

Bischoff-Ferrari, H. A., W. C. Willett, J. B. Wong, E. Giovannucci, T. Dietrich, and B. Dawson-Hughes. 2005. Fracture prevention with vitamin D supplementation: A meta-analysis of randomized controlled trials. *JAMA* 293:2257–2264.

Booth, S. L., F. Dallal, M. K. Shea, C. Gundberg, J. W. Peterson, and B. Dawson-Hughes. 2008. Effect of vitamin K supplementation on bone loss in elderly men and women. *J. Clin. Endocrinol. Metab.* 93:1217–1223.

Calvo, M. S., S. J. Whitting, and C. N. Barton. 2004. Vitamin D fortification in the United States and Canada: Current status and data needs. *Am. J. Clin. Nutr.* 80 (Suppl.):1710S–1716S.

Chapuy, M. C., M. E. Arlot, F. Duboeuf, J. Brun, B. Crouzet, S. Arnaud, P. D. Delmas, and P. J. Meunier. 1992. Vitamin D3 and calcium to prevent hip fractures in the elderly women. *N. Engl. J. Med.* 327:1637–1642.

Chen, Y. M., S. C. Ho, S. S. Lam, S. S. Ho, and J. L. Woo. 2003. Soy isoflavones have a favourable effect on bone loss in Chinese postmenopausal women with lower bone mass: A double-blind, randomized, controlled trial. *J. Clin. Endocrinol. Metab.* 88:4740–4747.

Cockayne, S., J. Adamson, S. Lanham-New, M. J. Shearer, S. Gilbody, and D. J. Torgerson. 2006. Vitamin K and the prevention of fractures. Vitamin K and the prevention of fractures. Systematic review and meta-analysis of randomized controlled trials. *Arch. Intern. Med.* 166:1256–1261.

Dahmer, S. and R. M. Schiller. 2008. Glucosamine. *Am. Fam. Physician* 78:471–476, 481.

Gartner, L.M. and F. R. Greer. 2003. Prevention of rickets and vitamin D deficiency: New guidelines for vitamin D intake. *Pediatrics* 111:908–910.

Gough, H., T. Goggin, A. Bissessar, M. Crowley, M. Baker, and N. Callaghan. 1986. A comparative study of the relative influence of different anticonvulsant drugs, UT exposure and diet on vitamin D and calcium metabolism in outpatients with epilepsy. Q. J. Med. 59:569–577.

Grant, A. M., A. Avenell, M. K. Campbell, A. M. McDonald, G. S. MacLennan, G. C. McPherson, F. H. Anderson, C. Cooper, R. M. Francis, C. Donaldson, W. J. Gillespie, C. M. Robinson, D. J. Torgerson, and W. A. Wallace; RECORD Trial Group. 2005. Oral vitamin D3 and calcium for secondary prevention of low-trauma fractures in elderly people (Randomized Evaluation of Calcium Or vitamin D, RECORD): A randomized, double-blind, placebo controlled trial. *Lancet* 365:1621–1628.

Gregory, P. J., M. Sperry, and A. Friedman Wilson. 2008. Dietary supplements for osteoarthritis. *Am. Fam. Physician* 77:177–184.

Haguenauer, D., V. Welch, B. Shea, and G. Wells. 2003. Fluoride for treating postmenopausal osteoporosis (Cochrane Review). In *The Cochrane Library*, Issue 4. Chichester, UK: John Wiley and Sons.

Hulley, S., C. Furberg, E. Barrett-Connor, J. Cauley, D. Grady, W. Haskell, R. Knopp, M. Lowery, S. Satterfield, H. Schrott, E. Vittinghoff, and D. Hunninghake; HERS Research Group. 2002. Noncardiovascular disease outcomes during 6.8 years of hormone therapy. Heart and Estrongen/Progestin Replacement Study follow-up (HERS II). *JAMA* 288:58–66.

Institute of Medicine, Food and Nutrition Board. 1997. Dietary reference intakes: Calcium, phosphorous, magnesium, vitamin D, and fluoride. Washington, DC: National Academy Press.

Iwamoto, I., S. Kosha, S. Noguchi, M. Murakami, T. Fujino, T. Douchi, and Y. Nagata. 1999. A longitudinal study of the effect of vitamin K_2 on bone mineral density in postmenopausal women: A comparative study with vitamin D_3 and estrogen-progestin therapy. Maturitas 31:161–164.

Iwamoto, J., T. Takeda, and S. Ichimura. 2001. Effect of menatetrenone on bone mineral density and incidence of vertebral fractures in postmenopausal women with osteoporosis: A comparison with the effect of etidronate. *J. Orthop. Sci.* 6:487–492.

Jackson, R. D., A. Z. LaCroix, M. Gass, R. B. Wallace, J. Robbins, C. E. Lewis, T. Bassford, S. A. A. Beresford, H. R. Black, P. Blanchette, D. E. Bonds, R. L. Brunner, R. G. Brzyski, B. Caan, J. A. Cauley, R. T. Chlebowski, S. R. Cummings, I. Granek, J. Hays, G. Heiss, S. L. Hendrix, B. V. Howard, J. Hsia, F. A. Hubbell, K. C. Johnson, H. Judd, J. M. Kotchen, L. H. Kuller, R. D. Langer, N. L. Lasser, M. C. Limacher, S. Ludlam, J. E. Manson, K. L. Margolis, J. McGowan, J. K. Ockene, M. J. O'Sullivan, L. Phillips, R. L. Prentice, G. E. Sarto, M. L. Stefanick, L. Van Horn, J. Wactawski-Wende, E. Whitlock, G. L. Anderson, A. R. Assaf, and D. Barad, for the Women's Health Initiative Investigators. 2006. Calcium plus vitamin D supplementation and the risk of fractures. *N. Engl. J. Med.* 354:669–683.

Jellin, J. M., ed. 2009. Natural medicines comprehensive database. http://www.naturaldatabase.com. Accessed January 8 and 12, 2009.

Kreijkamp-Kaspers, S., L. Kok, D. E. Grobbee, E. H. F. de Haan, A. Aleman, J. W. Lampe, and Y. T. van der Schouw. 2004. Effect of soy protein containing isoflavones on cognitive function, bone mineral density, and plasma lipids in postmenopausal women. JAMA 292:65–74.

Lawson, B., S. Rose, D. McLeod, A. Dowell. 2003. Changes in use of hormone replacement therapy after the report from the Women's Health Initiative: Cross-sectional survey of users. *BMJ* 327:845–846.

Lexicomp-online. http://www.crlonline.com. Accessed January 2, 2008.

Lohmander, L. S. and E. M. Roos. 2007. Clinical update: Treating osteoarthritis. *Lancet* 370:2082–2084.

Lydeking-Olsen, E., J. Beck-Jensen, K. D. R. Setchell, and T. Holm-Jensen. 2004. Soymilk or progesterone for prevention of bone loss: A 2 year randomized, placebo-controlled trial. *Eur. J. Nutr.* 43:246–257.

MacLean, C., S. Newberry, M. Maglione, M. McMahon, V. Ranganath, M. Suttorp, W. Mojica, M. Timmer, A. Alexander, M. McNamara, S. B. Desai, A. Zhou, S. Chen, J. Carter, C. Tringale, D. Valentine, B. Johnsen, and J. Grossman. 2008. Systematic review: Comparative effectiveness of treatments to prevent fractures in men and women with low bone density or osteoporosis. *Ann. Intern. Med.* 148:197–213.

Meunier, P. J., C. Roux, E. Seeman, S. Ortolani, J. E. Badurski, T. D. Spector, J. Cannata, A. Balogh, E. M. Lemmel, S. Pors-Nielsen, R. Rizzoli, H. K. Genant, and J. Y. Reginster. 2004. The effects of strontium ranelate on the risk of vertebral fracture in women with postmenopausal osteoporosis. *N. Engl. J. Med.* 350:459–468.

Micromedex Healthcare Series [Internet database]. Greenwood Village, CO: Thomson Healthcare, updated periodically. http://www.thomsonhc.com. Accessed January 12, 2009.

Morales, A. J., R. H. Haubrich, J. Y. Hwang, H. Asakura, and S. S. Yen. 1998. The effect of six months of treatment with a 100 mg daily dose of dehydroepiandrosterone (DHEA) on circulating sex steroids, body composition and muscle strength in age-advanced men and women. *Clin. Endocrinol.* 49:421–432.

Morelli, V., C. Naquin, and V. Weaver. 2003. Alternative therapies for traditional disease states: Osteoarthritis. *Am. Fam. Physician* 67:339–344.

Nair, K. S., R. A. Rizza, P. O'Brien, K. Dhatariya, K. R. Short, A. Nehra, J. L. Vittone, G. G. Klee, A. Basu, R. Basu, C. Cobelli, G. Toffolo, C. Dalla Man, D. J. Tindall, L. J. Melton 3rd, G. E. Smith, S. Khosla, and M. D. Jensen. 2006. DHEA in elderly women and DHEA or testosterone in elderly men. *N. Engl. J. Med.* 355:1647–1659.

National Institutes of Health Consensus Development Program. 2000. Osteoporosis Prevention, Diagnosis, and Therapy. National Institutes of Health, Bethesda, MD. http://consensus.nih.gov/2000/2000Osteoporosis111html.htm. Accessed January 2, 2009.

O'Connell, M. B. 1999. Prevention and treatment of osteoporosis in the elderly. *Pharmacotherapy* 19:7S–20S.

Office of Dietary Supplements. 2008. Dietary supplement fact sheet: Vitamin D. National Institutes of Health, Bethesda, MD. http://ods.od.nih.gov/factsheets/vitamind.asp. Accessed December 28, 2008.

Papadimitropoulos, E., G. Wells, B. Shea, W. Gillespie, B. Weaver, N. Zytaruk, A. Cranney, J. Adachi, P. Tugwell, R. Josse, C. Greenwood, and G. Guyatt; Osteoporosis Methodology Group and The Osteoporosis Research Advisory Group. 2002. Meta-analyses of therapies for postmenopausal osteoporosis. VIII. Meta-analysis of the efficacy of vitamin D treatment in preventing osteoporosis in postmenopausal women. *Endocr. Rev.* 23:560–569.

Prince, R. L., A. Devine, S. S. Dhaliwal, and I. M. Dick. 2006. Effects of calcium supplementation on clinical fracture and bone structure: Results of a 5-year, double-blind, placebo-controlled trial in elderly women. *Arch. Intern. Med.* 166:869–875.

Qaseem, A., V. Snow, P. Shekelle, R. Hopkins Jr., M. A. Forciea, and D. K. Owens; Clinical Efficacy Assessment Subcommittee of the American College of Physicians. 2008. Pharmacologic treatment of low bone density or osteoporosis to prevent fractures: A clinical practice guideline from the American College of Physicians. *Ann. Intern. Med.* 149:404–415.

Rossouw, J. E., G. L. Anderson, R. L. Prentice, A. Z. LaCroix, C. Kooperberg, M. L. Stefanick, R. D. Jackson, S. A. Beresford, B. V. Howard, K. C. Johnson, J. M. Kotchen, and J. Ockene; Writing Group for the Women's Health Initiative Investigators. 2002. Risks and benefits of estrogen plus progestin in healthy postmenopausal women: Principal results from the Women's Health Initiative randomized controlled trial. *JAMA* 228:321–333.

Sawitzke, A. D., H. Shi, M. F. Finco, D. D. Dunlop, C. O. Bingham 3rd, C. L. Harris, N. G. Singer, J. D. Bradley, D. Silver, C. G. Jackson, N. E. Lane, C. V. Oddis, F. Wolfe, J. Lisse, D. E. Furst, D. J. Reda, R. W. Moskowitz, H. J. Williams, and D. O. Clegg. 2008. The effect of glucosamine and/or chondroitin sulfate on the progression of knee osteoarthritis: A report from the Glucosamine/Chondroitin Arthritis Intervention Trial. *Arthritis Rheum.* 58:3183–3191.

Shea, B., G. Wells, A. Cranney, N. Zytaruk, V. Robinson, L. Griffith, Z. Ortiz, J. Peterson, J. Adachi, P. Tugwell, and G. Guyatt; Osteoporosis Methodology Group and The Osteoporosis Research Advisory Group. 2002. Meta-analyses of therapies for postmenopausal osteoporosis. VII. Meta-analysis of calcium supplementation for the prevention of postmenopausal osteoporosis. *Endocr. Rev.* 23:552–559.

Shiraki, M., Y. Shiraki, C. Aoki, and M. Miura. 2000.Vitamin K_2 (menatetrenone) effectively prevents fractures and sustains lumbar bone mineral density in osteoporosis. *J. Bone Miner. Res.* 15:515–521.

Slowing the progression of osteoarthritis: Do glucosamine and chondroitin help? Pharmacist's Letter 2008; 24:241111. http://www.pharmacistletter.com. Accessed January 12, 2009.

Sun, Y., M. Mao, L. Sun, Y. Feng, J. Yang, and P. Shen. 2002. Treatment of osteoporosis in men using dehydroepiandrosterone sulfate. *Chin. Med. J. (Engl.)* 115:402–404.

Tang, B. M., G. D. Eslick, C. Nowson, C. Smith, and A. Bensoussan. 2007. Use of calcium or calcium in combination with vitamin D supplementation to prevent fractures and bone loss in people aged 50 years and older: a meta-analysis. *Lancet* 370:657–666.

United States Bone and Joint Decade. Facts in Brief. http://www.usbjd.org/healthcare_pro/resources/Facts_in_Brief_2004.doc. Accessed January 16, 2009.

United States Department of Health and Human Services. 2004. Bone health and osteoporosis: A report of the surgeon general. Office of the Surgeon General. http://www.surgeongeneral.gov/library/bonehealth/content.html. Accessed January 10, 2009.

Wagner, C. L. and F. R. Greer. 2008. Prevention of rickets and vitamin D deficiency in infants, children, and adolescents. *Pediatrics* 122:1142–1152.

Wang, Y., L. F. Prentice, L. Vitetta, A. E. Wluka, and F. M. Cicuttini. 2004. The effect of nutritional supplements on osteoarthritis. *Altern. Med. Rev.* 9:275–296.

Whelan, A. M., T. M. Jurgens, and S. K. Bowles. 2006. Natural health products in the prevention and treatment of osteoporosis: Systematic review of randomized controlled trials. *Ann. Pharmacother.* 40:836–849.

Wu, C. H., Y. C. Yang, W. J. Yao, F. H. Lu, J. S. Wu, and C. J. Chang. 2002. Epidemiological evidence of increased bone mineral density in habitual tea drinkers. *Arch. Intern. Med.* 162:1001–1006.

Yao, F. A., A. S. Dobs, and T. T. Brown. 2006. Alternative therapies for osteoporosis. *Am. J. Chin. Med.* 34:721–730.

CHAPTER 17

Nutraceuticals for Skin Health

Raghunandan Yendapally

CONTENTS

INTRODUCTION

The skin is the most vital and largest organ in the human body. Skin is responsible for sense of touch. Skin is primarily composed of two layers: epidermis, the outer layer, and dermis, the inner layer. Skin primarily protects the human body from the damage caused by the external environmental agents such as UV rays, pathogens, and pollutants. In addition, vitamin D, a fat-soluble vitamin, is synthesized in the skin by the action of UV rays emitted by the sunlight. Apart from beneficial effects, UV rays emitted by the sunlight are associated with several adverse effects. For example, UV rays emitted by the sunlight are responsible for causing skin cancer. According to the National Cancer Institute, skin cancer is the most common type of cancer in United States, affecting approximately one million Americans every year [National Cancer Institute 2005]. UV rays have a wavelength ranging from 200 to 400 nm. UV rays based on their wavelengths are further classified into UV-A rays (320–400 nm), UV-B rays (280–320 nm), and UV-C rays (200–280 nm) [Katiyar 2007]. Sunlight primarily constitutes UV-A rays, to which we are mainly exposed, have a greater ability to penetrate the cutaneous layers, and are responsible for skin wrinkle, aging, and immune suppression [Katiyar 2007]. UV-B rays are mutagens and are shown to be responsible for melanoma (skin cancer that is formed in melanocytes) and non-melanoma types (skin cancer that is formed in basal cells and squamous cells) of skin cancers [Urbach 1978]. In addition, they are also responsible for erythema, sunburns, and immune suppression. UV-A and UV-B rays are responsible for about 90% of the estimated skin damage [Guercio-Hauer, MacFarlane, and Deleo 1994]. UV-C rays can potentially cause skin cancer and immunosuppression. Fortunately, we are not exposed to UV-C rays because the stratospheric ozone layer above the earth's surface blocks these rays [Katiyar 2007].

The unique nature of the skin facilitates permeation of various hydrophobic drugs. The delivery of the drug through the skin is referred to as topical drug delivery. Skin formulations are classified as creams, gels, lotions, and ointments. In addition, drugs are also delivered through the skin using patches, and this method of drug administration is referred to as transdermal drug delivery. Although the topical and transdermal drug delivery systems are extensively used in the treatment of various skin diseases, certain kinds of skin diseases are treated orally or systemically.

Skin infections have caused significant mortality and morbidity ever since the origination of the human race. In ancient times, many cultures have applied natural products such as soybean, curd, cheese, honey, and moldy bread to counteract skin infections. A wide variety of nutraceuticals, including, but not limited to, carotenoids, melatonin, proanthocyanidins, curcumin, ferulic acid, tea, linoleic acid, gingerol, curry leaf, CoQ_{10}, and silymarin, have been claimed to have beneficial effects in various skin diseases or disorders, such as melanoma and non-melanoma cancers, age-related skin diseases, acne, psoriasis, skin rash, inflammation, and immunomodulation. In addition, nutraceuticals are also used in skin care to improve skin texture, glow, and smoothness.

CAROTENOIDS

Carotenoids, such as beta-carotene, lutein, lycopene, and zeaxanthin, either alone or in combination, are commonly used as nutraceuticals in skin healthcare. In a recent study, it was shown that supplementation of carotenoid mixture containing lycopene (3 mg/day), lutein (3 mg/day), and beta-carotene (4.8 mg/day), along with α-tocopherol (10 mg/day) and selenium (75 μg/day), for a period of 12 weeks has improved skin density, thickness, scaling, smoothness, and wrinkling [Heinrich et al. 2006].

Prolonged exposure to UV rays significantly reduces the skin and plasma levels of carotenoids. It is evident that carotenoids exhibit *in vitro* antioxidant properties by scavenging reactive oxygen species. On the contrary, carotenoids at high concentrations under *in vitro* conditions possess pro-oxidant properties [Young and Lowe 2001; Eichler, Sies, and Stahl 2002; Offord et al. 2002]. Therefore, the dosage of carotenoids is very critical in achieving the beneficial effects.

A meta-analysis of effectiveness of beta-carotene in protection against sunburn supported that supplementation of beta-carotene protects the skin against sunburn in a time-dependent manner [Kopcke and Krutman 2008]. In a placebo-controlled, parallel study, it was found that supplementation of beta-carotene at 24 mg/day or carotenoid mix of beta-carotene, lutein, and lycopene at 8 mg/day each for 12 weeks significantly attenuated the UV-B-induced erythema in humans [Heinrich et al. 2003]. In another study, it was demonstrated that chronically UV-B-irradiated hairless mice when supplemented with lutein/zeaxanthin reduced the tumor volume size and decreased the multiplicity when compared with the animals fed with normal diet [Astner et al. 2007]. Therefore, from the above studies, it is evident that supplementation of dietary carotenoids has a positive effect on the skin health.

LUPEOL

Lupeol is a naturally occurring pentacyclic triterpine of plant origin present in several common fruits and vegetables, such as mangoes, figs, strawberries, and olives [Saleem et al. 2004]. It was demonstrated that lupeol, when administered topically at a dose level of 0.75 and 1.5 mg/animal 1 h before benzoyl peroxide, a free radical generator, treatment attenuated the early responses of tumors induced by benzoyl peroxide in mice skin [Sultana et al. 2003]. In another study, it was demonstrated that topical application of lupeol (1–2 mg/mouse) 30 min before the 12-*O*-tetradecanoyl-phorbol-13-acetate (TPA) (3.2 nmol/mouse) resulted in inhibition of TPA-induced skin cancer in CD-1 mice model [Saleem et al. 2004]. Recent studies suggest that lupeol has protective effects on 7,12-dimethylbenz[a] anthracene (DMBA)-induced DNA alkylation damage in mice skin. This particular study also implies the potential use of lupeol as a chemotherapeutic agent because DNA alkylation damage can lead to cancer and other genetic diseases [Nigam, Prasad, and Shukla 2007]. In addition, it is also observed that, when lupeol (and its natural and semi-synthetic derivatives) is administered topically, it exhibits anti-inflammatory properties with improved keratinocyte proliferation [Nikiema et al. 2001].

MELATONIN

Melatonin is a naturally occurring hormone secreted by the pineal gland. Melatonin (*N*-acetyl-5-methoxytryptamine) is biosynthesized by a series of reactions from indole amino acid tryptophan [Yu, Tsin, and Reiter 1992]. Melatonin is a highly potent antioxidant that acts by scavenging hydroxyl and lipid peroxidyl free radicals [Reiter et al. 1995]. Several studies investigated the photoprotective effects of melatonin in humans. The topical application of melatonin solution resulted in attenuation of UV-induced erythema in a dose-dependent manner [Bangha, Elsner, and Kistler 1997]. Interestingly, more pronounced results were obtained when melatonin was formulated in combination with vitamin C and vitamin E [Dreher et al. 1998].

Recently, Maldonado et al. [2007] reviewed the use of melatonin as a potential pharmacological support against thermal injury. It is evident that thermal injury results in lymphocytopenia and sleep deficiencies. It is believed that melatonin ameliorates burn injuries by inhibiting the proinflammatory cytokines and improving the sleep mechanisms.

It is evident from the body of research material published to date that melatonin exhibits tumorostatic properties, which includes melanomas and tumors of cutaneous origin [Cos and Sanchez-Barcelo 2000]. In a mice study, melatonin reduced the number of papillomas during the initiation as well as promotion stages of tumor induced by benzo[a]pyrene. This study also found that mice treated with melatonin prevent the binding benzo[a]pyrene and its metabolites to DNA [Kumar and Das 2000].

PROANTHOCYANIDINS

Oligomeric proanthocyanidins, when administered as a dietary supplement before UV radiation followed by topical application in the form of a lotion or a cream (Anthogenol), has attenuated the inflammation and improved the hydration properties of the skin in humans [Hughes-Formella, Wunderlich, and Williams 2007]. It was shown that supplementation of grape seed proanthocyanidins (GSPs) in the diet (0.2 and 0.5 % weight/weight [w/w]) to mice protected its skin from UV-B-induced oxidative stress by inhibiting mitogen-activated protein kinase and NF-κB cellular signal pathways [Sharma, Meeran, and Katiyar 2007]. In another study, the dietary supplementation of GSPs in mice attenuated UV-B-induced immunosuppression by inhibiting immunosuppressive cytokine IL-10 and inducing immunostimulatory cytokine IL-12 production [Sharma and Katiyar 2006].

Dietary administration of GSPs (0.2 and 0.5 % w/w) to hairless mice inhibited tumor incidence, tumor multiplicity, and tumor size in UV-B-induced initiation and promotion stages of mouse photocarcinogenesis [Mittal, Elmets, and Katiyar 2003]. In addition, it was also shown that dietary GSPs prevented the progression of UV-B-induced papillomas to carcinomas [Mittal, Elmets, and Katiyar 2003]. However, in a double-blind, placebo-controlled, randomized Phase II trial involving 66 volunteers, oral supplementation of GSP extract (100 mg three times a day) had no significant effect in patients with breast induration after radiotherapy for breast cancer [Brooker et al. 2006].

CURCUMIN

Curcumin is a yellow-colored phenolic compound obtained from turmeric. Turmeric is commonly used as a spice and coloring agent in many Asian foods for centuries. There are several reports that have demonstrated curcumin as an anti-inflammatory and antioxidant agent. In the past few years, there has been considerable interest in identifying the cellular and molecular mechanisms by which the isolates of turmeric cures the skin ailments. In a review, it was summarized that curcumin exhibits anti-inflammatory properties probably by inhibiting COX2 lipoxygenase, and iNOS [Srinivasan, Sudheer, and Menon 2007]. Because inflammation is closely associated with the tumor promotion, several studies have also demonstrated the beneficial effects of curcumin in attenuating tumor promotion.

One study found that topical application of curcumin inhibited benzo[a]pyrene-initiated and TPA-induced tumors in mouse skin in a dose-dependent manner [Huang et al. 1988]. The *in vitro* studies have shown that curcumin reduces human epidermal keratinocytes differentiation and proliferation and enhances apoptosis [Balasubramanian and Eckert 2007]. In another study, it was observed that curcumin at micromolar concentrations induces apoptosis and reduces proliferation in melanoma cells by suppressing nuclear factor κB inhibitor kinase and nuclear transcription factor NF-κB [Siwak et al. 2005].

In a randomized, controlled study in rats, it was found that oral administration of curcumin before and after the skin burns reduced the progression of unburned skin interspaces into full necrosis [Singer et al. 2007]. The oral administration of curcumin before γ-radiation has expedited the wound healing in mice by enhancing wound contraction and increasing the synthesis of collagen, hexosamine, NO, DNA, proliferation of fibroblasts, and vasculature [Jagetia and Rajanikant 2004]. Therefore, curcumin may be potentially used in the skin burn treatment and wound healing.

FERULIC ACID

Ferulic acid is a phenolic compound present in high concentrations in leaves, fruits, and vegetables. It possesses potent antioxidant properties, and the antioxidant potential of ferulic is explained by its unique structural characteristics. The phenoxy radical form of ferulic acid is resonance stabilized by the delocalization of electrons. Ferulic acid has been reported to scavenge hydroxyl, alkoxy, peroxy, and superoxide free radicals [Srinivasan, Sudheer, and Menon 2007].

Topical formulation of ferulic acid (0.5%), L-ascorbic acid (15%), and α-tocopherol (1%) has been shown to reduce skin erythema and sunburn cell formation [Lin et al. 2005]. This formulation has effectively reduced oxidative stress and thymidine dimer formation induced by UV radiation [Lin et al. 2005]. An *in vitro* study has shown that addition of ferulic acid at concentrations ranging from 1 to 10 μg/ml to human lymphocytes 30 min before UV-B irradiation has inhibited UV-B-induced lipid peroxidation and oxidative stress in a dose-dependent manner [Prasad et al.

2007]. In another study, it was shown that topical application of ferulic acid inhibited TPA-induced skin tumor formation [Huang et al. 1988].

TEA

Tea is obtained by the fermentation of fresh leaves of the plant *Camellia sinensis*. Polyphenols are the major constituents of the tea. During the past decade, several tests were conducted to evaluate the effect of black and green tea constituents on the skin health.

EGCG is one of the main constituents of green tea. In rats, the topical application of EGCG (2%) hydrophilic ointment 30 min before UV-A exposure has decreased skin damage caused by UV-A rays. However, the application of EGCG 30 min after the UV-A exposure has resulted in no beneficial effects [Sevin et al. 2007].

In an *in vivo* study, the aqueous extracts of black tea were formulated as a gel and tested for protection against a broad range of UV radiation (200–400 nm). In the subjects receiving black tea gel, no erythema was observed, indicating the potential application of black tea in sunscreens [Turkoglu and Cigirgil 2007].

A randomized, double-blind, three-arm parallel-group, vehicle-controlled, clinical study was conducted using Polyphenon E (MediGene AG, Munich, Germany), a proprietary extract of green tea leaves, for the treatment of extragenital and perianal warts. The topical application of Polyphenon E 10 and 15% ointment for 16 weeks has resulted in complete clearance of warts in 51 and 53% of the patients, respectively [Stockfleth et al. 2008].

Recently, it was reported that pretreatment with polymeric black tea polyphenolic fractions (PBP1–PBP5) or thearubigins has attenuated TPA-induced skin papillomas in mice. In this study, it was also found that PBP2 as the most potent polymeric phenolic fraction [Patel et al. 2008].

FATTY ACIDS

Studies have shown that fatty acids from animal and plant origins have a wide array of beneficial effects on skin health. Fish oils rich in omega-3 polyunsaturated fatty acids have been reported to significantly reduce the UV-B-induced erythema in humans [Orengo, Black, and Wolf 1992; Rhodes et al. 1994, 2003] and animals [Orengo et al. 1989]. EPA and DHA, the two omega-3 polyunsaturated fatty acids, were found to significantly inhibit the production of UV-B and TNF-α-induced IL-8, a proinflammatory cytokine, in keratinocytes [Storey et al. 2005]. In another study, EPA, when applied topically on human skin, reduced the UV-induced collagen and decreased the epidermal thickening. Therefore, EPA has potential application as photoprotective and anti-skin-aging agent [Kim et al. 2006].

A study was conducted in animals to determine whether the essential and nonessential fatty acids would affect the cutaneous wound healing after the surgery. Topical application of linolenic (omega-3), linoleic (omega-6), and oleic acids (omega-9) modulated the wound-healing process. It was also found that the application of oleic acids resulted in faster recovery when compared with linolenic, linoleic acids, and control [Cardoso et al. 2004].

Dietary supplementation with 1.0 or 1.5% (w/w) CLA has significantly inhibited skin papillomas in mice when compared with a diet without CLA and 0.5% (w/w) CLA [Belury et al. 1996]. It is hypothesized that CLA reduces skin papillomas by modulating peroxisome proliferators-activated receptor -δ [Belury 2007], which plays a major role in inflammatory responses and apoptosis.

Many studies were conducted to determine the efficacy of primrose oil on inflammatory diseases such as atopic eczema and psoriasis [Wright and Burton 1982; Berth-Jones and Graham-Brown 1993; Williams 2003]. Primrose oil constitutes approximately 70% linoleic and 10% gamma-linolenic acids. One study had 99 patients with atopic dermatitis treated with primrose oil or placebo for 12 weeks, and significant clinical improvement was shown at high doses [Wright and Burton 1982]. Another study was conducted to investigate the effectiveness of evening primrose oil for atopic dermatitis. In this study, 123 patients with atopic dermatitis were treated with primrose oil, combination of primrose oil and fish oil, or a placebo for 16 weeks [Berth-Jones and Graham-Brown 1993]. However, the treatment with these oils did not result in reduction of the skin surface area affected by atopic dermatitis. To conclude, some of the studies have demonstrated the beneficial effects of primrose oil for atopic dermatitis [Wright and Burton 1982], whereas the others did not [Berth-Jones and Graham-Brown 1993]. Therefore, the usage of primrose oil for atopic dermatitis remains questionable.

Cocoa butter is a natural fatty acid extracted from cocoa bean. It is widely used in many topical products such as creams, soaps, ointments, and lotions. Cocoa butter acts as an emollient and alleviates dry skin, inflammation, and irritation.

GINGER

Ginger rhizome is obtained from *Zingiber officinale*. The ginger rhizome is commonly used throughout the world as a spice and flavoring agent. The topical application of ethanol extract of ginger on SENCAR mouse skin has resulted in the significant protection against TPA-induced skin tumor [Katiyar, Agarwal, and Mukhtar 1996]. Subsequently, [6]-gingerol was determined as the principal component that is responsible for tumor protection [Park et al. 1998]. Furthermore, the topical application of [6]-gingerol has resulted in inhibition of DMBA-induced skin papillomagenesis [Park et al. 1998]. In another study, it was demonstrated that topical application of [6]-gingerol (30 μM) before UV-B radiation has resulted in the induction of COX2 mRNA and protein and NF-κB translocation in a hairless mice. This study is of particular significance, implying the potential application of [6]-gingerol against UV-B-induced skin disorders [J. K. Kim et al. 2007].

Topical application of ginger dry extract (DE) and gingerols enriched dry extract (EDE) in mice in the form of solution have exhibited dose-dependent anti-inflammatory properties: ID_{50} = 142 and 181 μg/cm [Minghetti et al. 2007]. However, the increase in [6]-gingerol concentration in the extract did not result in improved activity. In addition, application of DE and EDE in the form of medicated plasters resulted in the reduction of inflammation in mice [Minghetti et al. 2007].

ALOE VERA

Aloe vera is a small plant containing succulent leaves. The aloe extracts are commonly used in a wide variety of formulations such as skin creams, lotions, gels, and ointments. Topical application of *A. vera* gel (97.5%) in humans has significantly attenuated UV-induced erythema after 48 h [Reuter et al. 2008]. In a double-blind, placebo-controlled study involving 60 patients suffering from slight to moderate chronic plaque-type psoriasis found that topical application of *A. vera* extract (0.5% extract) in a hydrophilic cream (three times per day, five consecutive days per week) significantly cured lesions, erythema, and infiltration when compared with the placebo [Syed et al. 1996].

In mice, the topical administration of aloe gel (1 ml/9 cm^2 per day) and/or oral administration of aloe leaf extract (1000 mg/kg/day) for 16 weeks resulted in decreased number and size of the skin papillomas induced by DMBA [Chaudhary, Saini, and Goyal 2007]. The topical (25% *A. vera* in Eucerin cream) and oral (100 mg/kg/day) administration of *A. vera* in mice has significantly reduced the wound diameter by 50.8 and 62.5%, respectively, compared with controls [Davis et al. 1989]. *A. vera* tends to have some beneficial effects in preventing skin reactions during radiation therapy [Olsen et al. 2001], burn wound healing [Maenthaisong et al. 2007], desquamation, and pain related to radiation therapy [Heggie et al. 2002]. However, additional studies are warranted to unequivocally establish these wide arrays of beneficial effects of *A. vera.*

CoQ_{10}

CoQ_{10} is incorporated into a variety of skincare products because of its exceptional antioxidant properties. In an *in vitro* study involving human dermal fibroblasts, the combination of CoQ_{10} and carotenoids have resulted in attenuation of inflammation induced by UV radiation [Fuller et al. 2006]. In another study, CoQ_{10} nanoparticles were observed to attenuate the oxidative stress induced by UV-B-induced irradiation by increasing the manganese superoxide dismutase and glutathione peroxidase immunoreactivity and their protein levels in the hairless mouse skin [D. W. Kim et al. 2007].

SILYMARIN

Silymarin is the active constituent extracted from the fruits of *Silybum marianum* L. Gaertn. Chemically, silymarin is a mixture of flavolignanas primarily containing silybin

[Wagner, Horhammer, and Munster 1968]. Topical application of silymarin at a dose of 9 mg/application before UV-B exposure has been shown to reduce the tumor incidence, tumor multiplication, and tumor growth in mice [Katiyar et al. 1997]. In subsequent mice studies, it was reported that topical application of silymarin attenuates UV-B-induced immunosuppression and oxidative stress [Katiyar 2002] primarily through the inhibition of infiltration CD11b+ cells, a major source of oxidative stress [Katiyar, Meleth, and Sharma 2008]. In addition, it was reported that topical application of silybin and 2,3-dehydrosilybin on human keratinocytes at a dose of 1–50 μmol/L suppressed UV-A-induced oxidative damage [Svobodova et al. 2007]. Administration of silymarin either locally or locally and systemically has significantly protected rats from burn-induced oxidative damage [Toklu et al. 2007]. Together, these results imply the potential application of silymarin as a therapeutic agent for the treatment of skin cancers and burns.

CURRY LEAVES

Curry leaves are obtained as fresh leaves from *Murraya koenigii*. Curry leaves are commonly used as flavoring agents in Indian food preparations. The topical application of 10 and 20% curry leaf extract has been shown to reduce 29.05 and 43.75 % of DMBA-induced skin carcinoma in Swiss mice [Dasgupta, Rao, and Yadava 2003]. However, the principal active constituent of curry leaf that is responsible for anticarcinogenic properties is yet to be determined.

CONCLUSIONS

Many experimental studies have found that nutraceuticals, irrespective of the mode of administration (i.e., topical, oral, or systemic), exert the beneficial effects in skin disorders. This has led to the extensive usage of dietary supplements for the prevention and treatment of skin diseases. However, there are several contradictory reports challenging the effectiveness of nutraceuticals, which mandate large clinical studies in humans to validate the usefulness of nutraceuticals for skin care. Furthermore, pharmacological investigations are necessary to explore the effectiveness of supplementation of nutraceuticals with commonly used dermatological agents. Several experimental studies have demonstrated that nutraceuticals are safer than or have fewer side effects than prescription medications. This holds true only when a specific nutraceutical is supplemented during a particular period. However, in a real case scenario, people may take several drugs along with nutraceuticals for skin care. Therefore, it is necessary to study drug-nutraceutical interactions to understand their safety.

REFERENCES

Astner, S., A. Wu, J. Chen, N. Philips, F. Rius-Diaz, C. Parrado, M. C. Mihm, D. A. Goukassian, M. A. Pathak, and S. González. 2007. Dietary lutein/zeaxanthin partially reduces photoaging and photocarcinogenesis in chronically UVB-irradiated Skh-1 hairless mice. *Skin Pharmacol. Physiol.* 20:283–291.

Balasubramanian, S. and R. L. Eckert. 2007. Curcumin suppresses AP1 transcription factor-dependent differentiation and activates apoptosis in human epidermal keratinocytes. *J. Biol. Chem.* 282:6707–6715.

Bangha, E., P. Elsner, and G. S. Kistler. 1997. Suppression of UV-induced erythema by topical treatment with melatonin (N-acetyl-5-methoxytryptamine). Influence of the application time point. *Dermatology* 195:248–252.

Belury, M. A., K. P. Nickel, C. E. Bird, and Y. Wu. 1996. Dietary conjugated linoleic acid modulation of phorbol ester skin tumor promotion. *Nutr. Cancer* 26:149–157.

Belury, M. A., C. J. Kavanaugh, and K. L. Liu. 2007. Conjugated linoleic acid modulates phorbol ester-induced PPARdelta and K-FABP mRNA expression in mouse skin. *Nutr. Res* 27:48–55.

Berth-Jones, J. and R. A. Graham-Brown. 1993. Placebo-controlled trial of essential fatty acid supplementation in atopic dermatitis. *Lancet* 341:1557–1560.

Brooker, S., S. Martin, A. Pearson, D. Bagchi, J. Earl, L. Gothard, E. Hall, L. Porter, and J. Yarnold. 2006. Double-blind, placebo-controlled, randomised phase II trial of IH636 grape seed proanthocyanidin extract (GSPE) in patients with radiation-induced breast induration. *Radiother. Oncol.* 79:45–51.

Cardoso, C. R., M. A. Souza, E. A. Ferro, S. Favoreto Jr., and J. D. Pena. 2004. Influence of topical administration of n-3 and n-6 essential and n-9 nonessential fatty acids on the healing of cutaneous wounds. *Wound Repair Regen.* 12:235–243.

Chaudhary, G., M. R. Saini, and P. K. Goyal. 2007. Chemopreventive potential of *Aloe vera* against 7,12-dimethylbenz(a)anthracene induced skin papillomagenesis in mice. *Integr. Cancer Ther.* 6:405–412.

Cos, S. and E. J. Sanchez-Barcelo. 2000. Melatonin and mammary pathological growth. *Front. Neuroendocrinol.* 21:133–170.

Dasgupta, T., A. R. Rao, and P. K. Yadava. 2003. Chemomodulatory action of curry leaf (*Murraya koenigii*) extract on hepatic and extrahepatic xenobiotic metabolising enzymes, antioxidant levels, lipid peroxidation, skin and forestomach papillomagenesis. *Nutr. Res.* 23:1427–1446.

Davis, R. H., M. G. Leitner, J. M. Russo, and M. E. Byrne. 1989. Wound healing. Oral and topical activity of *Aloe vera. J. Am. Podiatr. Med. Assoc.* 79:559–562.

Dreher, F., B. Gabard, D. A. Schwindt, and H. I. Maibach. 1998. Topical melatonin in combination with vitamins E and C protects skin from ultraviolet-induced erythema: A human study in vivo. *Br. J. Dermatol.* 139:332–339.

Eichler, O., H. Sies, and W. Stahl. 2002. Divergent optimum levels of lycopene, beta-carotene and lutein protecting against UVB irradiation in human fibroblastst. *Photochem. Photobiol.* 75:503–506.

Fuller, B., D. Smith, A. Howerton, and D. Kern. 2006. Anti-inflammatory effects of CoQ10 and colorless carotenoids. *J. Cosmet. Dermatol.* 5:30–38.

Guercio-Hauer, C., D. F. Macfarlane, and V. A. Deleo. 1994. Photodamage, photoaging and photoprotection of the skin. *Am. Fam. Physician* 50:327–332, 334.

Heggie, S., G. P. Bryant, L. Tripcony, J. Keller, P. Rose, M. Glendenning, and J. Heath. 2002. A Phase III study on the efficacy of topical *Aloe vera* gel on irradiated breast tissue. *Cancer Nurs.* 25:442–451.

Heinrich, U., C. Gartner, M. Wiebusch, O. Eichler, H. Sies, H. Tronnier, and W. Stahl. 2003. Supplementation with beta-carotene or a similar amount of mixed carotenoids protects humans from UV-induced erythema. *J. Nutr.* 133:98–101.

Heinrich, U., H. Tronnier, W. Stahl, M. Bejot, and J. M. Maurette. 2006. Antioxidant supplements improve parameters related to skin structure in humans. *Skin Pharmacol. Physiol.* 19:224–231.

Huang, M. T., R. C. Smart, C. Q. Wong, and A. H. Conney. 1988. Inhibitory effect of curcumin, chlorogenic acid, caffeic acid, and ferulic acid on tumor promotion in mouse skin by 12-O-tetradecanoylphorbol-13-acetate. *Cancer Res.* 48:5941–5946.

Hughes-Formella, B., O. Wunderlich, and R. Williams. 2007. Anti-inflammatory and skin-hydrating properties of a dietary supplement and topical formulations containing oligomeric proanthocyanidins. *Skin Pharmacol. Physiol.* 20:43–49.

Jagetia, G. C. and G. K. Rajanikant. 2004. Role of curcumin, a naturally occurring phenolic compound of turmeric in accelerating the repair of excision wound, in mice whole-body exposed to various doses of gamma-radiation. *J. Surg. Res.* 120:127–138.

Katiyar, S. K. 2002. Treatment of silymarin, a plant flavonoid, prevents ultraviolet light-induced immune suppression and oxidative stress in mouse skin. *Int. J. Oncol.* 21:1213–1222.

Katiyar, S. K. 2007. UV-induced immune suppression and photocarcinogenesis: Chemoprevention by dietary botanical agents. *Cancer Lett.* 255:1–11.

Katiyar, S. K., R. Agarwal, and H. Mukhtar. 1996. Inhibition of tumor promotion in SENCAR mouse skin by ethanol extract of *Zingiber officinale* rhizome. *Cancer Res.* 56:1023–1030.

Katiyar, S. K., N. J. Korman, H. Mukhtar, and R. Agarwal. 1997. Protective effects of silymarin against photocarcinogenesis in a mouse skin model. *J. Natl. Cancer Inst.* 89:556–566.

Katiyar, S. K., S. Meleth, and S. D. Sharma. 2008. Silymarin, a flavonoid from milk thistle (*Silybum marianum* L.), inhibits UV-induced oxidative stress through targeting infiltrating CD11b+ cells in mouse skin. *Photochem. Photobiol.* 84:266–271.

Kim, D. W., I. K. Hwang, D. W. Kim, K. Y. Yoo, C. K. Won, W. K. Moon, and M. H. Won. 2007. Coenzyme Q_{10} effects on manganese superoxide dismutase and glutathione peroxidase in the hairless mouse skin induced by ultraviolet B irradiation. *Biofactors* 30:139–147.

Kim, H. H., S. Cho, S. Lee, K. H. Kim, K. H. Cho, H. C. Eun, and J. H. Chung. 2006. Photoprotective and anti-skin-aging effects of eicosapentaenoic acid in human skin in vivo. *J. Lipid Res.* 47:921–930.

Kim, J. K., Y. Kim, K. M. Na, Y. J. Surh, and T. Y. Kim. 2007. [6]-Gingerol prevents UVB-induced ROS production and COX-2 expression in vitro and in vivo. *Free Radic. Res.* 41:603–614.

Kopcke, W. and J. Krutmann. 2008. Protection from sunburn with beta-carotene: A meta-analysis. *Photochem. Photobiol.* 84:284–288.

Kumar, C. A. and U. N. Das. 2000. Effect of melatonin on two stage skin carcinogenesis in Swiss mice. *Med. Sci. Monit.* 6:471–475.

Lin, F. H., J. Y. Lin, R. D. Gupta, J. A. Tournas, J. A. Burch, M. A. Selim, N. A. Monteiro-Riviere, J. M. Grichnik, J. Zielinski, and S. R. Pinnell. 2005. Ferulic acid stabilizes a solution of vitamins C and E and doubles its photoprotection of skin. *J. Invest. Dermatol.* 125:826–832.

Maenthaisong, R., N. Chaiyakunapruk, S. Niruntraporn, and C. Kongkaew. 2007. The efficacy of *Aloe vera* used for burn wound healing: A systematic review. *Burns* 33:713–718.

Maldonado, M. D., F. Murillo-Cabezas, J. R. Calvo, P. J. Lardone, D. X. Tan, J. M. Guerrero, and R. J. Reiter. 2007. Melatonin as pharmacologic support in burn patients: a proposed solution to thermal injury-related lymphocytopenia and oxidative damage. *Crit. Care Med.* 35:1177–1185.

Minghetti, P., S. Sosa, F. Cilurzo, P. J. LardoneJ, D. X. Tan, J. M. Guerrero, and R. J. Reiter. 2007. Evaluation of the topical anti-inflammatory activity of ginger dry extracts from solutions and plasters. *Planta Med.* 73:1525–1530.

Mittal, A., C. A. Elmets, and S. K. Katiyar. 2003. Dietary feeding of proanthocyanidins from grape seeds prevents photocarcinogenesis in SKH-1 hairless mice: Relationship to decreased fat and lipid peroxidation. *Carcinogenesis* 24:1379–1388.

National Cancer Institute. 2005. What You Need To Know About™ Skin Cancer. http://www.cancer.gov/pdf/WYNTK/WYNTK_skin.pdf. Accessed August 11, 2008.

Nigam, N., S. Prasad, and Y. Shukla. 2007. Preventive effects of lupeol on DMBA induced DNA alkylation damage in mouse skin. *Food Chem. Toxicol.* 45:2331–2335.

Nikiema, J. B., R. Vanhaelen-Fastre, M. Vanhaelen, J. Fontaine, C. De Graef, and M. Heenen. 2001. Effects of antiinflammatory triterpenes isolated from *Leptadenia hastata* latex on keratinocyte proliferation. *Phytother. Res.* 15:131–134.

Offord, E. A., J. C. Gautier, O. Avanti, C. Scaletta, F. Runge, K. Krämer, and L. A. Applegate. 2002. Photoprotective potential of lycopene, beta-carotene, vitamin E, vitamin C and carnosic acid in UVA-irradiated human skin fibroblasts. *Free Radic. Biol. Med.* 32:1293–1303.

Olsen, D. L., W. Raub Jr., C. Bradley, M. Johnson, J. L. Macias, V. Love, and A. Markoe. 2001. The effect of *Aloe vera* gel/mild soap versus mild soap alone in preventing skin reactions in patients undergoing radiation therapy. *Oncol. Nurs. Forum* 28:543–547.

Orengo, I. F., H. S. Black, A. H. Kettler, and J. E. Wolf Jr. 1989. Influence of dietary menhaden oil upon carcinogenesis and various cutaneous responses to ultraviolet radiation. *Photochem. Photobiol.* 49:71–77.

Orengo, I. F., H. S. Black, and J. E. Wolf Jr. 1992. Influence of fish oil supplementation on the minimal erythema dose in humans. *Arch. Dermatol. Res.* 284:219–221.

Park, K. K., K. S. Chun, J. M. Lee, S. S. Lee, and Y. J. Surh. 1998. Inhibitory effects of [6]-gingerol, a major pungent principle of ginger, on phorbol ester-induced inflammation, epidermal ornithine decarboxylase activity and skin tumor promotion in ICR mice. *Cancer Lett.* 129:139–144.

Patel, R., R. Krishnan, A. Ramchandani, and G. Maru. 2008. Polymeric black tea polyphenols inhibit mouse skin chemical carcinogenesis by decreasing cell proliferation. *Cell Prolif.* 41:532–553.

Prasad, N., S. Ramachandran, K. Pugalendi, and V. Menon. 2007. Ferulic acid inhibits UV-B-induced oxidative stress in human lymphocytes. *Nutr. Res.* 27:559–564.

Reiter, R. J., D. Melchiorri, E. Sewerynek, B. Poeggeler, L. Barlow-Walden, J. Chuang, G. G. Ortiz, and D. Acuña-Castroviejo. 1995. A review of the evidence supporting melatonin's role as an antioxidant. *J. Pineal. Res.* 18:1–11.

Reuter, J., A. Jocher, J. Stump, B. Grossjohann, G. Franke, and C. M. Schempp. 2008. Investigation of the anti-inflammatory potential of *Aloe vera* gel (97.5%) in the ultraviolet erythema test. *Skin Pharmacol. Physiol.* 21:106–110.

Rhodes, L. E., S. O'Farrell, M. J. Jackson, and P. S. Friedmann. 1994. Dietary fish-oil supplementation in humans reduces UVB-erythemal sensitivity but increases epidermal lipid peroxidation. *J. Invest. Dermatol.* 103:151–154.

Rhodes, L. E., H. Shahbakhti, R. M. Azurdia, R. M. Moison, M. J. Steenwinkel, M. I. Homburg, M. P. Dean, F. McArdle, G. M. Beijersbergen van Henegouwen, B. Epe, and A. A. Vink. 2003. Effect of eicosapentaenoic acid, an omega-3 polyunsaturated fatty acid, on UVR-related cancer risk in humans. An assessment of early genotoxic markers. *Carcinogenesis* 24:919–925.

Saleem, M., F. Afaq, V. M. Adhami, and H. Mukhtar. 2004. Lupeol modulates NF-kappaB and PI3K/Akt pathways and inhibits skin cancer in CD-1 mice. *Oncogene* 23:5203–5214.

Sevin, A., P. Oztas, D. Senen, U. Han, C. Karaman, N. Tarimci, M. Kartal, and B. Erdoğan. 2007. Effects of polyphenols on skin damage due to ultraviolet A rays: an experimental study on rats. *J. Eur. Acad. Dermatol. Venereol.* 21:650–656.

Sharma, S. D. and S. K. Katiyar. 2006. Dietary grape-seed proanthocyanidin inhibition of ultraviolet B-induced immune suppression is associated with induction of IL-12. *Carcinogenesis* 27:95–102.

Sharma, S. D., S. M. Meeran, and S. K. Katiyar. 2007. Dietary grape seed proanthocyanidins inhibit UVB-induced oxidative stress and activation of mitogen-activated protein kinases and nuclear factor-kappaB signaling in in vivo SKH-1 hairless mice. *Mol. Cancer Ther.* 6:995–1005.

Singer, A. J., S. A. McClain, A. Romanov, J. Rooney, and T. Zimmerman. 2007. Curcumin reduces burn progression in rats. *Acad. Emerg. Med.* 14:1125–1129.

Siwak, D. R., S. Shishodia, B. B. Aggarwal, and R. Kurzrock. 2005. Curcumin-induced antiproliferative and proapoptotic effects in melanoma cells are associated with suppression of IkappaB kinase and nuclear factor kappaB activity and are independent of the B-Raf/mitogen-activated/extracellular signal-regulated protein kinase pathway and the Akt pathway. *Cancer* 104:879–890.

Srinivasan, M., A. R. Sudheer, and V. P. Menon. 2007. Ferulic acid: Therapeutic potential through its antioxidant property. *J. Clin. Biochem. Nutr.* 40:92–100.

Stockfleth, E., H. Beti, R. Orasan, F. Grigorian, A. Mescheder, H. Tawfik, and C. Thielert. 2008. Topical Polyphenon E in the treatment of external genital and perianal warts: a randomized controlled trial. *Br. J. Dermatol.* 158:1329–1338.

Storey, A., F. McArdle, P. S. Friedmann, M. J. Jackson, and L. E. Rhodes. 2005. Eicosapentaenoic acid and docosahexaenoic acid reduce UVB- and TNF-alpha-induced IL-8 secretion in keratinocytes and UVB-induced IL-8 in fibroblasts. *J. Invest. Dermatol.* 124:248–255.

Sultana, S., M. Saleem, S. Sharma, and N. Khan. 2003. Lupeol, a triterpene, prevents free radical mediated macromolecular damage and alleviates benzoyl peroxide induced biochemical alterations in murine skin. *Indian J. Exp. Biol.* 41:827–831.

Svobodova, A., A. Zdarilova, D. Walterova, and J. Vostalova. 2007. Flavonolignans from *Silybum marianum* moderate UVA-induced oxidative damage to HaCaT keratinocytes. *J. Dermatol. Sci.* 48:213–224.

Syed, T. A., S. A. Ahmad, A. H. Holt, S. A. Ahmad, S. H. Ahmad, and M. Afzal. 1996. Management of psoriasis with *Aloe vera* extract in a hydrophilic cream: a placebo-controlled, double-blind study. *Trop. Med. Int. Health* 1:505–509.

Toklu, H. Z., T. Tunali-Akbay, G. Erkanli, M. Yuksel, F. Ercan, and G. Sener. 2007. Silymarin, the antioxidant component of *Silybum marianum*, protects against burn-induced oxidative skin injury. *Burns* 33:908–916.

Turkoglu, M. and N. Cigirgil. 2007. Evaluation of black tea gel and its protection potential against UV. *Int. J. Cosmet. Sci.* 29:437–442.

Urbach, F. 1978. Evidence and epidemiology of UV-induced carcinogenesis in man. *Natl. Cancer Inst. Monogr.* 50:5–10.

Wagner, H., L. Horhammer, and R. Munster. 1968. On the chemistry of silymarin (silybin), the active principle of the fruits from *Silybum marianum* (L.) Gaertn. (*Carduus marianus* L.) (in German). *Arzneimittelforschung* 18:688–696.

Williams, H. C. 2003. Evening primrose oil for atopic dermatitis. *BMJ* 327:1358–1359.

Wright, S. and J. L. Burton. 1982. Oral evening-primrose-seed oil improves atopic eczema. *Lancet* 2:1120–1122.

Young, A. J. and G. M. Lowe. 2001. Antioxidant and prooxidant properties of carotenoids. *Arch. Biochem. Biophys.* 385:20–27.

Yu, H., A. T. C. Tsin, and R. J. Reiter. 1992. Melatonin: history, biosynthesis, and assay methodology. In *Melatonin: Biosynthesis, Physiological Effects, and Clinical Applications*. Edited by Yu, H. and R. J. Reiter. Boca Raton, FL: CRC Press, pp. 1–16.

CHAPTER 18

Tranquilizing Medicinal Plants: Their CNS Effects and Active Constituents—Our Experience

Mariel Marder and Cristina Wasowski

CONTENTS

INTRODUCTION

Mental disorder is a psychological or behavioral pattern that occurs in an individual and is thought to cause distress or disability that is not expected as part of normal development or culture. Mental and neurological disorders are highly prevalent worldwide, with 450 million people estimated to be suffering from them. They are responsible for about 1% of deaths, and they account for almost 11% of disease burden the world over. The magnitude of neurological disorders is huge, and these disorders are priority health problems around the world. The extension of life expectancy and the aging of the general populations in both developed and still-developing countries are likely to increase the prevalence of many chronic and progressive physical and mental conditions, including neurological disorders. The proportionate share of the total global burden of disease attributable to neuropsychiatric disorders is projected to rise to 14.7% by 2020 (WHO).

There are many different categories of mental disorders and many different facets of human behavior and personality that can become disordered. Mental illnesses are classified according to the symptoms that a patient experiences, as well as the clinical features of the illness. Some of the major categories of mental illness include anxiety disorders, cognitive disorders, developmental disorders, dissociative disorders, mood disorders, personality disorders, schizophrenia, and substance abuse disorders.

"Anxiety" is defined as a subjective emotional state of uneasiness, not pleasant, and even fearful. When the anxiety reaches pathological levels, the subject experiences conductual changes, apprehension, motor troubles, sweating, and hypertension.

The term "sedation" implies a general slowing down of cognitive functioning, whereas a "hypnotic" specifically means the induction of sleep itself. Conversely, "tranquilization" signifies emotional calming that may or may not lead to sleep but does not induce the feeling of drowsiness.

Traditional medicine has many cures for these ailments, most of them based on herbal preparations; however, modern medicinal chemistry has provided several drugs that are more or less effective, for the same purpose. The most spectacular success was achieved in 1957 with the synthesis of the benzodiazepines [Sternbach 1978], which still are, after 50 years of intense clinical research and use, the near-best medication to treat mental disorders.

Benzodiazepines, however, also produce several side effects, such as sedation, muscle relaxation, alcohol incompatibility, amnesia, and addiction [Woods et al. 1992]. These drawbacks have to be carefully considered in clinical therapeutical applications.

Although benzodiazepines are laboratory products, they were found also in nature, and, appropriately, their first detection was in the mammalian brain [Sangameswaran et al. 1986]. They were then identified in many other sources, such as foods, rumen, plasma, and cow's and human's milk [Medina and Paladini 1993].

When we attempted detection of benzodiazepines in several plants, including some used to prepare tranquilizing infusions, we unexpectedly discovered that some flavonoids present in them were ligands for the benzodiazepine binding site of the gamma aminobutyric acid receptor type A ($GABA_A$) [Medina et al. 1989, 1997, 1998; Paladini et al., 1999; Marder and Paladini 2002].

A search for novel pharmacotherapy from medicinal plants for psychiatric illnesses has progressed significantly in the past 20 years. This is reflected in the large number of herbal preparations for which psychotherapeutic potential has been evaluated in a variety of animal models. A considerable number of herbal constituents, whose behavioral effects and pharmacological actions have been well characterized, may be good candidates for additional investigations that may ultimately result in clinical use. Herbal remedies that have demonstrable psychotherapeutic activities have provided a potential to psychiatric pharmaceuticals and deserve increased attention in future studies.

This chapter deals with plants possessing CNS effects. However, because of the huge amount of plants belonging to this category, we decided to select a few plants and to focus our attention on them, mostly concerning the constituents that have significant therapeutic effects in animal models of CNS disorders.

ANIMAL MODELS

An animal model is a nonhuman animal that has a disease or injury that is similar to a human condition. These test conditions are often termed as animal models of disease. The use of animal models allows researchers to investigate disease states in ways that would be inaccessible in a human patient, performing procedures on the nonhuman animal that imply a level of harm that would not be considered ethical to inflict on a human.

To serve as a useful model, a modeled disease must be similar in etiology (mechanism of cause) and function to the human equivalent. Animal models are used to learn more about a disease, its diagnosis, and its treatment. For instance, behavioral analogs of anxiety or pain in laboratory animals can be used to screen and test new drugs for the treatment of these conditions in humans.

Housing, handling, and experimental procedures complied with the recommendations set forth by national and international committees for the care and use of laboratory animals, and all efforts are taken to minimize animal suffering. The number of animals used is always the minimum number consistent with obtaining significant data.

Radioreceptor Binding Assays (*In Vitro*)

Radioligand binding assays are used to evaluate the putative action of the extracts or their constituents on different brain receptors in brain homogenates (bovine, rat, and mouse).

Pharmacological Studies Conducted on Mice (*In Vivo*)

Figure 18.1 shows the pharmacological assays used in our laboratories to assess behavior in mice.

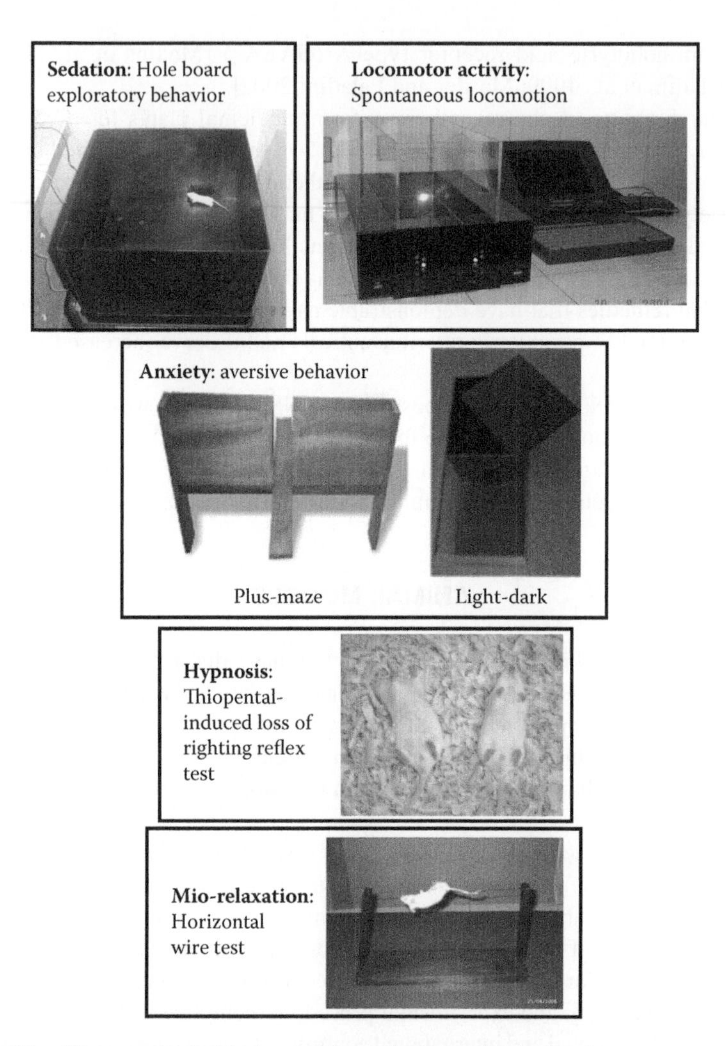

Figure 18.1 Pharmacological tests used for evaluating behavior in mice.

The Hole-Board Assay

The hole-board test, which was first introduced by Boissier and Simon [1962, 1964], offers a simple method for measuring the response of an animal to an unfamiliar environment. Previously, the hole-board test has been used to assess emotionality, anxiety, and/or responses to stress in animals [Rodriguez Echandia et al. 1987]. Some advantages of this test are that several behaviors can be readily observed and quantified, which makes possible a comprehensive description of the animals' behavior.

This assay is conducted in a walled, black Plexiglas arena with a floor of approximately 60 × 60 cm and 30-cm-high walls, with four centered and equally spaced holes in the floor, 2 cm in diameter each and illuminated by an indirect and dim light. Each animal, after the intraperitoneal injection of the vehicle or the drug, is placed in the center of the hole-board and allowed to freely explore the apparatus for 5 min; then, the number of holes explored, the time spent head dipping, and the number of rearings are measured. Changes in head-dipping behavior in the hole-board test reflect the anxiogenic and/or anxiolytic state in mice [Kliethermes and Crabbe 2006].

Assessment of Locomotor Activity

The spontaneous locomotion activity is measured in a box made of Plexiglas, with a floor of 30 × 15 cm and 15-cm-high walls. The locomotor activity is measured and expressed as total light beam counts during 5 min.

Sodium Thiopental-Induced Loss of Righting Reflex

A subhypnotic dose of sodium thiopental (35 mg/kg) is intraperitoneally injected to mice 20 min after a similar injection of the vehicle or the drug. The time of loss of righting reflex is determined as the interval between the loss and the recovery of the reflex [Ferrini et al. 1974]. The disappearance and the reappearance of the righting reflex are considered indications of duration of sleep.

The Elevated Plus-Maze Test

This test is based on the natural aversion of rodents for open spaces and uses a maze with two open and two closed arms (25 × 5 cm, each), with free access to all arms from the crossing point. The closed arms had walls 15 cm high all around. The maze is suspended 50 cm from the room floor. After administration of the drugs, mice are placed on the central part of the cross, facing an open arm. The number of entries and the time spent going into open and closed arms were counted during 5 min under red, dim light. A selective increase in the parameters corresponding to open arms reveals an anxiolytic effect. The total exploratory activity (number of entries in both arms) is also determined [Pellow et al. 1985].

Light/Dark Transition Test

The model is based on the innate aversion of rodents to brightly illuminated areas and on the spontaneous exploratory behavior in response to novel environment and light. The light/dark box consists of a Plexiglas box monitored with two compartments, distinguished by wall color, illumination, and size; one light area (30 × 21 × 21 cm, length × width × height) illuminated by a 60 W light in the ceiling of the compartment and with white walls, and a smaller dark compartment (14 × 21 × 21 cm, length × width × height) with black walls and not illuminated. An opening door (6 × 3 cm) located in the center of the partition at floor level connected the two compartments.

Animals are placed in the center of the dark or white area facing the wall opposite to the door. The following parameters were recorded during 5 min: (1) latency time of the first crossing to the light compartment, (2) the number of crossings between both compartments, (3) the total time spent in the illuminated zone of the cage, and (4) the overall movements in both areas [Bourin and Hascoët 2003]. A selective increase in the parameters corresponding to the light compartment and transitions reveals an anxiolytic effect.

Horizontal Wire Test

This assay is performed to evaluate myorelaxant effects of the drugs. The mice are lifted by the tail and allowed to grasp a horizontally strung wire (1 mm diameter, 15 cm long, and placed 20 cm above the table) with their forepaws, after which they are then released. The number of mice from each treatment group that do not grasp the wire with their forepaws within a 5 s period is recorded. A myorelaxant drug would impair the ability of the mice to grasp the wire. Muscle relaxation is commonly associated with sedation [Bonetti et al. 1982].

Seizure Testing

The pentylenetetrazol (PTZ) test is one of the models that represent the *in vivo* system most commonly used in the search for effective antiepileptic drugs.

In this assay, PTZ (200 mg/kg) is administered intraperitoneally to mice 15 min after injection of drug or vehicle. The number of mice showing clonic or tonic-clonic convulsions is determined.

SOME HERBS USED IN FOLKLORIC MEDICINE AS TRANQUILIZERS

Anxiolytics and sedatives essentially have the same underlying mechanisms of action: the stronger the agent, the greater the sedative effect, leading to coma in extreme cases. Four mechanisms of action have been implicated: (1) binding to GABA receptors leading to hyperpolarization of the cell membrane through increased influx of chlorine anions; (2) inhibition of excitatory amino acids, thereby also impairing the ability to form new memories; (3) sodium channel blockade, decreasing depolarization of the cell membrane; and (4) calcium channel blockade, decreasing the release of neurotransmitters into the synaptic cleft. Most complementary medicines prescribed for anxiolysis/sedation (e.g., kava kava, valerian, passion flower, and chamomile) could be GABAergic, although for some the mechanism of action remains unknown [Werneke et al. 2006].

Passiflora Species (Passion Flower)

The genus *Passiflora*, comprising about 500 species, is the largest in the family Passifloraceae (the passion flower family). The species of this genus are distributed in the warm temperate and tropical regions of America; they are much rarer in Asia, Australia, and tropical Africa. Several species are grown in the tropics for their edible fruits; the most widely grown is the *Passiflora edulis* Sims (passion fruit or purple granadilla). The discovery of several thousand year old seeds of *Passiflora* from the archaeological sites at Virginia and North America provides strong evidences of the prehistoric use of the fruits by the ancient "Red Indian" people.

The use of *Passiflora* as a medicine was lauded for the first time by a Spanish researcher Monardus in Peru in 1569, because the beautiful flowers of *Passiflora* appeared to him to be symbolic of the passion of Christ. Various species of *Passiflora* have been used extensively in the traditional system of therapeutics in many countries.

The extract of *Passiflora alata* (fragrant granadilla), with aloes, was reputed beneficial in atrophy of various parts. In Brazil, the said species, known as "Maracuja," has been put to use as an anxiolytic, sedative, diuretic, and an analgesic.

Passiflora capsularis is a reputed emmenagogue. *Passiflora contrayerva* is a counter-poison, deobstruent and cordial. *Passiflora edulis*, sometimes known as the "passionfruit," has been used as a sedative, diuretic, anthelmintic, antidiarrheal, stimulant, tonic, and also in the treatment of hypertension, menopausal symptoms, and colic of infants in South America. *Passiflora foetida* leaf infusion has been used to treat hysteria and insomnia in Nigeria. This plant is widely cultivated in India. The leaves are applied on the head for giddiness and headache; a decoction is given in biliousness and asthma.

The *Materia Medica Americana*, a Latin work published in Germany in 1787, mentions the use of *Passiflora incarnata* to treat epilepsy of the aged. An ancient report describes the use of this plant in spasmodic disorders and insomnia of infants and the old. *Passiflora incarnata* is a popular traditional European remedy as well as a homoeopathic medicine for insomnia and anxiety and has been used as a sedative tea in North America. The plant has been used (1) as an analgesic, antispasmodic, antiasthmatic, wormicidal, and sedative in Brazil, (2) as a sedative and narcoticin in Iraq, (3) in diseased conditions such as dysmenorrhea, epilepsy, insomnia, neurosis, and neuralgia in Turkey, (4) to cure hysteria and neurasthenia in Poland, and (5) to treat diarrhea, dysmenorrhea, neuralgia, burns, hemorrhoids, and insomnia in America.

Passiflora laurifolia Linn. (yellow granadilla, Jamaica honeysuckle) is used to treat nervous heart palpitations in Trinidad. The juice of *Passiflora maliformis* Linn. is used for intermittent fevers in Brazil. *Passiflora quadrangularis* Linn. (giant granadilla) is used throughout the Caribbean as a sedative and for headaches [Dhawan et al. 2004].

Passiflora coerulea L.

Passiflora coerulea (blue Passion flower), native to Brazil and introduced into Britain in the 17th century, is the most vigorous and tender species having traditional use of its fruit as a sedative and anxiolytic. In the West Indies, Mexico, the Netherlands, and South America, the root has been used as a sedative and vermifuge. In Italy, the plant has been used as an antispasmodic and sedative. In Mauritius, a

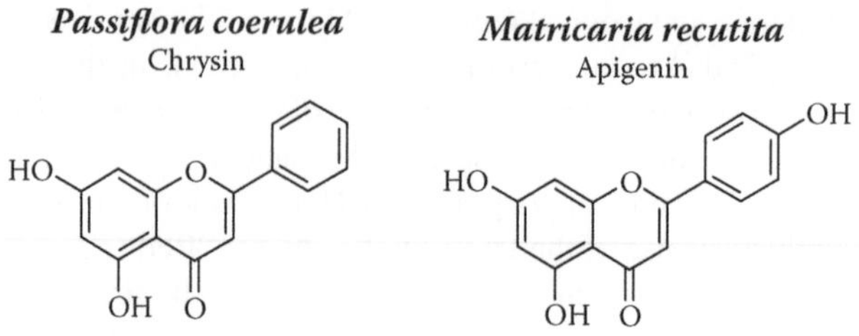

Passiflora coerulea
Chrysin

Matricaria recutita
Apigenin

HO O OH O

HO O OH OH O

Figure 18.2 Neuroactive flavones isolated from Pasionaria and chamomile.

tincture and an extract of the plant had been used as a remedy for insomnia caused by various nervous conditions but not caused by pain. The root has been used as a diuretic and a decoction of leaf as an emetic. In Argentina folk medicine, the aerial parts of *Passiflora coerulea* (where it is known as the Pasionaria or Mburucuyá in Guaraní) are used as mild antimicrobial agents in diseases, such as catarrh and pneumonia, and as a sedative.

Chrysin, 5,7-dihydroxyflavone (Figure 18.2), was isolated and identified from the dried branchlets of *Passiflora coerulea* L. (Passifloraceae). Chrysin was found to be a ligand for the benzodiazepine binding site in the $GABA_A$ receptor. Administered intraperitoneally to mice, chrysin was able to prevent the expression of tonic-clonic seizures induced by PTZ. Ro 15-1788, a central benzodiazepine receptor antagonist, abolished this effect. In addition, all of the treated mice lose the normal righting reflex, which suggests a depressant action of the flavonoid [Medina et al. 1990].

In the elevated plus-maze test, chrysin induced increases in the number of entries into the open arms and in the time spent on the open arms, consistent with an anxiolytic action. In the hole-board assay, chrysin increased the time spent head dipping. In the horizontal wire test, chrysin produced no effects. These data suggest that chrysin possesses anxiolytic actions without inducing sedation and muscle relaxation [Wolfman et al. 1994].

Matricaria recutita L. (Chamomile)

Chamomile has been used medicinally for thousands of years and is widely used in Europe. It is a popular treatment for numerous ailments, including sleep disorders, anxiety, digestion/intestinal conditions, skin infections/inflammation (including eczema), wound healing, infantile colic, teething pains, and diaper rash. In the United States, chamomile is best known as an ingredient in herbal tea preparations advertised for mild sedating effects.

German chamomile (*Matricaria recutita*) and Roman chamomile (*Chamaemelum nobile*) are the two major types of chamomile used for health conditions. They are believed to have similar effects on the body, whereas German chamomile may be slightly stronger. Although chamomile is widely used, there is not enough reliable research in humans to support its use for any condition.

The dried flower heads of *Matricaria recutita* L. (Asteraceae) are used in folk medicine to prepare a spasmolytic and sedative tea. The fractionation of the aqueous extract of this plant led to the detection of several fractions with significant affinity for the central benzodiazepine receptor and to the isolation and identification of 5,7,4′-trihydroxyflavone (apigenin) (Figure 18.2) in one of them. Apigenin competitively binds to the benzodiazepine binding site of the $GABA_A$ receptor, and it has no effect on muscarinic receptors, $\alpha 1$ adrenoceptors, or on the binding of muscimol to $GABA_A$ receptors. Apigenin has a clear anxiolytic activity in mice in the elevated plus maze without evidencing sedation or muscle relaxant effects at doses similar to those used for classical benzodiazepines, and no anticonvulsant action was detected. Electrophysiological studies performed on cultured cerebellar granule cells showed that apigenin reduced GABA-activated Cl^- currents in a dose-dependent manner. The effect was blocked by coapplication of Ro 15-1788, a specific benzodiazepine receptor antagonist [Viola et al. 1995; Avallone et al. 2000].

Other studies have also explored the behavioral effects of apigenin and chrysin in rats. These studies demonstrate that the two flavonoids were equally able to reduce locomotor activity when injected in rats. However, whereas chrysin exhibited a clear anxiolytic effect when injected at the dose of 1 mg/kg, apigenin has failed to exert this activity [Avallone et al. 2000; Zanoli et al. 2000].

Tilia Species (Linden)

Linden is the common name for the Tiliaceae, a family of chiefly woody shrubs and trees. Most genera are tropical, but the genus *Tilia*, or lime tree, in Europe and Asia, and basswood, in North America, is found throughout the North Temperate Zone. These deciduous trees are valued for ornament and shade. Their light, strong lumber, often called basswood, or whitewood, is variously used (e.g., for woodenware and cheap furniture and for beehives and honeycomb frames).

The dried flowers of these plants have been used widely in herbal teas, as a diuretic, stomachic, antineuralgic, sedative, and tranquilizer around the world. Despite the widespread use of the tea of linden in folk medicine, the number of scientific studies for the evaluation of its therapeutic use is limited.

In an attempt to add experimental confirmation to its popular medicinal use, a pharmacological profile of the chronic administration of the infusion of *Tilia petiolaris* DC., a deciduous tree native to South-East Europe and Western Asia, was performed.

The CNS-related effects of the infusion of *Tilia petiolaris* DC. inflorescences were evaluated in the hole-board, locomotor activity, and light-dark tests in mice. The results suggest that this infusion exerts an anxiolytic-like activity in mice [Loscalzo et al. 2008a]. The explanation of the activities noted must involve a knowledge of which compounds are present in the plant extract. Many studies have reported the CNS activity of *Tilia* extracts, but the isolation and identification of their bioactive principles is scarce. Some studies documented that aqueous extracts of linden produced sedative effects in mice [Coleta et al. 2001]. Viola et al. [1994] described the isolation of a pharmacologically active benzodiazepine binding site ligand from a fraction of the ethanolic extract of *Tilia tomentosa* and the anxiolytic effect exerted by a flavonoid fraction in mice. Recently, it was reported that the hexane, the methanol, and the aqueous extracts of *Tilia americana* var. *mexicana* demonstrated anxiolytic and sedative effects in mice [Aguirre-Hernandez et al. 2007a,b; Herrera-Ruiz et al. 2008; Pérez-Ortega et al. 2008].

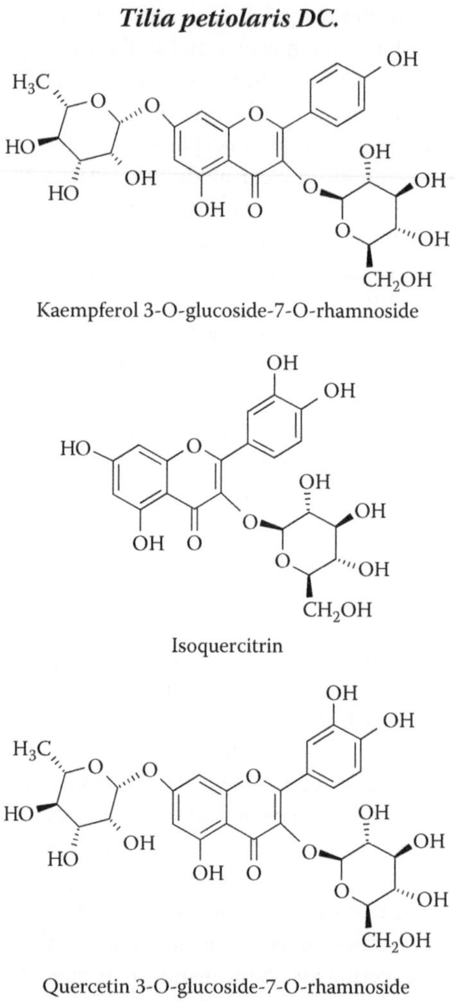

Tilia petiolaris DC.

Kaempferol 3-O-glucoside-7-O-rhamnoside

Isoquercitrin

Quercetin 3-O-glucoside-7-O-rhamnoside

Figure 18.3 Glycosilated flavonoids isolated from *Tilia*.

To identify the compounds responsible for the tranquilizing effects, pharmacological assay guided purification of a *Tilia petiolaris* DC. inflorescences ethanolic extract was performed. These studies resulted in the isolation and identification of three flavonoid glycosides: isoquercitrin, quercetin 3-*O*-glucoside-7-*O*-rhamnoside, and kaempferol 3-*O*-glucoside-7-*O*-rhamnoside (Figure 18.3). The behavioral actions of these compounds were examined in the hole-board, locomotor activity, and thiopental-induced loss of righting reflex tests in mice, showing clear depressant activities [Loscalzo et al. 2008b]. These results demonstrate the occurrence of neuroactive flavonoid glycosides in *Tilia*.

Salvia Species (Sage)

Salvia is a genus of plants in the mint family, Lamiaceae. It is one of three genera commonly referred to as sage. This genus includes approximately 700–900 species of shrubs, herbaceous perennials, and annuals with almost worldwide redistribution; the center of diversity and origin appears to be Central and South Western Asia.

The name *Salvia* derives from the Latin "salvere," which means "to heal." Indeed, this herb is highly regarded for its healing qualities.

Several types of *Salvia* are used medicinally, such as aromatic varieties (usually strongly scented leaves, also used as herbs), non-aromatic varieties (not considered medicinal, but many still have a scent), Chia sages, and Divinorum (Diviner's sage) that contains a diterpenoid used for spiritual and recreational purposes and Alzheimer's disease (research has shown that it improves cognitive function over a period of several months) [Grundmann et al. 2007]. The aromatic sages strengthen the lungs and can be used in teas or tinctures to prevent coughs.

Salvia species, such as *S. officinalis* L., *S. lavandulaefolia* Vahl., and *S. miltiorrhiza* Bung. are prominent for their reputed beneficial effects on memory disorders, depression, and cerebral ischemia [Perry et al. 2003].

S. elegans Vahl (Lamiaceae), popularly known as "mirto," is a shrub that has been used widely in Mexican traditional medicine for the treatment of different CNS diseases, principally anxiety. The antidepressant and anxiolytic-like effects of hydroalcoholic (60%) extract of *S. elegans* (leaves and flowers) were demonstrated in mice [Herrera-Ruiz et al. 2006].

Salvia guaranítica St. Hil.

S. guaranítica St. Hil., is also sometimes called Blue anise sage, anise-scented sage, Brazilian sage, giant blue sage, sapphire sage, or various other common names. It is a species of sage native to South America, including Brazil, Paraguay, Uruguay, and Argentina. It is a popular ornamental plant in mild areas where its leaves purportedly were used by the Guarani Indians of Brazil as a sedative.

Cirsiliol (5,3′, 4′-trihydroxi, 6, 7-dimethoxiflavone) (Figure 18.4) was isolated and identified by a bioguided purification of the ethanolic extract of the aerial parts

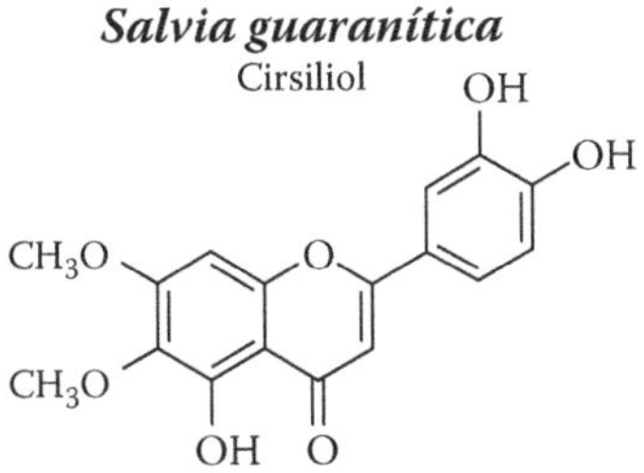

Figure 18.4 Sedative flavone isolated from *Salvia*.

of *S. guaranítica* St. Hil. Cirsiliol is a competitive low-affinity ligand for the benzodiazepine binding site of the $GABA_A$ and exerted sedative and hypnotic effects in mice without inducing anxiolysis, muscle relaxation, and prevention of seizures [Marder et al. 1996; Viola et al. 1997].

Aloysia polystachia (Griseb.) Moldenke ("Burrito")

The family Verbenaceae comprises about 175 genus and 2,300 species, distributed in the tropics and subtropics, mainly in the temperate zone of Southern Hemisphere.

Aloysia polystachya (Griseb.) Moldenke (Verbenaceae), popularly named "burrito" in Paraguayan folk medicine, is a well-known medicinal plant that has been used for a wide variety of indications, including digestive and respiratory tract disorders. Leaves of *A. polystachya* are used, in Argentina, for respiratory diseases (colds and cough), GI pain, as antiemetic and sedative remedy, or to treat "nervous diseases." Therapeutic actions of other species of *Aloysia* (i.e., *A. triphilla*) include febrifuge, sedative, stomachic, diuretic, and antispasmodic activities. However, no scientific references or experimental evaluation regarding its CNS activity or toxicity was found.

The study to analyze the behavioral effects of the crude hydroethanolic extract of the aerial parts of *A. polystachya* showed that it exhibits low toxicity, no lethality, did not induce any significant changes in several behavioral and physiological parameters, showed a slight decrease in spontaneous locomotor activity, and showed an increase in breath frequency. The extracts of this plant also evidenced anxiolytic and antidepressant effects in rodents, and these activities are mediated by another mechanism than the benzodiazepine binding site modulation of the $GABA_A$ receptor [Hellión-Ibarrola et al. 2005, 2008].

Valeriana Species

The underground organs of members of the genus *Valeriana* (Valerianaceae) as well as related genera such as *Nardostachys* are used in the traditional medicine of many cultures as mild sedatives and tranquilizers and to aid in the induction of sleep. *V. officinalis*, *V. wallichii*, *V. edulis*, and *V. fauriei* are the species most commonly used. It is remarkable that all these species are used for much the same purposes. This plant is still the subject of considerable research aimed at establishing the chemical and pharmacological basis of the activity that has been shown clearly in a number of animal and clinical studies [Houghton 1999].

Valerian is a good example of both the negative and positive aspects of herbal drugs. The considerable variation in its composition and content, as well as the instability of some of its constituents, cause serious problems for standardization, but the range of components that contribute to its overall activity suggest that it may correct a variety of underlying causes of conditions that necessitate a general sedative or tranquilizing effect.

Valerian preparations contained in commercially available products are extracted with water, water/methanol, or water/ethanol mixtures. In several placebo-controlled clinical studies, an improvement of sleep-related parameters after treatment with aqueous or ethanolic extracts was demonstrated.

In the search for the active substances of *Valeriana*, many compounds have been isolated and identified during the past 120 years, but it is as yet uncertain as to which of them are responsible for the recorded actions [Bos et al. 1996; Houghton 1999].

The most popular compounds, in this connection, are the valepotriates, the baldrinals, and valerenic acid derivatives, as well as some other members of the essential oil.

The presence of a volatile oil in *Valeriana* has been known for a long time, although its characteristic odor, which many find very unpleasant, is attributable to the release of isovaleric acid from some volatile oil components and other constituents by enzyme activity rather than to the oil itself. The composition of the volatile oil is very variable and depends on climate and other ecological factors. The oil contains monoterpenes, chiefly consisting of borneol and its acetyl and isovaleryl esters, but the sesquiterpene components are distinctive and have received most attention regarding their biological activity. Three major types of sesquiterpene skeleton are found, and these are exemplified by valerenic acid, valeranone, and kessyl glycol. The valerenic acid and kessyl ring systems are unique to the Valerianaceae. Valerenic acid has so far been found in no other organism apart from *V. officinalis*, whereas valeranone is found as the major component of the oil of *V. wallichii* and the related plant *N. jatamansii*. Compounds with the kessyl ring system are the major constituents in the volatile oil of Japanese valerian *V. fauriei* but are also found in *V. officinalis* oil [Houghton 1999].

There is certain evidence that valerenic acid may contribute to central effects of extracts derived from *V. officinalis*. It was shown to possess anticonvulsant properties [Hiller and Zetler 1996]. In animal experiments, valerenic acid showed tranquilizing and/or sedative activity [Bent et al. 2006]. This compound also modulates or, at high concentrations, activates $GABA_A$ receptors as shown for recombinant receptors expressed in *Xenopus* oocytes [Khom et al. 2007] or neonatal brainstem neurons [Yuan et al. 2004], and its specific binding site on $GABA_A$ receptors was identified [Benke et al. 2009].

The valepotriates consist of the furanopyranoid monoterpene skeleton commonly found in the glycosylated forms known as iridoids. They decompose rapidly to give homobaldrinal and related products. Valepotriates were also discussed to be determinant for central actions of valerian [Andreatini et al. 2002]. They are detectable at trace levels or not at all in recent drug preparations [European Scientific Cooperative on Phytotherapy 2003; Schulz and Hänsel 2004]. The presence of an epoxide group and its alkylating potential, in many of the valepotriates, has raised concerns about their cytotoxicity and consequent potential carcinogenicity.

An approach to detect new active substances in *Valeriana* extracts consists in searching for the presence of ligands for the principal brain receptors predominantly associated with anxiolytic, sedative, and/or sleep-enhancing properties [Marder and Paladini 2002]. Bodesheim and Hölzl [1997] found that the lignan (+) hydroxypinoresinol, present in *Valeriana* extracts, is a medium- to low-affinity ligand for the serotonin receptor, but its *in vivo* effects were not investigated.

In addition, the adenosine system may also account for the central action of valerian extracts because, in some studies, an interaction of the extracts, as well as of an olivil derivate at the adenosine A1 receptor, was observed [Müller et al. 2002; Schumacher et al. 2002].

However, several facts have cast doubts on the relevance of the described compounds to explain *Valeriana* extract effects. The principal of them are as follows: (1) the central depressant action of valepotriates, valeranone, and of the essential oil of *Valeriana* could not be demonstrated by a reduction of the glucose turnover in rat brain [Hölzl 1997]; (2) the sedative potency of these compounds is rather low (>30 mg/kg, in mice) [Hölzl 1997]; (3) the valepotriates rapidly decompose, and the baldrinals are chemically reactive and may form polymers [Bos et al. 1996]; hence both valepotriates and baldrinals disappear rapidly from the extracts; and (4) the roots and rhizomes of different *Valeriana* species show large differences with regard to their constituents.

Valeriana wallichii **DC. and** ***Valeriana officinalis*** **L.**

In our laboratory, we have applied the "ligand-searching approach" using, as far as possible, purified extracts, and we were able to report the presence of 6-methylapigenin (Figure 18.5) in *V. wallichii* and *V. officinalis* and to prove that it is a benzodiazepine binding site ligand. We have also made the first report of the presence of 2 S(−)hesperidin (Figure 18.5) in *V. wallichii* and in *V. officinalis* and of linarin (Figure 18.5) in *V. officinalis* and found that they have sedative and sleep-enhancing properties in mice. 6-Methylapigenin, in turn, had anxiolytic activity [Wasowski et al. 2002; Marder et al. 2003, 2005; Fernández et al. 2004]. The effective doses of these compounds are commensurable with their concentrations in the plant extracts and with the doses used in folkloric medicine.

Although isolating and identifying individual chemical constituents with relevant bioactivity provides a rational scientific basis for the medicinal use of a plant, synergistic effects in crude extracts are common. Synergistic interactions are of vital importance in phytomedicines and explain the difficulty in isolating a single active ingredient or to explain the efficacy of low concentrations of active constituents in an herbal product. This concept, that a whole or partially purified extract of a plant offers advantages over a single isolated ingredient, also underpins the philosophy in herbal medicine [Williamson 2001].

CONCLUSION

We have demonstrated that 6-methylapigenin was able to potentiate the sleep-enhancing properties of hesperidin together with a potentiation of the sedative effects by simultaneous administration of linarin with valerenic acid. It was also demonstrated that the potentiated effect of hesperidin is shared with various $GABA_A$

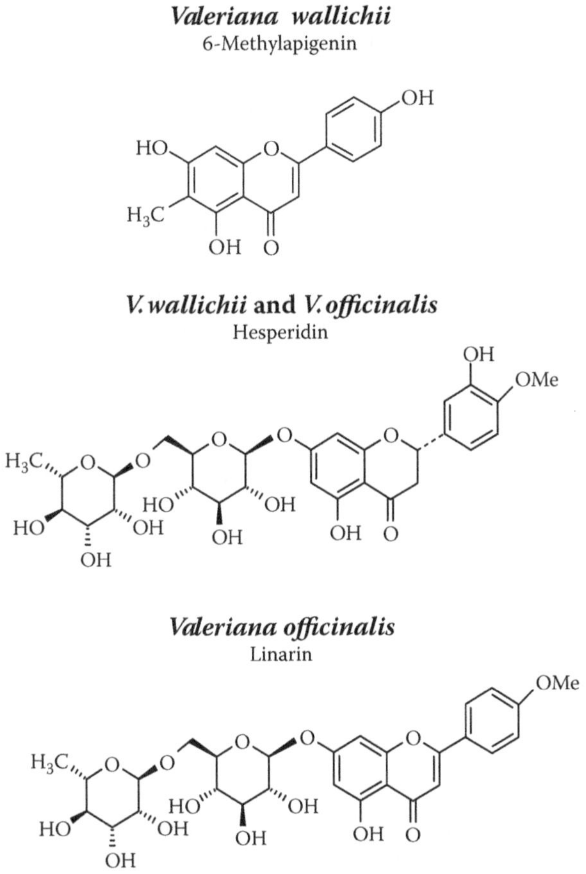

Valeriana wallichii

6-Methylapigenin

V. wallichii* and *V. officinalis

Hesperidin

Valeriana officinalis

Linarin

Figure 18.5 Neuroactive compounds isolated from *Valeriana*.

receptor ligands, among them various benzodiazepines widely used in human therapy (alprazolam, bromazepam, midazolam, and flunitrazepam) and with the classical agonist diazepam, in which a synergistic interaction was proved using an isobolar analysis. All the reported data up to now strongly suggest that the behavioral effects induced by hesperidin do not involve classical $GABA_A$ receptors, at least not directly [Fernández et al. 2005, 2006; Loscalzo et al. 2008].

We explored the participation of various other brain receptors besides $GABA_A$, namely opioid, serotonin (5-hydroxytryptamine type 2, $5\text{-}HT_2$) and α1 adrenoceptors, on hesperidin-sedative actions. The endogenous opioid system is critical for many physiological and behavioral effects. They play a major role in pain-controlling systems and also modulate affective behaviors. The serotoninergic system is known to modulate mood, sleep, and appetite and, thus, is implicated in the control of numerous behavioral and physiological functions. The α1 adrenoceptors are involved in locomotion, cognitive functions, and the control of motor activity.

The results obtained provide the first pharmacological evidence about the involvement of opioid receptors in the sedative and antinociceptive effects of hesperidin. Our results suggest a possible beneficial use of the association of hesperidin with benzodiazepines not only to improve human sedative therapy but also in the management of pain.

REFERENCES

Aguirre-Hernandez, E., A. L. Martinez, M. E. Gonzalez-Trujano, J. Moreno, H. Vibrans, and M. Soto-Hernandez. 2007a. Pharmacological evaluation of the anxiolytic and sedative effects of *Tilia americana L.* var. *mexicana* in mice. *J. Ethnopharmacol.* 109:140–145.

Aguirre-Hernández, E., H. Rosas-Acevedo, M. Soto-Hernández, A. L. Martínez, J. Moreno, M. E. González-Trujano. 2007b. Bioactivity-guided isolation of β-sitosterol and some fatty acids as active compounds in the anxiolytic and sedative effects of *Tilia americana* var. *mexicana. Planta Med.* 73:1148–1155.

Andreatini, R., V. A. Sartori, M. L. Seabra, and J. R. Leite. 2002. Effect of valepotriates (valerian extract) in generalized anxiety disorder: A randomized placebo-controlled pilot study. *Phytother. Res.* 16:650–654.

Avallone, R., P. Zanoli, G. Puia, M. Kleinschnitz, P. Schreier, and M. Baraldi. 2000. Pharmacological profile of apigenin, a flavonoid isolated from *Matricaria chamomilla. Biochem. Pharmacol.* 59:1387–1394.

Benke, D., A. Barberis, S. Kopp, K. Altmann, M. Schubiger, K. E. Vogt, U. Rudolph, and H. Möhler. 2009. $GABA_A$ receptors as in vivo substrate for the anxiolytic action of valerenic acid, a major constituent of valerian root extracts. *Neuropharmacology* 56:174–181.

Bent, S., A. Padula, D. Moore, M. Patterson, and W. Mehling. 2006. Valerian for sleep: A systematic review and metaanalysis. *Am. J. Med.* 119:1005–1012.

Bodesheim, U. and J. Hölzl. 1997. Isolierung, Strukturaufklörung und Radiorezeptorassays von Alkaloiden und Lignanen aus *Valeriana officinalis. L. Pharmazie* 52:387–391.

Boissier, J. R. and P. Simon. 1962. La reaction dexploration chez la souris. *Therapie* 17:1225–1232.

Boisser, J. R. and P. Simon. 1964. Dissociation de deux composantes dans le compartment dinvestigation de la souris. *Arch. Int. Pharmacodyn.* 147:372–387.

Bonetti, E. P., E. L. Pierri, R. Cumin, M. Schaffner, E. R. Gamzu, R. Muller, and W. Haefely. 1982. Benzodiazepine antagonist RO 15-1788: Neurological and behavioral effects. *Psychopharmacology* 78:8–18.

Bos, R., H. Hendriks, J. J. C. Scheffer, and H. J. Woerdenbag. 1998. Cytotoxic potential of valerian constituents and valerian tinctures. *Phytomedicine* 5:219–225.

Bourin, M. and M. Hascoët. 2003. The mouse light/dark box test. *Eur. J. Pharmacol.* 463:55–65.

Coleta, M., M. G. Campos, M. D. Cotrim, and A. Proenca da Cunha. 2001. Comparative evaluation of *Melissa officinalis* L., *Tilia europaea* L., *Passiflora edulis* Sims. and *Hypericum perforatum* L. in the elevated plus maze anxiety test. *Pharmacopsychiatry* 34 (Suppl. 1):S20–S21.

Dhawan, K., S. Dhawan, and A. Sharma. 2004. *Passiflora*: A review update. *J. Ethnopharmacol.* 94:1–23.

European Scientific Cooperative on Phytotherapy. 2003. ESCOP Monographs. Stuttgart, Germany: Thieme.

Fernández, S. P., C. Wasowski, L. M. Loscalzo, R. E. Granger, G. A. R. Johnston, A. C. Paladini, and M. Marder. 2006. Central nervous system depressant action of flavonoid glycosides. *Eur. J. Pharmacol.* 539:168–176.

Fernández, S. P., C. Wasowski, A. C. Paladini, and M. Marder. 2004. Sedative and sleep-enhancing properties of linarin, a flavonoid-isolated from *Valeriana officinalis*. *Pharmacol. Biochem. Behav.* 77:399–404.

Fernández, S. P., C. Wasowski, A. C. Paladini, and M. Marder. 2005. Synergistic interaction between hesperidin, a natural flavonoid, and diazepam. *Eur. J. Pharmacol.* 512:189–198.

Ferrini, R., G. Miragoli, and B. Taccardi. 1974. Neuro-pharmacological studies on SB 5833, a new psychotherapeutic agent of the benzodiazepine class. *Arzneim.-Forsch. (Drug Res.)* 24:2029–2032.

Grundmann, O., S. M. Phipps, I. Zadezensky, and V. Butterweck. 2007. *Salvia divinorum* and salvinorin A: An update on pharmacology and analytical methodology. *Planta Med.* 73:1039–1046.

Hellión-Ibarrola, M. C., D. A. Ibarrola, Y. Montalbetti, M. L. Kennedy, O. Heinichen, M. Campuzano, J. Tortoriello, S. Fernández, C. Wasowski, M. Marder, T. C. De Lima, and S. Mora. 2006. The anxiolytic-like effects of *Aloysia polystachya* (Griseb.) Moldenke (Verbenaceae) in mice. *J. Ethnopharmacol.* 105:400–408.

Hellión-Ibarrola, M. C., D. A. Ibarrola, Y. Montalbetti, M. L. Kennedy, O. Heinichen, M. Campuzano, E. A. Ferro, N. Alvarenga, J. Tortoriello, T. C. De Lima, and S. Mora. 2008. The antidepressant-like effects of *Aloysia polystachya* (Griseb.) Moldenke (Verbenaceae) in mice. *Phytomedicine* 15:478–483.

Herrera-Ruiz, M., Y. García-Beltrán, S. Mora, G. Díaz-Véliz, G. S. Viana, J. Tortoriello, and G. Ramírez. 2006. Antidepressant and anxiolytic effects of hydroalcoholic extract from *Salvia elegans*. *J. Ethnopharmacol.* 107:53–58.

Herrera-Ruiz, M., R. Román-Ramos, A. Zamilpa, J. Tortoriello, and J. E. Jiménez-Ferrer. 2008. Flavonoids from *Tilia americana* with anxiolytic activity in plus-maze test. *J. Ethnopharmacol.* 118:312–317.

Hiller, K. O. and G. Zetler. 1996. Neuropharmacological studies on ethanol extracts of *Valeriana officinalis* L.: Behavioural and anticonvulsant properties. *Phytother. Res.* 10:145–151.

Hölzl, J. 1997. The pharmacology and therapeutics of valerian. In *Valerian: The genus Valeriana: medicinal and aromatic plants. Industrial Profiles*. Edited by Houghton, P. J. Amsterdam, The Netherlands: Harwood Academic Publishers, pp. 55–75.

Houghton, P. J. 1999. The scientific basis for the reputed activity of Valerian. *J. Pharm. Pharmacol.* 51:505–512.

Khom, S., I. Barburin, E. Timin, A. Hohaus, G. Tanner, B. Kopp, and S. Hering. 2007. Valerenic acid potentiates and inhibits $GABA_A$ receptors: Molecular mechanism and subunit specificity. *Neuropharmacology* 53:178–187.

Kliethermes, C. L. and J. C. Crabbe. 2006. Pharmacological and genetic influences on hole board behaviors in mice. *Pharmacol. Biochem. Behav.* 85:57–65.

Loscalzo, L. M., C. Wasowski, and M. Marder. 2008a. The truth about Tilia teas: Their CNS effects. III. Argentinean Medicinal Chemistry Workshop, Argentinean Association of Chemistry (AQA), Los Cocos, Córdoba, Argentina, November 2008.

Loscalzo, L. M., C. Wasowski, and M. Marder. 2008b. Neuroactive flavonoid glycosides from *Tilia petiolaris* DC. extracts. From Cell to Society 6, ASN event, Blue Mountain, Australia, November 2008.

Marder, M., J. H. Medina, A. C. Paladini, H. Viola, and C. Wasowski. 2005. Sedative materials and treatments. Patent WO/2003/061678. EP 01.12.2004 2003701586. Published February 9, 2005. Issued July 23, 2008.

Marder, M. and A. C. Paladini. 2002. $GABA_A$-Receptor ligands of flavonoid structure. *Curr. Topics Med. Chem.* 2:853–867.

Marder, M., H. Viola, C. Wasowski, S. Fernandez, J. H. Medina, and A. C. Paladini. 2003. 6-methylapigenin and hesperidin: New valeriana flavonoids with activity on the CNS. *Pharmacol. Biochem. Behav.* 75:537–545.

Marder, M., C. Wasowski, H. Viola, C. Wolfman, P. Waterman, J. H. Medina, and A. C. Paladini. 1996. Cirsiliol and caffeic acid ethyl ester, isolated from *Salvia guaranitica*, are competitive ligand for the central benzodiazepine receptor. *Phytomedicine* 3:29–31.

Medina, J. H. and A. C. Paladini. 1993. *Naturally occurring benzodiazepines.* Edited by Izquierdo, I. and J. H. Medina. New York, NY: Ellis Horwood, pp. 28–43.

Medina, J. H., A. C. Paladini, C. Wolfman, M. Levi de Stein, D. Calvo, L. E. Diaz, and C. Peña. 1990. Chrysin (5,7-di-OH-flavone), a naturally-occurring ligand for benzodiazepine receptors, with anticonvulsant properties. *Biochem. Pharmacol.* 40:2227–2231.

Medina, J. H., C. Peña, M. Levi de Stein, C. Wolfman, and A. C. Paladini. 1989. Benzodiazepine-like molecules as well as other ligands for the brain benzodiazepine receptor are relatively common constituents of plants. *Biochem. Biophys. Res. Comm.* 165:547–553.

Medina, J. H., H. Viola, C. Wolfman, M. Marder, C. Wasowski, D. Calvo, and A. C. Paladini. 1997. Overview: Flavonoids. A new family of benzodiazepine receptor ligands. *Neurochem. Res.* 22:419–425.

Medina, J. H., H. Viola, C. Wolfman, M. Marder, C. Wasowski, D. Calvo, and A. C. Paladini. 1998. Neuroactive flavonoids: New ligands for the benzodiazepine receptors. *Phytomedicine* 5:235–243.

Müller, C., B. Schumacher, A. Brattström, E. A. Abourashed, and U. Koetter. 2002. Interaction of valerian extracts and a fixed valerian-hop extract combination with adenosine receptors. *Life Sci.* 71:1939–1949.

Paladini, A. C., M. Marder, H. Viola, C. Wolfman, C. Wasowski, and J. H. Medina. 1999. Flavonoids and the central nervous system: From forgotten factors to potent anxiolytic compounds. *J. Pharm. Pharmacol.* 51:519–526.

Pellow, S., P. Chopin, S. E. File, and M. Briley. 1985. Validation of open:closed arm entries in an elevated plus-maze as a measure of anxiety in the rat. *J. Neurosci. Methods* 14:149–167.

Pérez-Ortega, G., P. Guevara-Fefer, M. Chávez, J. Herrera, A. Martínez, A. L. Martínez, and M. E. González-Trujano. 2008. Sedative and anxiolytic efficacy of *Tilia americana* var. *mexicana* inflorescences used traditionally by communities of State of Michoacan, Mexico. *J. Ethnopharmacol.* 116:461–468.

Perry, N. S. L., C. Bollen, E. K. Perry, and C. Ballard. 2003. Salvia for dementia therapy: Review of pharmacological activity and pilot tolerability clinical trial. *Pharmacol. Biochem. Behav.* 75:651–659.

Rodriguez Echandia E. L., S. T. Broitman, and M. R. Foscolo. 1987. Effect of the chronic ingestion of chlorimipramine and desipramine on the hole board response to acute stresses in male rats. *Pharmacol. Biochem. Behav.* 26:207–210.

Sangameswaran, L., H. M. Fales, P. Friedrich, and A. L. De Blas. 1986. Purification of a benzodiazepine from bovine brain and detection of benzodiazepine-like immunoreactivity in human brain. *Proc. Natl. Acad. Sci. USA* 83:9236–9240.

Schulz, V. and R. Hänsel. 2004. *Rationale Phytotherapie*. Berlin, Germany: Springer.

Schumacher, B., S. Scholle, J. Hölzl, N. Khudeir, S. Hess, and C. E. Müller. 2002. Lignans isolated from Valerian: Identification and characterization of a new olivil derivative with partial agonistic activity at A1 adenosine receptors. *J. Nat. Prod.* 65:1479–1485.

Sternbach, L. H. 1978. The benzodiazepine story. *Prog. Drug Res.* 22:229–266.

Viola, H., C. Wasowski, M. Levi de Stein, C. Wolfman, R. Silveira, F. Dajas, J. H. Medina, and A. C. Paladini. 1995. Apigenin, a component of *Matricaria recutita* flowers, is a central benzodiazepine receptors-ligand with anxiolytic effects. *Planta Med.* 61:213–216.

Viola, H., C. Wasowski, M. Marder, C. Wolfman, A. C. Paladini, and J. H. Medina. 1997. Sedative and hypnotic properties of *Salvia guaranítica* St. Hill. and of its active principle, Cirsiliol. *Phytomedicine* 4:45–50.

Viola, H., C. Wolfman, M. Levi de Stein, C. Wasowski, C. Peña, J. H. Medina, and A. C. Paladini. 1994. Isolation of pharmacologically active benzodiazepine receptor ligands from *Tilia tomentosa* (Tiliaceae). *J. Ethnopharmacol.* 44:47–53.

Wasowski, C., M. Marder, H. Viola, J. H. Medina, and A. C. Paladini. 2002. Isolation and identification of 6-methylapigenin, a competitive ligand for the brain GABA(A) receptors, from *Valeriana wallichii. Planta Med.* 68:934–936.

Werneke, U., T. Turner, and S. Priebe. 2006. British complementary medicines in psychiatry. Review of effectiveness and safety. *J. Psychiatry* 188:109–121.

Williamson, E.M.. 2001. Synergy and other interactions in phytomedicines. *Phytomedicine* 8:401–409.

Wolfman, C., H. Viola, A. C. Paladini, F. Dajas, and J. H. Medina. 199-4. Possible anxiolytic effects of chrysin, a central benzodiazepine receptor ligand isolated from *Passiflora coerulea. Pharmacology Biochem. Behav.* 47:1–4.

Woods, J. H., Katz J. L., G. Winger. 1992. Benzodiazepines: Use, abuse and consequences. *Pharmacol. Rev.* 44:151–347.

Yuan, C. S., S. Mehendale, Y. Xiao, H. H. Aung, J. T. Xie, and M. K. AngLee. 2004. The GABAergic effects of valerian and valerenic acid on rat brain stem neuronal activity. *Anesth. Analg.* 98:353–358.

Zanoli, P., R. Avallone, and M. Baraldi. 2000. Behavioral characterisation of the flavonoids apigenin and chrysin. *Fitoterapia* 71 (Suppl 1.): S117–S123.

CHAPTER 19

Dietary Foods

Pamela Mason

CONTENTS

INTRODUCTION

Nutraceuticals find therapeutic application in a wide range of food products. These include foods and drinks enriched or fortified with vitamins, minerals, fatty acids, proteins and amino acids, fiber, plant phytochemicals, probiotic bacteria, fruit extracts, traditional spices, and herbal ingredients. These products could have a key role to play in health enhancement and/or disease risk reduction in the 21st century. In the developed world, a high proportion of the diet is derived from processed and convenience foods. In this context, specific dietary foods and drinks appropriate for an individual's lifestyle, health status, and genotype could make an important contribution to a healthy diet.

Moreover, demographic change, an increasing incidence of degenerative diseases, many as the result of the global obesity epidemic, and the concomitant rise in the costs of healthcare are increasing the pressure to shift from treatment to prevention. This change in emphasis could include the substitution of dietary foods for drugs to provide specific health benefits. Examples of currently available food ingredients that have shown some promise and could potentially be used as substitutes or partial replacements for drugs (depending on the generation of an adequate evidence base) include the following:

- Beta-glucan for controlling blood glucose as a substitute for oral hypoglycemic drugs and insulin
- Probiotic bacteria (instead of antidiarrheal drugs) for diarrhea
- Probiotics as potential substitutes for aminosalicylates and corticosteroids in the management of IBD (e.g., Crohn's disease, ulcerative colitis)
- Probiotics to assist in the eradication of Helicobacter pylori together with acid suppressant and antibiotic therapy
- Prebiotics instead of laxatives in the management of infrequent bowel function
- Bioactive peptides as substitutes for antihypertensive medication in the management of high blood pressure
- Phytosterols, soy proteins, and soluble fiber as substitutes for statins in cholesterol lowering
- Chitosan or CLA as substitutes for orlistat and other antiobesity drugs
- Omega-3 fatty acids for mental health conditions (e.g., depression, schizophrenia) instead of antidepressants and antipsychotics
- Melatonin for promotion of sleep instead of benzodiazepines

A dietary approach is financially logical for healthcare systems in that foods are not only cheaper than medicines but are also purchased by consumers rather than by the government or the insurer [Institute of Grocery Distribution 2003]. This trend toward self care is already apparent, partly as a reaction against conventional medicine but also because many individuals feel good about taking control of their health. Dietary change, including the consumption of dietary foods, can represent one form of self care. However, this shift to self care requires educated and informed consumers. Individuals need to know what dietary actions they should be taking, as do the professionals who advise them. This requires an effective evidence base around the benefits of specific foods for specific individuals and population groups. There is therefore a need for the food industry to demonstrate that individual foods consumed in normal amounts can improve health or influence disease occurrence.

THE MARKET

The market for dietary foods and supplements is large, although the year-to-year increase appears to have slowed somewhat in recent years. A recent report found that the so-called "functional foods" sector increased worldwide by 8.3% in value terms in the year to September 2007, a substantial slowing of growth from the 22.1% rise shown in the year to September 2006 [Key Note Publications 2008]. This was accounted for by a decline in sales of probiotic yogurt drinks, whereas sales of cholesterol-lowering spreads increased only slightly. However, there was positive growth in probiotic yogurts, soya milks, and fortified breakfast cereals. A recent Mintel report noted that the U.K. functional foods sector grew by only 3% in 2007 compared with more than 20% in 2006 [Mintel 2008], yet Mintel still predicted a U.K. growth of 72% between 2007 and 2012, with sales increasing from £613 million to £1 billion over this five year period [Mintel 2008].

DEFINITIONS

Dietary foods encompass an enormous range of products, usually produced by adding nutraceuticals, including nutrients and other potentially health-promoting ingredients, and/or by removing or reducing potentially less healthy ingredients (e.g., saturated fat, sugar, salt). Increasingly, this term is also used to include those conventional foods that are promoted for their favorable nutritional properties but have no added ingredients. Oats, for example, are marketed on the basis of their soluble fiber content, whereas certain breakfast cereals are promoted for their whole-grain and fiber content.

Such foods can be regarded as products eligible for health claims. This helps to improve product marketability and communication of potential health and therapeutic benefits to the consumer. Dietary foods produced for this purpose are variously categorized as "functional foods," "designer foods," "vita foods," "medifoods," "alicaments," and "pharmafoods."

There is no consensus on the definition of any these terms, but collectively they refer to foods and beverages constructed to confer health and/or therapeutic benefits beyond the nutritional value of the foods themselves. In the early 1990s, the term "nutraceutical" might have been included in this category because it was originally defined as a "food (or part of a food) that provides medical or health benefits, including the prevention and/or treatment of a disease" [DeFelice 2002]. In 1999, however, whole foods were distinguished from the natural bioactive compounds derived from them by using the term functional foods to describe the former and nutraceuticals to describe the latter [Zeose 1999]. Under this newer definition, nutraceuticals are functional ingredients derived from foods and are formulated as powders, pills, and other pharmaceutical forms not generally considered to be foods. However, this definition does not exclude the possibility that nutraceuticals can be extracted from some conventional foods (e.g., berry extracts, fish oils) and then added to other food products to produce so-called functional foods. In such cases, the food product could be described as a functional food, the added ingredient as a nutraceutical.

None of these terms has any regulatory definition. The term nutraceutical, for example, is not recognized by the FDA or the European Scientific Committee on Food. Moreover, some nutraceuticals or functional ingredients are sold as supplements or ingredients for potential addition to foods in some countries, whereas in other countries these substances are sold as drugs requiring medical prescription.

In contrast to a nutraceutical or dietary supplement, a functional food is a food or drink product consumed as part of the daily diet [Scientific concepts of functional foods in Europe 1999; Halstead 2003]. Functional foods cannot easily be defined because they are not single, well-characterized entities, although many definitions have been produced. These definitions range from simple statements, such as "foods that may provide health benefits beyond basic nutrition" and "foods marketed with the message of benefit to health" to more complex definitions, such as "foods derived from naturally occurring substances consumed as part of the daily diet and possessing particular physiological benefits when ingested" or "foods that encompass potentially helpful products, including any modified food or food ingredient that may provide a health benefit beyond that of the traditional nutrient it contains" [Roberfroid 2002]. The term functional food covers a wide variety of food products, with a variety of components, both nutrients and non-nutrients, influencing a range of body functions associated with health and/or reduction in disease risk. It has been argued that functional food should therefore be understood as a concept rather than a term to be defined [Roberfroid 2002].

The concept of functional foods was developed in Japan. During the early 1980s, the Japanese government funded research studies to evaluate the functionalities of various foods. This was followed in 1991 by the establishment of a Foods for Specified Health Uses (FOSHU) system, designed to introduce a legal category of foods with potential benefits as part of a national effort to control the escalating costs of healthcare [Omaha, Ikeda, and Moriayama 2006]. According to the Japanese Ministry of Health, Labour, and Welfare, FOSHU include the following:

- Foods that are expected to have a specific health effect attributable to relevant constituents or foods from which allergens have been removed are considered FOSHU.
- Also included are foods in which the effect of such addition or removal has been scientifically evaluated and permission is granted to make claims regarding their specific beneficial effects on health.

- To be identified as FOSHU, evidence is required that the final food product, but not isolated individual components, is likely to exert a health or physiological effect when consumed as part of an ordinary diet. FOSHU foods should be in the form of ordinary foods (i.e., not tablets or capsules).

The functional food concept has been further developed in the United States and in Europe. In the 1990s, the International Life Science Institute in Europe developed a functional food project designed to assess the state of the art in functional foods. Known as Functional Food Science in Europe (FUFOSE), this European Commission concerted action involved large numbers of European experts in nutrition and related sciences, who elaborated, for the first time, a global framework that included a strategy for the identification and development of functional foods and for the scientific substantiation of their effects to justify health-related claims [Scientific concepts of functional foods in Europe 1999].

According to this European consensus, a food can be regarded as "functional" if it is satisfactorily demonstrated to affect beneficially one or more "target functions" in the body, beyond adequate nutritional effects in a way that is relevant to either an improved state of health and well being and/or to the reduction of risk of a disease. In this context, target function refers to genomic, biochemical, physiological, or behavioral functions [Roberfroid 2002]. According to this consensus, the unique features of functional foods are as follows [Scientific concepts of functional foods in Europe 1999]:

- Are conventional or everyday food
- Are consumed as part of the normal diet
- Are composed of naturally occurring component(s), sometimes in unnatural proportions
- Have a positive effect on the target function beyond nutritive value
- May enhance well-being and health and/or reduce the risk of disease to provide health benefits so as to improve quality of life, including physical, psychological, and behavioral performances
- Have authorized and scientifically substantiated health claims

The FUFOSE document emphasizes the food nature of functional foods. They are not tablets or capsules or any form of dietary supplement but are consumed as part of a normal dietary pattern and must demonstrate their effects in amounts that can normally be expected to be consumed in the diet. Beneficial effects must be demonstrated to the satisfaction of the scientific community and can be used to justify health claims, an enhanced function claim, or a disease risk reduction claim [Roberfroid 2002]. In essence, therefore, functional foods can be regarded as food products eligible for health claims.

LEGISLATION AND CLAIMS

Legislation throughout much of the world does not recognize functional foods as a distinct category of foods, as for example in Japan. This means that functional foods must comply with all relevant food legislation with respect to composition, labeling, and claims. Claims fall into two categories: medicinal claims and health claims. A medicinal claim states or implies that a food has the capacity to treat, prevent, or cure human disease or makes reference to such a property. Medicinal claims are currently prohibited worldwide for food and drink products. However, in a few cases, medicinal licenses have been granted for certain food supplements (e.g., folic acid may prevent neural tube defects).

A health claim is a direct, indirect, or implied claim in food labeling, advertising, and promotion indicating that consumption of a food carries a specific health benefit (e.g., dietary fiber can help maintain a healthy gut) or reduces the risk of a specific health detriment or disease risk factor (e.g., soy protein can help reduce LDL cholesterol). Health claims have been allowed in the United States since 1993 on certain foods. These contain components for which the FDA has accepted there is objective evidence for a correlation between nutrients or foods in the diet and specific disease risks on the basis of "the totality of publicly available scientific evidence" and in which there is substantial agreement among qualified experts that the claims were supported by the evidence. Twelve types of food components are currently approved by the FDA to make health claims [U.S. Food and Drug Administration 2008].

In Europe, until recently, there has been no harmonized legislation on health claims, and claims have been dealt with at a national level. However, health claims legislation has recently been passed in the European Union, and foods outside of the scope of the medicinal law will be able to make health claims and reduction of disease risk claims, following scientific evaluation by the European Food Safety Authority. An approved list of claims is expected by 2010.

DEVELOPMENT OF FUNCTIONAL FOODS

The following can be considered a functional food:

- A food to which a component(s) has been added during the processing stage to provide benefits. This component(s) can be an officially recognized nutrient (e.g., the addition of folic acid to reduce a woman's risk of having an infant with a neural tube defect or calcium to contribute to bone health), a non-nutrient (e.g., soluble fiber to promote heart health, phytosterol to reduce serum cholesterol, probiotic bacteria to improve GI health), or an herb (e.g., *Ginkgo biloba* or ginseng to improve alertness, *Echinacea* to support the immune system).
- A food from which a component has been removed so that the food has reduced adverse health effects (e.g., the reduction of *trans* fatty acids by the removal of partially hydrogenated vegetable oil, or the reduction in saturated fatty acids by substitution with monounsaturated or polyunsaturated fatty acids or reduction in total fat)
- A natural food in which one of the components has been naturally enhanced through special growing conditions. For example, can the antioxidant content of plant foods be increased through genetic modification?
- A natural food in which a new component has been introduced through special growing conditions. For example, the European Lipgene project includes work on

the introduction of genes from marine algae into rapeseed plants to enable the synthesis and accumulation of EPA and DHA [Napier and Sayanova 2005]. If this project is successful, this technique could provide an alternative for incorporating long-chain omega-3 fatty acids into functional foods.
- A food in which the bioavailability or stability of one or more components has been increased to provide greater bioavailability of a beneficial component. Selection of raw materials and optimization of processing conditions needs to be carefully studied to enhance the retention of minerals, the bioactivity of proteins and peptides, and beneficial effects of components such as nondigestible carbohydrates, resistant starch, and fat replacers. The study of fermentation processes in both food components and the GI tract is key to the development of probiotics with increased resistance to the environment within the GI tract. The development of effective prebiotics (e.g., nondigestible oligosaccharides) with optimal colonic fermentation rates for health benefits will depend on the evaluation of bioconversion processes and the monitoring of their fermentation in the gut.
- A food in which the nature of one or more components has been chemically modified to improve health (e.g., infant formulas containing hydrolyzed protein to reduce the risk of allergenicity)
- Any combination of the above.

Development of a functional food demands a thorough understanding of basic body functions and the identification of one or more food components that could target a function (i.e., genomic, biochemical, physiological, or behavioral) to improve or maintain health and/or reduce risk of disease. Such potential interactions between the food component and the target function should be plausible and a possible mechanism of action proposed. Biomarkers relevant to the target functions being considered also need to be identified and validated. These could be metabolites, specific proteins, hormones, or enzymes, physiological parameters (e.g., blood pressure, heart rate, GI transit time, etc.), or changes in physical and intellectual performance using objective parameters [Roberfroid 2002]. The proposed functional effect then needs to be demonstrated in carefully designed studies, using these validated markers. Studies should include establishment of effective dose and a safety assessment. The functional food should then be fully developed and, finally, evaluated in a controlled clinical study.

THERAPEUTIC USES FOR FUNCTIONAL FOODS

Functional foods (of which some examples are shown in Table 19.1) are available for use in several specific health concerns, of which the following are key:

- Cardiovascular health
- GI health
- Oxidative stress
- Weight management
- Diabetes
- Cognitive/mental health
- Joint and bone health

Cardiovascular Health

Cardiovascular diseases (CVD) are a group of degenerative diseases affecting the whole cardiovascular system, including coronary heart disease, peripheral artery disease, and stroke. To understand the role that functional foods could play in these conditions, it is essential to identify the risk factors for CVD. These include elevated serum cholesterol levels, particularly LDL cholesterol, oxidative modification of LDL cholesterol, low concentration of HDL cholesterol, high blood pressure, high homocysteine levels, damage to the artery lining, compromised endothelial function, and increased blood clot formation.

Fatty Acids

Altering the amounts and proportions of fatty acids in the diet can influence the levels of blood lipids and warrants particular attention in preventing and treating CVD. Saturated fatty acids with chain lengths up to 16 carbon atoms increase plasma LDL cholesterol; they also increase plasma HDL concentration but only to a small extent [Lichtenstein et al. 1998]. Unsaturated fatty acids in the *trans* configuration that are formed during some manufacturing processes (rather than those *trans* fatty acids formed from hydrogenation in ruminant animals and found in dairy products and meat) can increase plasma LDL and reduce HDL cholesterol [Erkkila et al. 2008]. Diets low in saturated fatty acids and *trans* fatty acids could therefore reduce the risk of CVD.

The *cis*-unsaturated fatty acids, including the mono-unsaturates (e.g., oleic acid found in olive oil) and the poly-unsaturates (e.g., linoleic and ALA) reduce plasma LDL concentrations, and some do this without significantly lowering the beneficial HDL cholesterol [Erkkila et al. 2008]. Olive oil has been the subject of many research studies for its antioxidant, anti-inflammatory, and antithrombic properties in reducing CVD risk [Perez-Jimenez et al. 2007]. The long-chain omega-3 fatty acids (EPA and DHA) found in fish oils can reduce plasma triacylyglycerols, counteract blood clotting, and promote improvements in arterial and endothelial integrity. Functional foods enriched in these unsaturated fatty acids could reduce CVD risk.

Soluble Fiber

Soluble fiber appears to influence both hepatic cholesterol and lipoprotein metabolism, thus increasing bile loss. Soluble fiber can reduce LDL concentrations, particularly in people with high levels. Other suggested benefits of soluble fiber include inhibition of hepatic fatty acid synthesis by colonic fermentation products, increase in intestinal motility, and a slowing of macronutrient absorption leading to lowering of the glycemic response and improved insulin sensitivity [Salas-Salvado et al. 2006; Theuwissen and Mensink 2008]. Soluble fiber sources currently used in functional foods include psyllium and dietary fructans (e.g., inulin, oligofructose), both of which have beneficial effects on cardiovascular risk factors [Brighenti 2007;

Petchetti et al. 2007]. However, new functional fiber sources could potentially be generated by upgrading raw materials and by products that are rich in carbohydrates, using extraction and fractionation techniques. The challenge exists to modify the cell wall matrix of raw plant materials to alter the binding of water, bile salts, and macronutrients to optimize the glycemic index

Phytosterols

Phytosterols (plant sterols and stanols) are present in many fruits, vegetables, nuts, seeds, legumes, cereals, vegetable oils, and other plant sources. These substances have been shown in numerous studies to reduce plasma LDL concentration by an average of 10% and are thought to act by reducing intestinal cholesterol absorption [Ortega, Palencia, and Lopez-Sobaler 2006]. Most studies have investigated the effect of phytosterols administered in a fat spread vehicle. However, giving phytosterols in other vehicles such as low-fat milk [Clifton et al. 2004; Thomsen et al. 2004; Noakes et al. 2005], yogurt [Mensink et al. 2002; Noakes et al. 2005; Korpela et al. 2006], low-fat cheese [Korpela et al. 2006], orange juice [Devaraj, Sutret, and Jialal 2006], and a lemon-flavored drink or egg white has also been shown to reduce serum cholesterol [Spilburg et al. 2003]. In addition, combining phytosterols with statins has been shown to have an additive effect on cholesterol lowering [Normen, Holmes, and Frohlich 2005].

Flavonoids

Flavonoids represent a diverse range of polyphenolic compounds, including flavonols, flavones, flavanones, flavan-3-ols, isoflavones, anthocyanins, and proanthocyanidins, found naturally in plant foods (e.g., fruits, vegetables, grains, herbs, and beverages). Diets high in plant foods and rich in polyphenols have been inversely associated with risk for CVD and other chronic diseases, which may be explained by their range of biological activities observed *in vitro*, including anti-inflammatory, vasodilatory, antiplatelet and antioxidant effects, and induction of apoptosis [Hooper et al. 2008].

Depending on processing, flavonoids can be particularly abundant in cocoa and chocolate, tea, soy, and red wine, foods and beverages that are increasingly promoted as functional foods in their own right. Dark chocolate is formulated with a higher percentage of cocoa bean than milk chocolate and therefore often contains greater quantities of flavonoids. Dark chocolate has been associated with antioxidant effects, such as reduced susceptibility of LDL cholesterol to oxidation, and with improvements in endothelial dysfunction and lowering of blood pressure [Keen et al. 2005; Engler and Engler 2006; Erdman et al. 2008], but studies have not consistently demonstrated a favorable effect on cholesterol levels. In addition, there are no epidemiological studies specifically evaluating the effect of chocolate and CVD risk. There is a need for prospective cohort studies and additional randomized clinical trials to investigate the long-term impact of chocolate on CVD risk and outcomes. Tea (green, black, and oolong tea) has also been associated with reduced cardiovascular risk, particularly with improved endothelial function [Stangl, Lorenz, and Stangl 2006; Hodgson et al. 2008].

Soy Protein

Products containing soya protein and soya-enriched diets have been shown to reduce both total and LDL cholesterol in many studies in animals and humans with raised cholesterol levels. However, recent studies have suggested that effects on serum cholesterol are much smaller than previously thought [Sacks et al. 2006]. Nevertheless, many soy products should be beneficial to cardiovascular and overall health because of their high content of polyunsaturated fats, fiber, vitamins, and minerals and low content of saturated fat.

B Vitamins

High plasma homocysteine is associated with increased risk of CVD. Folic acid and vitamins B_6 and B_{12} have the potential to lower homocysteine and therefore reduce cardiovascular drink. However, clinical trials evaluating the effect of these B vitamins on cardiovascular outcomes have not been promising [Clarke et al. 2007].

Gastrointestinal Disease

The GI tract plays a major role in maintaining health and reducing disease risk. This is achieved by the complex microbial environment that, when in healthy balance, helps to prevent invasion of pathogenic bacteria and maintain the integrity of the immune system so reducing the risk of infection and severe allergic reaction. However, the gut microflora can play an important role not only in infection and allergy but also in constipation, irritable bowel syndrome, IBDs, and, possibly, colorectal cancer. The main groups of GI health-promoting bacteria are the *Bifidobacteria* and the *Lactobacilli*.

Probiotics

Probiotics are defined as live microbial food ingredients that, when ingested in sufficient amounts, exert health benefits to the consumer. The main bacteria used as probiotics are various species of *Lactobacilli* and *Bifidobacteria*. They are commonly used in yogurts and fermented dairy drinks. Major health benefits include reduction in the incidence or severity of GI infections (particularly antibiotic associated diarrhea, rotavirus, and traveler's diarrhea) and alleviation of lactose intolerance [Zuccotti et al. 2008]. There is some evidence, which requires confirmation in additional trials, that probiotics help in the management of inflammatory bowel conditions, such as ulcerative colitis, Crohn's disease, and irritable bowel syndrome [Hedin, Whelan, and Lindsay 2007]. Regular probiotic consumption may also afford some protection against developing bowel cancer. Optimizing gut microflora with probiotics may also reduce the risk of allergic disease by improving the barrier to antigen penetration and/or by stimulating anti-allergenic immunological processes. There is evidence that probiotics can reduce the risk of development of atopic eczema in infants [Furrie 2005], but evidence for a protective effect in asthma and allergic rhinitis is weaker. Probiotics may also be beneficial in cholesterol lowering [Zhao and Yang 2005] and reduction in *Helicobacter pylori* infection [Franceschi et al. 2007]. Survival of the bacteria during transit through the stomach and GI tract is essential for achieving these benefits, and, because the probiotic bacteria do not become part of the host's gut microflora, they must be taken regularly to sustain their beneficial effects.

Prebiotics

Prebiotics are indigestible oligosaccharides (small carbohydrate polymers) that enter the large bowel and selectively enhance the growth of certain bacteria within the bowel. Like probiotics, they can favorably alter the microbial balance in the bowel. Key examples include fructose, oligofructose, and inulin, which can be extracted commercially from chicory root, but are also present in other foods, such as artichokes, asparagus, and bananas. Although these substances are not digested in the small intestine, they are fermented by the colonic microflora and selectively stimulate the growth of *Bifidobacteria*. By promoting the growth of beneficial gut bacteria, prebiotics could have similar effects to probiotics. Evidence also suggests that they increase the absorption of certain minerals such as calcium and magnesium [Cashman 2003] and may inhibit pre-cancerous lesions in the colon [Geier, Butler, and Howarth 2006]. The major applications for prebiotics are in dairy products, baked goods and breads, spreads, meat products, salad dressings, and confectionery.

Prebiotics can be mixed with probiotics to produce a symbiotic. The aim is to improve the colonization of the bowel with the probiotic bacteria by use of the prebiotic and/or to stimulate growth of endogenous *Bifidobacteria*. There is some evidence that addition of prebiotics may prolong the colonization by *Bifidobacteria* after consumption of the probiotic is stopped. However, if the effect of the probiotic alone is large, amplification of *Bifidobacterial* colonization by the prebiotic may be difficult to demonstrate.

Oxidative Stress

Oxidative stress, which is considered to be attributable to the formation of reactive oxygen species, is believed to be a contributor to aging and many of the diseases associated with aging (e.g., CVD, cancer, cataract, age-related macular degeneration, Parkinson's disease, Alzheimer's disease, and osteoarthritis). The body has several defenses against oxidative stress, including antioxidant enzymes and the minerals and trace elements that are involved in the activity of these enzymes, vitamins (e.g., vitamins C and E), and glutathione. If exposure to oxidants is high and the body's defenses are unable to cope, oxidative stress develops. Antioxidants naturally present in foods (e.g., vitamins C and E, carotenoids, flavonoids, and other polyphenols) are potentially useful candidates for functional ingredients. Plant foods that contain these substances, such as berries (e.g., cranberry, blueberry, goji, and acai), mangosteen, pomegranate, tomato, and grape, are increasingly finding application in the functional food industry as potential antioxidants.

Weight Management

During recent decades, the scale of obesity has increased worldwide to epidemic proportions. Obesity develops when energy intake is greater than energy expenditure, the excess energy being stored mainly as fat in the adipose tissue. A number of proposed functional food ingredients have been shown to act pre-absorptively to bind dietary fat in the GI tract or post-absorptively to influence substrate utilization or thermogenesis. These include chitosan, CLA, diglycerides, MCTs, green tea, caffeine, calcium, and capsaicin.

Chitosan

Chitosan is a polysaccharide extracted from chitin, which is a structural component of crustacean shells, crabs, shrimps, and lobsters. Chitosan binds fat molecules as a result of its ionic nature. When taken orally, chitosan has been reported to be able to bind 8–10 times its own weight in fat from food that has been consumed. This prevents fat from being absorbed and the body then has to burn stored fat, which may lead to reductions in body fat and body weight. Human trials with chitosan in obese subjects have produced conflicting results, and any effect of chitosan is likely to be small and of no clinical significance [Jull et al. 2008].

Conjugated Linoleic Acid

CLA is a collective term for a number of naturally occurring isomers of linoleic acid containing conjugated double bonds, both *cis* and *trans*. The two most abundant isomers are *cis*-9, *trans*-11 CLA and *trans*-10, *cis*-12 CLA. CLA is found naturally in food derived from ruminant animals (e.g., cows and sheep) and can also be produced synthetically. CLA is promoted for body weight loss, and there is evidence from animal studies that CLA supplementation could be beneficial in weight management. However, human data are conflicting. CLA appears to affect body composition rather than body weight and might be effective in offsetting lean body mass loss occurring during periods of strict energy control or in elderly people [Silveira et al. 2007]. The *trans*-10, *cis*-12 isomer has been suggested to be responsible for these effects and also for the insulin resistance that has been observed in some studies. A mixture of the two isomers, although less potent, may be more suitable for safety reasons, but this has not been proven in human studies.

Diglycerides

Diglycerides are present as minor constituents of various oils. When used in place of triglycerides, which predominate in food fats and oils, diglycerides show some potentially promising effects for weight control. They have a similar energy value and bioavailability to triglycerides but may not be taken up by the adipose tissue to the same extent as triglycerides because of differences in post-absorptive handling of the diacylglycerol components. In human studies, the magnitude of postprandial triglyceridaemia has been shown to be significantly lower with diglycerides than with triglycerides [Tomonobu, Hase, and Tokimitsu 2006]. Consumption of diglycerides has been associated with greater reduction in body weight and abdominal fat mass compared with triglycerides [Maki et al. 2002]. These effects are thought to be attributable to enhanced postprandial fat oxidation [Saito et al. 2006], which may suppress appetite. However, more studies are required to define the optimal dose, mechanism of action, and magnitude of effect on body weight that can be expected in practice.

Medium Chain Triglycerides

MCTs are triglycerides with fatty acids having a chain length of 6–12 carbon atoms. MCTs occur naturally, particularly in coconut and palm oil. They differ from LCTs in that their fatty acids are absorbed directly into the portal circulation and transported to the liver for rapid oxidation. The exact mechanism by which MCTs may influence energy balance is not clear, although production of ketone bodies may be involved. Postprandial energy expenditure increases in humans after consumption of MCT [St-Onge et al. 2003], and there is some evidence for reduced energy intake when meals are supplemented with MCTs. Data on weight loss are inconsistent, however, with some studies showing weight loss [St-Onge et al. 2008] and body fat loss, whereas others found no effects of MCTs on body weight or body composition [Roynette et al. 2008]. Doses of 10 g/day or more seem to be required for meaningful efficacy, but such high doses limit product quality and palatability and there is a potential for GI adverse effects.

Green Tea

Green tea contains high quantities of catechin polyphenols, such as epicatechin, epicatechin gallate, epigallocatechin, and EGCG, the latter being the most abundant and probably the most pharmacologically active. Green tea also contains catechin. Tea catechins have been shown to inhibit catechol-*O*- methyl-transferase, the enzyme that degrades noradrenaline, whereas caffeine inhibits phosphodiesterase, an enzyme that degrades intracellular cyclic AMP and by antagonizing the inhibitory effect of adenosine on increasing noradrenaline release. Both tea catechins and caffeine are therefore likely to increase the stimulatory effects of noradrenaline on energy and lipid metabolism. In short-term studies, green tea has been shown to stimulate thermogenesis and fat oxidation in some studies [Shixian et al. 2006; Boschmann and Thielecke 2007] but not others [Diepvens et al. 2005]. Long-term studies with green tea constituents have reported decreased body weight and body fat. It is not clear whether discrepancies in the data are attributable to caffeine intake, tea catechins, or both [Westerterp-Plantenga, Lejeune, and Kovacs 2005]. Green tea finds application as a functional ingredient in drinks.

Caffeine

Caffeine has both thermogenic and anorectic properties, but long-term administration of caffeine in doses of up to 600 mg/day has not been associated with reduced weight loss.

Calcium

Dietary calcium plays a pivotal role in the regulation of energy metabolism. High calcium diets reduce adipose tissue accretion and weight gain during periods of overconsumption and increase fat breakdown to preserve thermogenesis during energy restriction, thereby accelerating weight loss [Zemel 2002]. A review analyzing data from six observational studies and three controlled trials has shown that high calcium intakes are associated with lower weight gain at mid-life [Heaney, Davies, and Barger-Lux 2002]. However, these effects are difficult to separate from other dietary and lifestyle factors, most notably the consumption of dairy products that are low in energy density and high in protein.

Capsaicin

Capsaicin and other pungent spices have attracted attention as functional ingredients because of their enhancement of fat oxidation and thermogenesis. However, long-term data in humans are lacking, and the use of capsaicin as a functional ingredient may be limited by its pungency and burning effect in the mouth and stomach.

Diabetes Mellitus

Overweight and lack of physical activity have been consistently associated with increased risk of type 2 diabetes. However, dietary composition also appears to be important, and the diet for the management of type 2 diabetes is not significantly different from that recommended for diabetes prevention. Available evidence supports the use of whole-grain foods, vegetables, fruits, foods low in saturated fat, and starchy foods with a low glycemic index. Given that compliance with dietary recommendations in diabetes is often poor, functional foods may be valuable in both treatment and prevention. Low glycemic index starchy foods are of particular interest because of their potential beneficial effects on glucose metabolism and insulin sensitivity. Oral amino acids in the form of snacks have also been studied to positive benefits in blood glucose control and insulin sensitivity in a recent trial in patients with type 2 diabetes [Solerte et al. 2008]. Spices such as cinnamon, coriander, garlic, and turmeric may also be beneficial antidiabetic food adjuncts [Srinivasan 2005].

Numerous studies suggest that chromium, particularly NBC or chromium-nicotinate, may be effective in attenuating insulin resistance and lowering plasma cholesterol levels. Genetics appear to have an influence on these effects, and nutrigenomic studies may help to shed light on the individuals who could benefit from additional chromium [Lau et al. 2008].

Cognitive and Mental Health

A number of functional food ingredients could benefit cognitive and mental function. These include ingredients that are associated with immediate effects such as caffeine, guarana, and ginseng, which can lead to improvement in measures of cognitive performance (e.g., reaction time, attention, vigilance, and psychomotor performance). Carbohydrates exert beneficial effects on various aspects of mental performance, such as faster information processing, better word recall, and improvements in decision time and working memory, but high carbohydrate meals will eventually produce drowsiness and sleepiness whereas the amino acid tryptophan promotes drowsiness and fatigue and reduces the time to sleep.

Other ingredients are associated with longer-term effects, such as reduction in depression, changes in memory, and mental performance in aging with the possibility of reducing the risk of dementia, including Alzheimer's disease. Omega-3 fatty acids, SAMe, and folic acid have attracted attention as potential functional ingredients for depression, whereas *Ginkgo biloba* and omega-3 fatty acids represent potential functional ingredients for the prevention of age-related mental changes.

Joint Health

Osteoarthritis is one of the most prevalent and debilitating chronic conditions affecting older people. Current recommendations for management include nonpharmacological measures such as weight loss and increased physical activity and pharmacological interventions (e.g., NSAIDs). Serious adverse effects are associated with the use of NSAIDs, which creates a need for safe and alternative therapies. In addition, the absence of any cure for osteoarthritis makes prevention important. Nutraceuticals such as glucosamine and chondroitin are used as food supplements but are starting to find application as functional ingredients in foods. Evidence is also emerging for collagen hydrolysate, methylsulfonylmethane, SAMe, and soybean unsaponifiables, all of which could be used as functional ingredients in foods [Ameye and Chee 2006; Bello and Oesser 2006; Frech and Clegg 2007; Clark et al. 2008].

Specific functional functions with their key ingredients and claimed health-promoting benefits are shown in Table 19.1.

DIETETIC FOODS

Another type of "dietary food" is the so-called "dietetic food." In Europe, these are defined as foodstuffs intended to satisfy the nutritional requirements of specific groups of the population and are intended for individuals with a specific disease or condition. In contrast to functional foods, they are legally defined. They are also marketed directly to health professionals, whereas functional foods are marketed to consumers. Examples of dietetic foods include the following:

Table 19.1 Examples of Functional Foods and Drinks on the U.K. Market in 2008

Product	Examples of Claims on Product Packaging and/or Product Website	Key Ingredient Content
Products fortified with vitamins and minerals		
Fortified breakfast cereals (e.g., Kellogg's, Nestle, Weetabix, grocery stores' own brands)		Range of vitamins and minerals
Tropicana Essentials Multivitamins	"With 6 additional vitamins to keep you fighting fit all day long. Vitamin C to help iron absorption; vitamins B1, B2 and B6 to help turn food into energy; provitamin A to help support healthy skin and eyes; vitamin E, as an antioxidant, to help protect your cells."	Range of vitamins and minerals
Conventional food products with no added nutrients		
Optivita (Kellogg's) cereals and cereal bars	"Oats as part of a balanced diet low in saturated fat and healthy active lifestyle can help reduce blood cholesterol. Optivita contains 0.8g of oat beta glucan per 30g serving."	Beta-glucan
Quaker Oats	"Oats contain a soluble fibre which soaks up cholesterol. Reducing your cholesterol can help maintain a healthy heart."	Beta-glucan
Shredded Wheat (Nestle)	"100 per cent whole grain wheat. Whole grain foods contain a combination of protein, fibre, vitamins, minerals, antioxidants and carbohydrates. Together these help keep your heart healthy and help maintain a healthy body."	Whole-grain wheat
Tropicana Orange Juice	"Just 150ml of Tropicana Pure Premium equals one of your 5 daily portions of fruit and vegetables as recommended by health and medical experts. But a 250ml glass of delicious Tropicana Orange Juice contains a full day's supply of vitamin C, provides a good source of folic acid (33%RDA), is naturally sodium and fat free, provides a tasty way to promote healthy blood pressure."	Vitamin C naturally present in the orange juice
Products with added calcium		
Tropicana Essentials Calcium	Calcium is renowned for its vital role in keeping your bones strong and healthy. Tropicana Essentials Calcium contains the same level of calcium as milk—meaning it's a great source of this essential nutrient for anyone who is lactose-intolerant, or doesn't like milk.	Calcium

Products with added fibre		
Tropicana Essentials Fibre	Extra fibre to keep your digestive system regular	
Products with added omega-3 fatty acids		
Brainstorm iQ3 cereal bars	"with Omega-3 and Omega-6, which may help concentration and brain and eye function"	Omega-3 and omega-6 fatty acids
Columbus eggs	"Columbus eggs contain twice as much polyunsaturated fatty acids as standard eggs"	850 mg of long-chain omega-3 fatty acids/egg
Flora Omega 3 Plus spread	"An average serving will provide you with a third of your recommended daily intake of fish sourced Omega 3."	One serving (2 × 10 g) = 600 mg of plant omega-3 and 135 mg of fish omega-3
Flora Omega 3 Probiotic Plus Mini Drink	"Packed full of Omega 3—which helps to maintain heart health—and friendly bacteria, making it great for your digestive system."	EPA/DHA, 80 mg per drink
St. Ivel Fresh Milk with Omega 3	"Contains at least 20× more long chain omega 3 (EPA and DHA) than any other standard whole or semi-skimmed milk." St Ivel Fresh Milk with Omega 3 contains the most important long chain Omega 3s derived from fish oils. These are the most effective and their health benefits are widely acknowledged.	113 mg of long-chain omega-3 per 250 ml serving
Products with added phytosterols		
Benecol (spreads, dairy free drinks, yogurt drinks, yogurts, cream cheese)	"Helps block the uptake of cholesterol in the gut. Therefore, with Benecol less cholesterol enters the blood stream."	Plant stanols
Flora Pro-Activ (spreads, mini drinks, yogurts, semi-skimmed milk drinks)	"Plant sterols can lower LDL by 10–15% in just three weeks, when moving to a healthy diet and lifestyle."	Plant sterols
Minicol cheese	Extensive clinical trials, undertaken in a U.K. research Institute, have demonstrated that the newly developed product reduces overall cholesterol levels by an average of 5.7%, bad (LDL) cholesterol levels falling by 17%.	Plant sterols

(*Continued*)

Table 19.1 *(Continued)*

Product	Examples of Claims on Product Packaging and/or Product Website	Key Ingredient Content
Products containing prebiotics		
Danone Actimel yogurt drink	"More resistant to the digestive process than some other cultures (which means more of the bacteria make it into your intestine) *L. casei* Imunitass helps support your body's defences by topping up the levels of good bacteria found in your gut."	*L. casei* immunitas
Danone Activia yogurt	"Activia helps keep your digestive system ticking away nicely by helping to improve slower digestive transit. A slower transit may make you occasionally feel bloated which can make you feel uncomfortable in yourself. Activia is the only yogurt in the UK that is scientifically proven to help improve slower digestive transit"	*Bifidus Actiregularis*
Muller Vitality (yogurts and yogurt drinks)	"Each bottle contains pre and probiotics to help maintain a healthy digestive system. Keeps your tummy working like clockwork."	
Yakult fermented drink	"Self defence for your gut"	Live probiotic strain of *Lactobacillus casei* Shirota
Products with added prebiotics		
Muller Vitality Yogurts and Yogurt Drinks	"Each bottle contains pre and probiotics to help maintain a healthy digestive system. Keeps your tummy working like clockwork."	
Warburton's Healthy Inside Bread	Inulin "works by feeding or stimulating your own good bacteria to grow and multiply, thus increasing the amount of good bacteria in your body."	Inulin
Products with added peptides		
Flora Pro-Activ Blood Pressure Mini Drink	"Control blood pressure"	Ameal peptide containing 6.6 mg of dairy peptides

Products containing soya		
Alpro soya products (alternatives to milk, yogurt and cream), desserts, tofu	"Can help to lower your cholesterol (25g of soya protein per day as part of a diet which is low in saturated fat has been proven to lower your cholesterol levels)."	Soya
Burgen Soya and Linseed Bread	"Soya and linseeds are good sources of plant oestrogens which may be beneficial for women's health."	Soya
So Good soya milk	"All So Good products contribute to developing these 5 important health benefits: healthy heart, strong bones, lowers LDL (bad) cholesterol, low GI, protective antioxidants."	Soya
Products promoted for antioxidant content		
Delvaux-Acticoa chocolate	"Acticoa dark chocolate contains 2.33 times the antioxidants you'll find in standard dark chocolate; 9g a day delivers the RDA of antioxidants."	Website states that Acticoa cocoa drink (2.5% cocoa powder) per serving contains 540 mg of total polyphenol and 342 mg of total flavanol (for an equivalent standard cocoa drink, the figures stated are 210 and 188 mg, respectively)
Diet Coke Plus Antioxidant	"Antioxidant"	Green tea powder, vitamin C
Innocent Superfood Smoothies (various blends)	Contain various blends of fruits for which antioxidant claims are made, e.g. "*goji berries*, "full of antioxidants and the richest source of beta-carotene of all known foods on earth"; *pomegranates, blueberries and acai,* "all of which contain high levels of antioxidants."	
Lipton Green Tea (soft drink)	"Source of green tea antioxidants"	Green tea extracts, 10.8%
Ocean Spray (various juice drinks and juices containing cranberry with and without other fruits such as blueberry, blackcurrant, raspberry, mango)	"Rich in vitamin C and antioxidants." 100ml of Ocean Spray Cranberry and Pomegranate will typically contain 40mg proanthocyanidins.	Vitamin C, 75 mg/250 ml serving

(*Continued*)

Table 19.1 *(Continued)*

Product	Examples of Claims on Product Packaging and/or Product Website	Key Ingredient Content
PomWonderful Pomegranate Juice	"There are antioxidants, and then there are *antioxidants*. Only POM Wonderful 11% Pomegranate Juice is backed by over £10million of initial scientific research that's shown encouraging results for prostate and cardiovascular health. And every sip helps guard your body against free radicals."	Antioxidant activity quoted as 6.1 mM polyphenols
Pomegreat (various drinks, e.g., Pomegreat Original; Blueberry; Raspberry; Pomegreat 100; Pomegreat 100 Blueberry; Pomegreat Acai)	A 250 ml glass of Pomegreat provides half an adult's RDA of antioxidant vitamins A, C and E. "Bursting with antioxidants, they (pomegranates) also have cholesterol lowering properties"; "Packed with antioxidant vitamins" "Pomegranates are bursting with antioxidants"; "Fantastic source of antioxidants, which naturally defend and protect your body against free radicals"; "Acai berries are brimming with antioxidants";	Vitamin C, 30 mg/250 ml serving; vitamin E, 5 mg/250 ml serving
Rubicon (pomegranate, blueberry and cranberry exotic blend)	"Pomegranates, blueberries and cranberries are known as superfoods and are crammed full of antioxidants."	Vitamin C, 60 mg/200 ml serving
Tropicana (blueberry blend; pomegranate blend) juices	"Rich in antioxidants"	
Welch's Purple Grape Juice, Purple Grape and Raspberry Blend, Purple Grape and Strawberry Blend	"Natural antioxidant power"; "Packed full of antioxidant power"; "Welch's Concord Purple Grape has more naturally occurring antioxidant power than many other popular fruit juices"; "The Concord Purple Grape is higher in antioxidants than more common red and white grapes."	Antioxidant activity measured using ORAC; absolute values not quoted on the label

- Foods for infants and young children, including infant formulas and follow-on formulas, processed cereal-based foods, and baby foods (weaning foods)
- Foods intended for use in energy-restricted diets for weight reduction
- Foods for sports people and athletes
- Foods for special medical purposes (e.g., enteral/tube feeds, gluten-free foods for celiac disease)

FUTURE DEVELOPMENTS

The trend for increasing numbers of functional ingredients and functional foods has increased during recent years. Whether and how much this will continue is at present unclear. Factors that will encourage future development of dietary foods include the increasing global burden of obesity and the health conditions associated with it (e.g., CVD and diabetes) and an increasingly older population and the health problems common in that group (e.g., decline in bone mass, joint function, vision, and cognitive function). Pressures of cost containment for health systems and an emphasis on prevention, well-being, and self care could also help to drive this market forward.

Rapid developments in nutritional science and food technology will likely create opportunities for new food products. An understanding of components beyond traditional nutrients (e.g., phytochemicals) will likely lead to the development of new active ingredients. Furthermore, knowledge of mechanisms and development of valid biomarkers should help in the development of more target and disease-specific foods. New formulations, such as nutritional patches and sprays, will be introduced to deliver functional ingredients. New technologies with validated biomarkers could enable consumers to measure their own nutritional status and to assess the effects of functional foods on their health.

Functional foods of the future are likely to be targeted at specific subgroups of the population for very specific health benefits. To this end, the influence of genes on metabolic processes, the influence of foods on genetic expression, and individual susceptibility to diet-related disease will help in the production of specific functional foods for specific population groups.

However, future challenges for functional foods lie in the possibility of a return to "fundamental eating patterns" and increasing consumer skepticism of manufactured food products carrying health claims. Regulation, both nationally and internationally, particularly for product claims and labeling, is also likely to increase. Restrictions on genetically modified ingredients will also shape the future of functional foods. Education will be of paramount importance if consumers and the professionals who advise them are to understand the value of these foods and consume them.

REFERENCES

Ameye, L. G. and W. S. Chee. 2006. Osteoarthritis and nutrition. From nutraceuticals to functional foods: A systematic review of the scientific evidence. *Arthritis Res. Ther* 8:R127.

Bello, A. E. and S. Oesser. 2006. Collagen hydrolysate for the treatment of osteoarthritis and other joint disorders: A review of the literature. *Curr. Med. Res. Opin.* 22:2221–2232.

Boschmann, M. and F. Thielecke. 2007. The effects of epigallocatechin-3-gallate on thermogenesis and fat oxidation in obese men: A pilot study. *J. Am. Coll. Nutr.* 26:389S–395S.

Brighenti, F. 2007. Dietary fructans and serum triacylglycerols: A meta-analysis of randomized controlled trials. *J. Nutr.* 137 (Suppl. 11):2552S–2556S.

Cashman, K. 2003. Prebiotics and calcium bioavailability. *Curr. Issues Intest. Microbiol.* 4:21–32.

Clark, K. L., W. Sebastianelli, K. R. Flechsenhar, D. F. Aukermann, F. Meza, R. L. Millard, J. R. Deitch, P. S. Sherbondy, and A. Albert. 2008. 24-Week study on the use of collagen hydrolysate as a dietary supplement in athletes with activity-related joint pain. *Curr. Med. Res. Opin.* 24:1485–1496.

Clarke, R., S. Lewington, P. Sherliker, and J. Armitage. 2007. Effect of B vitamins on plasma homocysteine concentrations and on risk of cardiovascular disease and dementia. *Curr Opin Clin Nutr Metab Care* 10(1):32–39.

Clifton, P. M., M. Noakes, D. Sullivan, N. Erichsen, D. Ross, G. Annison, A. Fassoulakis, M. Cehun, and P. Nestel. 2004. Cholesterol-lowering effects of plant sterol esters differ in milk, yoghurt, bread and cereal. *Eur. J. Clin. Nutr.* 58:503–509.

DeFelice, S. 2002. FIM (Foundation for Innovation in Medicine) rationale and proposed guidelines for the Nutraceutical Research and Education Act, 2002: NREA. Presented at FIM's 10th Nutraceutical Conference, November 10–11, 2002. http://www.fimdefelice.org/archives/arc.researchact.html. Accessed August 6, 2008.

Devaraj, S., B. C. Autret, and I. Jialal. 2006. Reduced-calorie orange juice beverage with plant sterols lowers C-reactive protein concentrations and improves the lipid profile in human volunteers. *Am. J. Clin. Nutr.* 84:756–761.

Diepvens, K., E. M. Kovacs, I. M. Nijs, N. Vogels, and M. S. Westerterp-Plantenga. 2005. Effect of green tea on resting energy expenditure and substrate oxidation during weight loss in overweight females. *Br. J. Nutr.* 94:1026–1034.

Engler, M. B. and M. M. Engler. 2006. The emerging role of flavonoid-rich cocoa and chocolate in cardiovascular health and disease. *Nutr. Rev.* 64:109–118.

Erdman, J. W. Jr., L. Carson, C. Kwik-Uribe, E. M. Evans, and R. R. Allen. 2008. Effects of cocoa flavanols on risk factors for cardiovascular disease. *Asia Pac. J. Clin. Nutr.* 17 (Suppl. 1):284–287.

Erkkila, A., V. D. de Mello, U. Riserus, D. E. Laaksonen. 2008. Dietary fatty acids and cardiovascular disease: An epidemiological approach. *Prog. Lipid Res.* 47:172–187.

Franceschi, F., A. Cazzato, E. C. Nista, E. Scarpellini, D. Roccarina, G. Gigante et al. 2007. Role of probiotics in patients with Helicobacter pylori infection. *Helicobacter* 12 (Suppl 2):59–63.

Frech, T. M. and D. O. Clegg. 2007. The utility of nutraceuticals in the treatment of osteoarthritis. *Curr. Rheumatol. Rep.* 9:25–30.

Furrie, E. 2005. Probiotics and allergy. *Proc. Nutr. Soc.* 64:465–469.

Geier, M. S., R. N. Butler, and G. S. Howarth. 2006. Probiotics, prebiotics and synbiotics: A role in chemoprevention for colorectal cancer? *Cancer Biol. Ther.* 5:1265–1269.

Halsted, C. H. 2003. Dietary supplements and functional foods: 2 sides of a coin? *Am. J. Clin. Nutr.* 77(Suppl. 4):1001S–1007S.

Harris, W. S., M. Miller, A. P. Tighe, M. H. Davidson, and E. J. Schaefer. 2008. Omega-3 fatty acids and coronary heart disease risk: Clinical and mechanistic perspectives. *Atherosclerosis* 197:12–24.

Heaney, R. P., K. M. Davies, and M. J. Barger-Lux. 2002. Calcium and weight: Clinical studies. *J. Am. Coll. Nutr.* 21:152S–155S.

Hedin, C., K. Whelan, and J. O. Lindsay. 2007. Evidence for the use of probiotics and prebiotics in inflammatory bowel disease: A review of clinical trials. *Proc. Nutr. Soc.* 66:307–315.

Hodgson, J. M. 2008. Tea flavonoids and cardiovascular disease. *Asia Pac. J. Clin. Nutr.* 17 (Suppl. 1):288–290.

Hooper, L., P. A. Kroon, E. B. Rimm, J. S. Cohn, I. Harvey, K. A. Le Cornu, J. J. Ryder, W. L. Hall, and A. Cassidy. 2008. Flavonoids, flavonoid-rich foods, and cardiovascular risk: A meta-analysis of randomized controlled trials. *Am. J. Clin. Nutr.* 88:38–50.

Institute of Grocery Distribution. 2003. Working Group Report. Future foods for wellbeing: An expert panel's view of the next 25 years. http://www.igd.com. Accessed August 8, 2008.

Jull, A. B., C. Ni Mhurchu, D. A. Bennett, C. A. Dunshea-Mooij, and A. Rodgers. 2008. Chitosan for overweight or obesity. *Cochrane Database Syst. Rev.* 2008:CD003892.

Keen, C. L., R. R. Holt, P. I. Oteiza, C. G. Fraga, and H. H. Schmitz. 2005. Cocoa antioxidants and cardiovascular health. *Am. J. Clin. Nutr.* 81 (Suppl. 1):298S–303S.

Key Note Publications. 2008. Nutraceuticals. http://www.nutraingredients.com. Accessed July 8, 2008.

Korpela, R., J. Tuomilehto, P. Hogstrom, L. Seppo, V. Piironen, P. Salo-Vaananen, J. Toivo, C. Lamberg-Allardt, M. Kärkkäinen, T. Outila, J. Sundvall, S. Vilkkilä, and M. J. Tikkanen. 2006. Safety aspects and cholesterol-lowering efficacy of low fat dairy products containing plant sterols. *Eur. J. Clin. Nutr.* 60:633–642.

Lau, F. C., M. Bagchi, C. K. Sen, and D. Bagchi. 2008. Nutrigenomic basis of beneficial effects of chromium (III) on obesity and diabetes. *Mol. Cell. Biochem.* 317:1–10.

Lichtenstein, A. H., E. Kennedy, P. Barrier, D. Danford, N. D. Ernst, S. M. Grundy et al. 1998. Dietary fat consumption and health. *Nutr. Rev.* 56:S3–S19; discussion S19–S28.

Maki, K. C., M. H. Davidson, R. Tsushima, N. Matsuo, I. Tokimitsu, D. M. Umporowicz, M. R. Dicklin, G. S. Foster, K. A. Ingram, B. D. Anderson, S. D. Frost, and M. Bell. 2002. Consumption of diacylglycerol oil as part of a reduced-energy diet enhances loss of body weight and fat in comparison with consumption of a triacylglycerol control oil. *Am. J. Clin. Nutr.* 76:1230–1236.

Mensink, R. P., S. Ebbing, M. Lindhout, J. Plat, M. M. van Heugten. 2002. Effects of plant stanol esters supplied in low-fat yoghurt on serum lipids and lipoproteins, non-cholesterol sterols and fat soluble antioxidant concentrations. *Atherosclerosis* 160:205–213.

Mintel. 2008. Functional foods. http://www.nutraingredients.com. Accessed July 8, 2008.

Napier J. and A. Sayanova. 2005. The production of very long chain PUFA biosynthesis in transgenic plants: Towards a sustainable source of fish oils. *Proc. Nutr. Soc.* 64:387–393.

Noakes, M., P. M. Clifton, A. M. Doornbos, and E. A. Trautwein. 2005. Plant sterol ester-enriched milk and yoghurt effectively reduce serum cholesterol in modestly hypercholesterolemic subjects. *Eur. J. Nutr.* 44:214–222.

Normen, L., D. Holmes, and J. Frohlich. 2005. Plant sterols and their role in combined use with statins for lipid lowering. *Curr. Opin. Investig. Drugs* 6:307–316.

Ohama, H., H. Ikeda, and H. Moriyama. 2006. Health foods and foods with health claims in Japan. *Toxicology* 221:95–111.

Ortega, R. M., A. Palencia, and A. M. Lopez-Sobaler. 2006. Improvement of cholesterol levels and reduction of cardiovascular risk via the consumption of phytosterols. *Br. J. Nutr.* 96 (Suppl. 1):S89–S93.

Perez-Jimenez, F., J. Ruano, P. Perez-Martinez, F. Lopez-Segura, and J. Lopez-Miranda. 2007. The influence of olive oil on human health: Not a question of fat alone. *Mol. Nutr. Food Res.* 51:1199–1208.

Petchetti, L., W. H. Frishman, R. Petrillo, and K. Raju. 2007. Nutriceuticals in cardiovascular disease: Psyllium. *Cardiol. Rev.* 15:116–122.

Regulation (EC) No 1924. 2006 of the European Parliament and of the Council of 20 December 2006 on nutrition and health claims made on foods. *Official Journal of the European Union* 2006.

Roberfroid, M. B. 2002. Global view on functional foods: European perspectives. *Br. J. Nutr.* 88 (Suppl 2):S133–S138.

Roynette, C. E., I. Rudkowska, D. K. Nakhasi, and P. J. Jones. 2008. Structured medium and long chain triglycerides show short-term increases in fat oxidation, but no changes in adiposity in men. *Nutr. Metab. Cardiovasc. Dis.* 18:298–305.

Sacks, F. M., A. Lichtenstein, L. Van Horn, W. Harris, P. Kris-Etherton, and M. Winston. 2006. Soy protein, isoflavones, and cardiovascular health: An American Heart Association Science Advisory for professionals from the Nutrition Committee. *Circulation* 113:1034–1044.

Saito, S., K. Tomonobu, T. Hase, and I. Tokimitsu. 2006. Effects of diacylglycerol on post-prandial energy expenditure and respiratory quotient in healthy subjects. *Nutrition* 22:30–35.

Salas-Salvado, J., M. Bullo, A. Perez-Heras, and E. Ros. 2006. Dietary fiber, nuts and cardio-vascular diseases. *Br. J. Nutr.* 96 (Suppl. 2):S46–S51.

Scientific concepts of functional foods in Europe. Consensus document [No authors listed]. 1999. *Br. J. Nutr.* 81 (Suppl. 1):S1–S27.

Shixian, Q., B. VanCrey, J. Shi, Y. Kakuda, and Y. Jiang. 2006. Green tea extract thermogen-esis-induced weight loss by epigallocatechin gallate inhibition of catechol-O-methyl-transferase. *J. Med. Food.* 9:451–458.

Silveira, M. B., R. Carraro, S. Monereo, and J. Tebar. 2007. Conjugated linoleic acid (CLA) and obesity. *Public Health Nutr.* 10:1181–1186.

Solerte, S. B., M. Fioravanti, E. Locatelli, R. Bonacasa, M. Zamboni, C. Basso, A. Mazzoleni, V. Mansi, N. Geroutis, and C. Gazzaruso. 2008. Improvement of blood glucose control and insulin sensitivity during a long-term (60 weeks) randomized study with amino acid dietary supplements in elderly subjects with type 2 diabetes mellitus. *Am. J. Cardiol.* 101:82E–88E.

Spilburg, C. A., A. C. Goldberg, J. B. McGill, W. F. Stenson, S. B. Racette, and J. Bateman et al. 2003. Fat-free foods supplemented with soy stanol-lecithin powder reduce choles-terol absorption and LDL cholesterol. *J. Am. Diet. Assoc.* 103:577–581.

Srinivasan, K. 2005. Plant foods in the management of diabetes mellitus: Spices as beneficial antidiabetic food adjuncts. *Int. J. Food Sci. Nutr.* 56:399–414.

Stangl, V., M. Lorenz, and K. Stangl. 2006. The role of tea and tea flavonoids in cardiovascular health. *Mol. Nutr. Food Res.* 50:218–228.

St-Onge, M. P. and A. Bosarge. 2008. Weight-loss diet that includes consumption of medium-chain triacylglycerol oil leads to a greater rate of weight and fat mass loss than does olive oil. *Am. J. Clin. Nutr.* 87:621–626.

St-Onge, M. P., C. Bourque, P. J. Jones, R. Ross, and W. E. Parsons. 2003. Medium- versus long-chain triglycerides for 27 days increases fat oxidation and energy expenditure without resulting in changes in body composition in overweight women. *Int. J. Obes. Relat. Metab. Disord.* 27:95–102.

Theuwissen, E. and R. P. Mensink. 2008. Water-soluble dietary fibers and cardiovascular dis-ease. *Physiol. Behav.* 94:285–292.

Thomsen, A. B., H. B. Hansen, C. Christiansen, H. Green, and A. Berger. 2004. Effect of free plant sterols in low-fat milk on serum lipid profile in hypercholesterolemic subjects. *Eur. J. Clin. Nutr.* 58:860–870.

Tomonobu, K., T. Hase, and I. Tokimitsu. 2006. Dietary diacylglycerol in a typical meal suppresses postprandial increases in serum lipid levels compared with dietary triacylglycerol. *Nutrition* 22:128–135.

U.S. Food and Drug Administration, Center for Food Safety and Applied Nutrition. Health claims that meet significant scientific agreement (SSA). http://www.cfsan.fda.gov/~dms/lab-ssa.html. Accessed November 8, 2008.

Westerterp-Plantenga, M. S., M. P. Lejeune, and E. M. Kovacs. 2005. Body weight loss and weight maintenance in relation to habitual caffeine intake and green tea supplementation. *Obes. Res.* 13:1195–1204.

Zeisel, S. H. 1999. Regulation of "nutraceuticals." *Science* 285:1853–1855.

Zemel, M. B. 2002. Regulation of adiposity and obesity risk by dietary calcium: Mechanisms and implications. *J. Am. Coll. Nutr.* 21:146S–151S.

Zhao, J. R.and H. Yang 2005. Progress in the effect of probiotics on cholesterol and its mechanism (in Chinese). *Wei Sheng Wu Xue Bao* 45:315–319.

Zuccotti, G. V., F. Meneghin, C. Raimondi, D. Dilillo, C. Agostoni, and E. Riva et al. 2008. Probiotics in clinical practice: An overview. *J. Int. Med. Res.* 36 (Suppl. 1):1A–53A.

CHAPTER 20

Antiviral Nutraceuticals from Pomegranate (*Punica granatum*) Juice

Girish J. Kotwal

CONTENTS

INTRODUCTION

Pomegranate juice has been considered to have multiple health benefit for centuries [Jurenka 2008]. It is currently being marketed/promoted as an antioxidant [Khan et al. 2008] with disease-preventive, prophylactic, and therapeutic health benefits in aging [Afaq and Muktar 2006], prostate cancers [Siddiqui et al. 2004; Adhami and Muktar 2006, 2007; Bemi et al. 2006; Santilo and Lowe 2006; Bell and Hawthorne 2008; Syed et al. 2008] and cancers in general [Aggarwal and Shisodia 2004, 2006; Klass and Shin 2007; Syed et al. 2007; Heber 2008], cardiovascular health [Aviram et al. 2002], diabetes [Saxena and Vikram 2004; Katz et al. 2007; Li et al. 2008], dermatological conditions [Baumann 2007], and inflammation [Lansky and Neumann 2006]. Pomegranate (*Punica granatum*) juice from America was first shown by Neurath et al. [2005], to be the only fruit juice

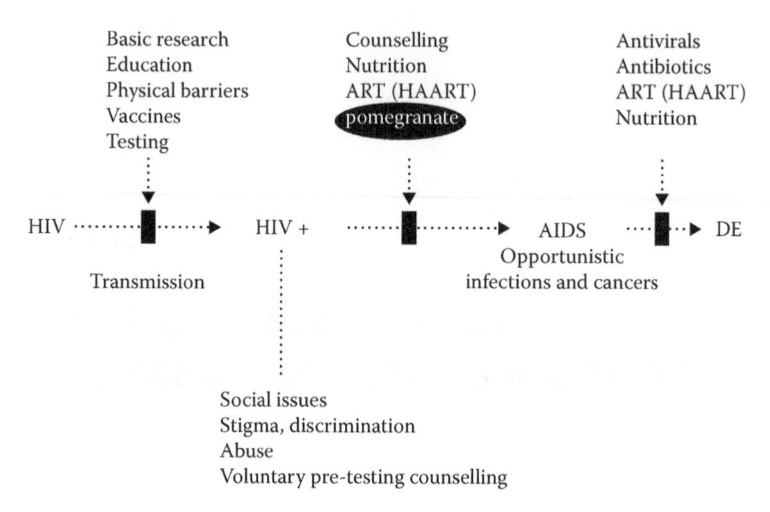

Figure 20.1 A modified multipronged approach to HIV infection. (Modified from Kotwal, J., *Journal of the Royal Society of Medicine*, 97, 2004.)

from a number of different fruit juices tested to possess an HIV-1 entry inhibitor activity; it was proposed as a candidate topical vaginal microbicide against all clades of HIV for prevention of HIV infection, which would afford a control to the females during a heterosexual encounter. This anti-HIV activity of pomegranate juice has yet to find itself to be a key agent in the proposed multipronged approach to target progression of HIV-infected person to AIDS (Figure 20.1). Considering the severe side effects of established anti-retrovirals and the predisposition to drug resistance, there is considerable scope for antiviral nutraceuticals from natural sources such as pomegranate juice to be used in the fight against HIV and AIDS. Subsequent studies described here have shown that the pomegranate juice has a far broader antiviral activity against enveloped viruses such as influenza, including the potentially pandemic H5N1 [Kotwal 2006], poxviruses, and herpes viruses; it can be also be ultrafiltered to separate the low-molecular-weight nutraceuticals termed enveloped virus neutralizing compounds (EVNCs) [Kotwal 2007].

DISTRIBUTION OF POMEGRANATE CROPS

Pomegranate trees grow in almost every continent. They may have originated in Arabia, and nowhere are they found in such abundance as in the markets and street corners as in Amman, Jordan, which can be described as the pomegranate capital of the world [G. J. Kotwal, unpublished observation]. These trees grow in the Mediterranean region in Europe. In the continent of Africa, the trees can be found in the southwestern-most tip of Africa in Cape Town, as well as in Egypt and Morocco and most likely across regions in Africa. The pomegranate tree is found in several parts of Asia and Australia and the warmer regions of North America, especially in California, Texas, and Florida. They were brought to North America from Spain [Welch 2002]. The antiviral activity from the pomegranate juice is quite comparable, whether it is derived tested from the commercially available POM, from California, or from homemade juice in Cape Town, South Africa [Kotwal, unpublished observation]. Pomegranate tree can grow as a highly branched, multi-stemmed tree ranging from 15 to 30 feet. If not harvested at the right time, a pomegranate tree without nets can be inhabited by birds, and pomegranate trees in the wild often get moldy and damaged. There is, therefore, a fairly narrow window during which pomegranate needs to be harvested. One needs to pluck the fruit from the trees one at a time and quickly separate from the damaged ones and store in a refrigerator. Generally, there is only one pomegranate crop per year. Pomegranate trees grow in places with lots of sunshine, range of acidic and alkaline pH, and in deep, well-drained clay soil. It can be propagated by planting seeds or cutting of a tree that bears good fruit.

POMEGRANATE JUICE EXTRACTION

After selection of undamaged or uninfected pomegranates, the pomegranates are processed by hand by first making a groove along the sides with a knife and then pulling out the calyx lobes of the pomegranate. Each pomegranate grain, or granatum, with its seeds is then removed and pooled in a clean beaker. Although each part of the pomegranate has different nutraceuticals, the process described here was developed to generate the pure juice. The contents of the beaker are then emptied in the inner compartment of a special mixer/grinder, which has a filter grid that retains the seeds and allows the juice to pass through during the grinding/juicing process. The filtered juice from the grinder is then sterile filtered and can be stored in a refrigerator. All testing for antiviral activity is then performed in a Biosafety Level 2 laminar flow hood if it is testing against laboratory strain of influenza, pox viruses, or herpes. Anti-HIV or hepatitis C virus, or testing against H5N1, is performed in a Biosafety Level 3 laminar flow hood. The sterile filtered pomegranate juice container has to be opened only in a sterile environment; if not, it can be easily contaminated with fungus and bacteria because it is rich in sugars and nutrients. Additional ultrafiltration is used to separate the larger than 3,000 Da proteins, for example, the lipid transfer proteins of the size 7–9 kDa from the EVNCs, which are less than 3,000 and most likely in the range of 500–1,000 Da. The EVNCs can be further purified by column chromatography, but the active ingredients remain a mixture that has potent and stable antiviral activity.

POMEGRANATE COMPOSITION

The pomegranate tree and its fruit can be considered as a bounty of unique compartmentalized nutraceuticals. As such, the composition can be considered within each of these anatomical areas (roots, stem/bark, leaves, flower, seeds, and juicy pulp) surrounding the seeds and the pericarp (peel and rind) and has been reviewed by Jurenka [2008]. The roots and the bark have a number of piperidine alkaloids and ellagitannins (punicalin, punicalagin). The flower has triterpenoids (maslinic acid, Asiatic acid), ursolic acid, and gallic acid. The pomegranate seeds are a powerful source of nutraceuticals with beneficial effects in cancer, lipidemia, and as antioxidants. The oils in the pomegranate seeds include elagic acid, fatty acids, predominantly linolenic acid (CLA, punicic acid), and phytosterols (beta-sitosterol, campesterol and stigmasterol) [Kaufmann and Wiesman 2007]. The pomegranate juicy pulp/granatum, which is mostly water, is where we found the antiviral activity. We have not tested the other areas of the fruit or the tree and cannot at this time state whether there is any antiviral activity. The juice has a number of constituents, including red, brown, or pink color pigments such as anthocyanins, which can stain clothes, dishes, or sinks. The juice contains small molecules such as sugars (glucose), amino acids, iron and many other minerals, ascorbic acid, caffeic acid, ellagic acid, gallic acid, cathechin, EGCG, quercitin, and rutin. Although it is not at this time known as to which of these compounds are responsible for the antiviral activity, one could speculate that it could be attributable to a mixture of compounds termed EVNCs and not any single or solo performer that can be associated with the antiviral activity of the pomegranate juice, and such EVNCs are not present in other fruit juices tested. Thus, ascorbic acid, iron, amino acids, and glucose could be considered not to influence the antiviral activity. It is possible that the other acids could, in concert, have a role in penetrating the viral lipid envelope and thus causing viral neutralization. In this context, it is noteworthy that Lanasky [2006] emphasizes that nutraceutical products, standardized to 40% ellagic acid, may be dangerous because the health benefits may be attributable to the synergy among the different pomegranate constituents and not restricted to the ellagic acid alone. Pomegranate juice also has fenhexamid [Hengel et al. 2003], but, because it is also present in juices from cranberry and blueberry, it is unlikely that the antiviral activity is attributable to fenhexamid, especially because neither blueberry nor cranberry was found to have antiviral activity. In addition to the smaller than 3,000 Da compounds, the pomegranate juice has two lipid transfer proteins (LTP1a and LTP1b) of a low molecular weight around 8 kDa [Zoccatelli et al. 2007]. These proteins are currently being investigated as elicitors of allergy and hence should be removed from the juice when considering any human antiviral trials. The peel/rind of the pomegranate has constituents similar to the juice with the exception of much greater quantities of flavonols, flavones, flavonones, and phenolic punicalagins.

OBJECTIVES OF THE STUDY

To determine whether the antiviral activity is specifically against all the clades of HIV or that it is also against other enveloped viruses, we have since tested the juice for antiviral activity against poxviruses, a number of influenza strains (including

H5N1), and herpes simplex viruses 1 and 2. Also, to determine whether the antiviral activity is found in pomegranate juice across the planet, we have tested and compared the antiviral activity from two different continents: North America and Africa.

METHOD USED

The overall design of a safe antiviral assay developed by us is illustrated in Figure 20.2. Essentially, we mixed a million virus particles of attenuated recombinant vaccinia virus, vGK5 [described previously by Kotwal et al. 1989], with varying amounts of the juice for 5 min to 1 h at 37°C and then obtained a virus titer on African Green Monkey cells called BSC-1 cells, with and without treatment. Similarly, for influenza strains, we quantitated the virus with and without treatment using an hemagglutinin inhibition assay as shown in Figure 20.3. To determine whether the pomegranate juice produced from a different continent can reproducibly have antiviral activity, we tested the pomegranate juice from the southwestern-most tip of Africa and compared it with the commercially available juice called POM. To identify the specific compounds and determine the structure of the bioactive antiviral compounds, we have separated

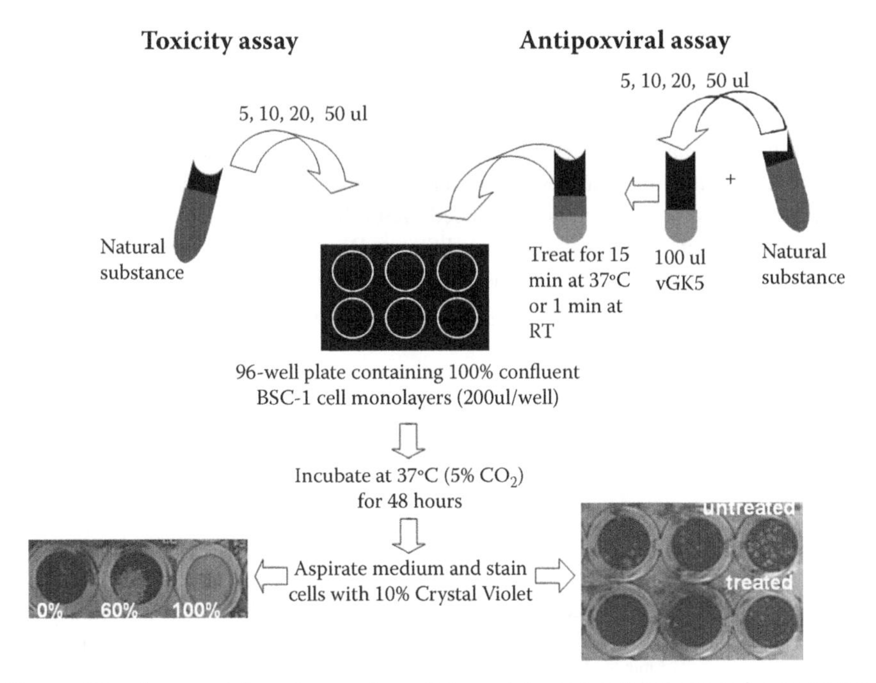

Figure 20.2 A general flow diagram for cellular toxicity and antiviral assay for evaluating toxicity and antiviral activity of candidate agents.

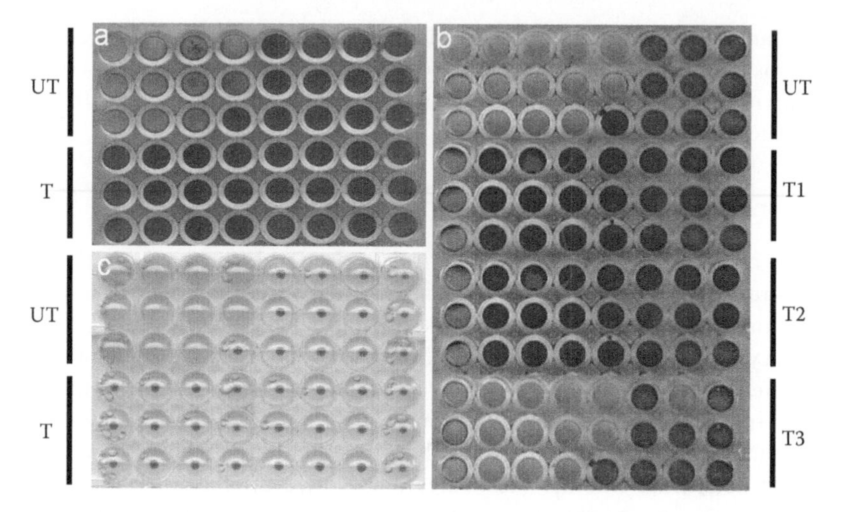

Figure 20.3 In-vitro: Inhibition of infectivity (a and b) and hemagglutinin activity (c) of influenza virus strains X31 and H5N1.

the juice containing only the less than 3,000 Da compounds and also tested it for antiviral activity.

RESULTS AND CONCLUSION

The pomegranate juice tested against vGK5 (Figure 20.4) as well as influenza strains and both the major herpes viruses had significant antiviral activity (Figure 20.5). Also, the clear pomegranate juice produced in South Africa had activity similar to that produced by a manufacturer in the United States. Besides possible benefits in preventing and treating cancers, heart diseases, diabetes, inflammatory diseases, and aging, pomegranate juice consumption in significant amounts or as the nutraceuticals termed

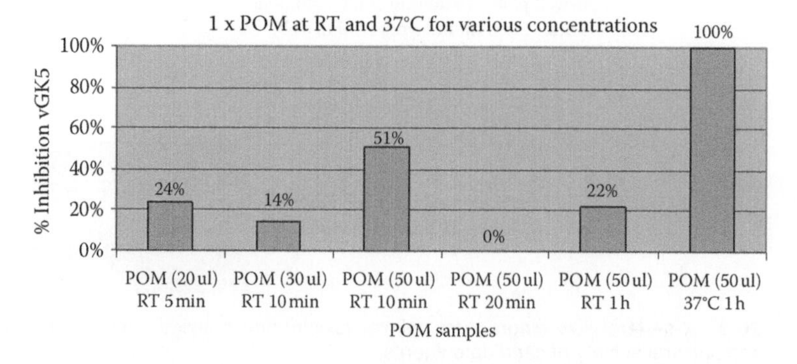

Figure 20.4 POM inhibits replication of poxviruses. No toxicity to BSC-1 cells detected.

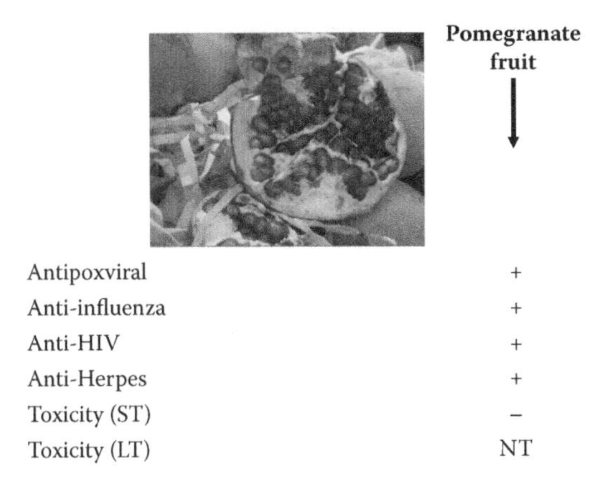

Figure 20.5 Antiviral activity of pomegranate juice.

EVNCs could contribute to lowering viral infections attributable to influenza, hepatitis viruses, herpes viruses, and slowing down the progression of HIV-infected persons to AIDS. Growing pomegranate trees worldwide would contribute to sustain the growing demand of the pomegranate tree and to better health of people around the world.

ACKNOWLEDGEMENTS

The author would like to acknowledge the technical assistance of Ms. Melissa Abrahams for her role in the assay of the antipoxviral activity. The authors would also like to acknowledge the expert assistance of the groups of Dr. Mark Sangster of the University of Tennessee at Knoxville and Dr. Richard Webby of St. Jude Children's Hospital Research Center for assistance with the antiviral assay of flu viruses, including H5N1, using pomegranate juice.

REFERENCES

Adhami, V. M. and H. Mukhtar. 2006. Polyphenols from green tea and pomegranate for prevention of prostate cancer. *Free Radic. Res.*40:1095–1104.

Adhami, V. M. and H. Mukhtar. 2007. Anti-oxidants from green tea and pomegranate for chemoprevention of prostate cancer. *Mol. Biotechnol.* 37:52–57.

Afaq, F. and Mukhtar H. 2006. Botanical antioxidants in the prevention of photocarcinogenesis and photoaging. *Exp. Dermatol.* 15:678–684.

Aggarwal, B. B. and S. Shishodia. 2004. Suppression of the nuclear factor-kappaB activation pathway by spice-derived phytochemicals: reasoning for seasoning. *Ann. N.Y. Acad. Sci.* 1030:434–441.

Aggarwal, B. B. and S. Shishodia. 2006. Molecular targets of dietary agents for prevention and therapy of cancer. *Biochem. Pharmacol.* 71:1397–1421.

Aviram, M., L. Dornfeld, M. Kaplan et al. 2002. Pomegranate juice flavonoids inhibit low-density lipoprotein oxidation and cardiovascular diseases: Studies in atherosclerotic mice and in humans. *Drugs Exp. Clin. Res.* 28:49–62.

Baumann, L. 2007. Botanical ingredients in cosmeceuticals. *J. Drugs Dermatol.* 6:1084–1088.

Bell, C. and S. Hawthorne. 2008. Ellagic acid, pomegranate and prostate cancer: A mini review. *J. Pharm. Pharmacol.* 60:139–144.

Bemis, D. L., A. E. Katz, and R. Buttyan. 2006. Clinical trials of natural products as chemopreventive agents for prostate cancer. *Expert Opin. Investig. Drugs.* 15:1191–1200.

Ghadirian, P., J. M. Ekoé, and J. P. Thouez.1992. Food habits and esophageal cancer: An overview. *Cancer Detect. Prev.* 16:163–168.

Heber, D. 2008. Multitargeted therapy of cancer by ellagitannins. *Cancer Lett.* 269:262–268.

Hengel, M., B. Hung, J. Engerbreton, and T. Shibamoto. 2003. Analysis of hexamid in cranberry, blueberry and pomegranate by liquid chromatography-tandem mass spectrometry. *J. Agric. Food Chem.* 51:6635–6639.

Jurenka, J. S. 2008. Therapeutic applications of pomegranate (*Punica granatum* L.). *Altern. Med. Rev.* 13:128–144.

Katz, S. R., R. A. Newman, and E. P. Lansky. 2007. *Punica granatum*: heuristic treatment for diabetes mellitus. *J. Med. Food.* 10:213–217.

Kaufman, M. and Z. Weisman. 2007. Pomegranate oil analysis with emphasis on MALD-TOF/MS triacylglycerol fingerprinting. *J. Agric. Food Chem.* 55:10405–10413.

Khan, N., F. Afaq, and H. Mukhtar. 2008. Cancer chemoprevention through dietary antioxidants: progress and promise. *Antioxid. Redox. Signal* 10:475–510.

Klass, C. M. and D. M. Shin. 2007. Current status and future perspectives of chemoprevention in head and neck cancer. *Curr. Cancer Drug. Targets.* 7:623–632.

Kotwal, G. J. 2006. Avian influenza in humans: First Annual Conference. Latest advances on prevention, therapies and protective measures, Paris, France. *IDrugs* 9:625–626.

Kotwal, G. J. 2007. Genetic diversity-independent neutralization of pandemic viruses (e.g. HIV), potentially pandemic (e.g. H5N1 strain of influenza) and carcinogenic (e.g. HBV and HCV) viruses and possible agents of bioterrorism (variola) by enveloped virus neutralizing compounds (EVNCs). *Vaccine* 26:3055–3058.

Kotwal, G. J., A. Hugin, and B. Moss. 1989. Mapping and insertional mutagenesis of a vaccinia virus gene encoding a 13,800 Da secreted protein. *Virology* 171:579–587.

Lansky, E. P. and R.A. Newman. 2006. *Punica granatum* (pomegranate) and its potential for prevention and treatment of inflammation and cancer. *J. Ethnopharmacol.* 109:177–206.

Lansky, E. P. 2006. Beware of pomegranates bearing 40% ellagic Acid. *J. Med. Food.* 9:119–122.

Li, Y., Y. Qi, T. H. Huang, J. Yamahara, B. D. Roufogalis. 2008. Pomegranate flower: A unique traditional antidiabetic medicine with dual PPAR-alpha/-gamma activator properties. *Diabetes Obes. Metab.* 10:10–17.

Neurath, A. R., N. Strick, Y. Y. Li, and A. K. Debnath. 2005. *Punica granatum* (pomegranate) juice provides an HIV-1 entry inhibitor and candidate topical microbicide. *Ann. N. Y. Acad Sci.* 1056:311–327.

Reed, J. D., C. G. Krueger, and M. M. Vestling. 2005. MALDI-TOF mass spectrometry of oligomeric food polyphenols. *Phytochemistry* 66:2248–2263.

Santillo, V. M. and F. C. Lowe. 2006. Role of vitamins, minerals and supplements in the prevention and management of prostate cancer. *Int. Braz. J. Urol.* 32:3–14.

Saxena, A. and N. K. Vikram. 2004. Role of selected Indian plants in management of type 2 diabetes: A review. *J. Altern. Complement. Med.* 10:369–378.

Siddiqui, I. A., F. Afaq, V. M. Adhami, and H. Mukhtar. 2008. Prevention of prostate cancer through custom tailoring of chemopreventive regimen. *Chem. Biol. Interact.* 171:122–132.

Syed, D. N., F. Afaq, and H. Mukhtar. 2007. Pomegranate derived products for cancer chemoprevention. *Semin. Cancer Biol.* 17:377–385.

Syed, D. N., Y. Suh, F. Afaq, and H. Mukhtar. 2008. Dietary agents for chemoprevention of prostate cancer. *Cancer Lett.* 265:167–176.

Welch, W. C. 2002. Pomegranate, *Punica granatum*. Horticulture update. http://aggiehorticulture.tamu.edu/extension/newsletters/hortupdate/sep02/art1sep.html. Accessed October 2008.

CHAPTER 21

Herbal Remedy: Safe or Not Safe? How to Use Them?

Hieu T. Tran

CONTENTS

INTRODUCTION

Over the years, herbals have gained popularity in the Western world as a potential alternative to contemporary medicine. With its use, patients do not think to report or list their herbals as potential medication to their physicians, pharmacists, and healthcare providers. Being used as medication to treat different types of illnesses in traditional medicine, one would think that these products will also carry with them a danger of side effects, toxicity, and interactions with other medication if used in excess or not under supervision of an expert in traditional medicine.

Numerous reports on herbal side effects were reported. Therefore, we would like to explore the various factors that need to be discussed in the realm of herbal usage and its safety.

First of all, herbal safety depends on the product itself, its manufacturer, and its conditions during the culture, the harvest, the production, the storage, and its distribution. Second, herbals contain several types of chemical components that traditional medicine has learned how to use judiciously to arrive at the final result. The conditions can influence factors such as content and the quality of the ingredients of the plant. This could explain the variety of potency and/or therapeutic effects of certain plants.

The following paragraphs will try to present the above-named factors of this potential contribution of traditional medicine to the contemporary treatment of illnesses.

THE QUALITY

The quality includes the following factors:

1. Habitats:
   a. Tame versus wild plants (wild blackberries), "nature intended," sun (ginseng, fur), rain (water-soluble substances), and quality of the soil (*Atropa belladonna*)
2. Harvesting conditions:
   a. Dry (flowers, seeds, barks best in spring with exceptions such as wild cherry bark in autumn), before the plant blossoms (leaves), ripeness (fruits)
3. Parts of the plant (harvest time):
   a. Barks exposed: spring
   b. Root barks: almost any time of the year
   c. Whole root: in June as long as we can see the dead top; ginseng in July
   d. Entire herb, leaves, and flowers: July, August, and September (bloom starts)
4. Drying and storage:
   a. Herbs, leaves, and flowers: dry in the shade to avoid loss of activity
   b. Large root: to be cut to ensure drying
   c. Storage: dry and low temperature

PLANT PARTS

Plant parts include the following:

1. Barks: portion of the woody exogenous stem or root
   - Measurements, external color, markings, fractures, color
2. Woods: part of the woody exogenous stem or root that lies inside the cambium ring

- Sapwood: still functions in the vegetative process of the plant
- Heartwood: no more function in the transportation of sap
- External surfaces: fiber or porous
- External colors: variable
- Fracture: tough and fibrous
- Internal color: same as external color

3. Leaves and flowers:
   - Size, shape, external marking, the feel, color, fracture (usually no importance for the identification)
4. Fruits and seeds:
   - Shape, marking, colors, fracture (no great importance)
5. Odor and taste:
   - Odor: depends on the amount of volatile constituents (aromatic, balsamic, spicy, alliacious, camphoraceous…)
   - Taste: the quality or flavor of the substance perceived
     - True taste: acid, sour, saline, saccharine, alkaline, bitter
     - Tasteless: insoluble in the saliva
     - Imparting a sensation to the tongue: mucilagenous, oily, astringent, pungent, acrid, nauseous.

PLANT SECONDARY METABOLITES

Secondary Metabolites

These substances that are commonly of limited molecular weight of <3,000 Da. Secondary metabolites are not waste or detoxification products, "ballast," reservoir of potential future properties, for the function of primary metabolism, or just waiting to become functional at some point in life. Secondary metabolites existed for the survival of the plants and possess specific physiological responses dependent on structure-receptor fitness.

Molecular Basis

These are traditionally similar in their structures (and biosynthesis) or based on their biological functions:

- DNA-binding antibiotic (i.e., bleomycins with specific action to cleave DNA)
- Alkylating and crosslinking DNA (e.g., mitomycin C)
- Bacterial cell wall synthesis (e.g., penicillin, vancomycin)

Types of Secondary Metabolites

The plants can possess the secondary metabolites that have physiological effects on the body systems (e.g., adrenal glands and lipids), or the secondary metabolites can also act as a medication with its side effects:

- Alkaloids
- Iridoid
- Lactones
- Flavonoids
- Steroidal
- Terpenes

STRESS COMPOUNDS

Stress compounds are products of either primary or secondary metabolism. Accumulation in plants in a concentration higher than normal, secondary to injury or insult to the plant metabolism:

- Mechanical wounding
- UV rays
- Dehydration
- Chemicals
- Infection

Their Roles?

Stress compounds can be potential pharmaceutical compounds (e.g., phytoalexins as antifungal) or toxicity.

Types

Their types include phenols, resins, carbohydrates, hydroxycinnamic acid derivatives, coumarins, and steroidal compounds.

RECOMMENDATIONS FOR HERBAL USAGE

One should consider herbals as a medication for the following reasons:

- Its excessive use might lead to toxicity (e.g., in Western medicine, acetaminophen known as Tylenol is a popular over-the-counter drug and has been reported as unsafe by the government because cases of liver failure were reported with excessive doses of the medication).
- Patients might want to seek advices or recommendations from a certified expert in this new modality of treatment.
- Products should be from a reputable manufacture company with all the ingredients listed as complete similarly to the listing of Western medicine.
- Any herbal usage will need to be reported or discussed with your physician or pharmacist or other healthcare providers.

Setup for an Herb-Drug Interaction to Occur

The following can lead to an herb-drug interaction:

- Herbal products usage not mentioned to the physician
- Easily accessible
- No mechanism of controlling, surveillance, and reporting the event
- Lack of education to the public and the healthcare practitioners

Common Herbal Drug Interactions

Common herb-drug interactions include the following:

1. Chamomile
    - Mild sedation, antispasmodic, antiseptic
    - Anaphylactic reaction with cross-allergy with ragweed
    - Anticoagulant effect
2. *Echinacea*
    - Anti-infective with immunologic active polysaccharides
    - Possible hepatotoxicity
3. Feverfew
    - Migraine
    - Cross-allergy with ragweed, chamomile, yarrow
    - Antiplatelet activity
    - Post-feverfew syndrome
4. Garlic
    - Antispasmodic, antiseptic, antiviral, promotion of leukocytosis
    - Antihypertension
    - Antihypercholesterolemia
5. Ginger
    - Antinausea and antispasmodic agent
    - Prolong bleeding time
6. *Ginkgo*
    - Free radical scavenger
    - Spontaneous bilateral subdural hematoma
    - May decrease the effectiveness of oral anticonvulsants
7. Ginseng
    - Interfere with digoxin assay
    - Increase adrenal steroid genesis
    - Immune modulatory effects in mice
    - Hypoglycemic agent
    - Hypertension, insomnia, headache, vomiting, epistaxis
    - Antiplatelet effect
    - Avoid in patient with manic-depressive disorders
8. St. John's Wort
    - Diuretic, urinary antiseptic
    - Benign prostate hypertrophy

Table 21.1 A Summary of the Most Common Herb Usage and Their Clinical Implications

Herbs	Uses	Administration	Adverse drug reaction	Clinical issues
Aloe	Burn, sunburn, moisturizer, laxative, general healing	Gel, juice	Contact dermatitis, intestinal cramping	Delay wound healing, loss of GI potassium; not recommended during pregnancy
Bilberry	Cataract, macular degeneration, diabetic retinopathy	20–40 mg orally three times a day (as anthocyanidin)	None reported	Inhibits platelet aggregation, lower blood glucose
Cayenne	Pain (shingles, headache, diabetic neuropathy, osteo-rheumatoid arthritis; thermogenesis, stomach protectant	0.025–0.075% capsaicin four times a day, Diet	Local burning, stomach upset, diarrhea	Reduce platelet aggregation, increase fibrinolytic activity, remove cayenne from hands with vinegar
Chamomile	Antispasmodic, sedative, anti-inflammatory	Tea (fresh or dried orally, 3–4 cups daily), compress	Hypersensitivity, GI upset	Delayed effect, may reduce drug absorption
Dong Quai	Amenorrhea, dysmenorrhea, menopause	Powdered root or tea: orally, 1–2 g; tincture (1:5): 1 teaspoon; fluid extract: 1 ml	Sunburn, reduce blood pressure, CNS stimulation	Phototoxic, inhibit platelet aggregation, ? synergism with calcium channel blocker
Echinacea	Viral, bacteria, candida infection (upper respiratory infection, urinary tract infections, snake bite)	Fluid extract: orally (1:1), 1–2 ml three times a day; solid extract: (6.5:1), 300 mg	Tingling sensation on tongue, fever from freshly pressed juice, cross sensitivity with sunflower seeds	Avoid in patient with autoimmune diseases (lupus, rheumatoid arthritis, leukemia)
Feverfew	Prophylaxis migraine headache, relieves fever, arthritis	Dry powdered orally, leaf cap 25–100 mg daily or 2–3 leaves daily	Aphthous ulcer from chewing leaves	Reduce platelet aggregation, increase fibrinolytic activity, uterine stimulant (avoid in pregnancy), not for children <2 years of age, delayed effect (4–6 months)
Garlic	Broad-spectrum antimicrobial, lower blood pressure, cholesterol	10 mg of alliin, or one clove of fresh garlic orally daily	GI irritation	Inhibits platelet aggregation and increases fibrinolytic activity, increase insulin production

Ginger	Motion, morning sickness, postoperative nausea, arthritis, migraine headache, muscular pain	Powdered ginger root: nausea/vomiting (N/V) 250 mg orally four times a day; arthritis, 125–1,000 mg four times a day	GI upset (>6 g daily)	Inhibits platelet aggregation, may decrease calcium channel blocker effect
Ginkgo	Vascular deficiency, diabetic retinopathy	40 mg three to four times orally leaf extract	GI discomfort, headache	Take 2–3 weeks for response, 12 weeks for positive
Ginseng (Chinese, Korean, American)	Adaptogen, antistress, antifatigue, regulate blood pressure, menopausal symptoms	10 mg of ginsenoside, orally three times a day	Nervousness, breast tenderness, low toxicity profile	Reduce platelet aggregation, reduce blood sugar, high dose may reduce the immune system in early stage of infection 2 weeks on and 2 weeks off, avoid during pregnancy
Ginseng (Siberian)	Same, with reduce anginal symptoms, immune system booster	Fluid extract: 2–4 ml once three times per day; solid extract: 100–200 mg daily	Insomnia with high dose, irritability, anxiety, diarrhea, headache, hypertension, pericardial pain	Estrogenic effect, not to use during pregnancy, avoid in manic-depressive condition
Goldenseal	Bacterial, fungal infection of mucous membrane, GI infection with diarrhea, cirrhosis, inflamed gallbladder, eye infection	250–500 mg orally three times a day	N/V, CNS stimulant, interfere with GI vitamin B manufacture	Hypoglycemic effect, prophylactic use for traveler's diarrhea (1 week before and 1 week after), do not exceed 2 months usage at a time, not during pregnancy
Hawthorn	Atherosclerosis, HTN, angina, congestive heart failure, rheumatoid arthritis, periodontal disease	Fluid extract (1:1) orally: 1–2 ml three times a day; freeze-dried hawthorn berries, 1–1.5 g	High dose may induce hypotension and sedation	Inhibit angiotension-converting enzyme, potentiate cardiac glycosides, increase use of vitamin C, do not discontinue abruptly
LaPacho	Bacterial, viral, fungal, and parasitic infection, intestinal and vaginal candidiasis	Tea or extract: orally 1.5–2 g daily	None reported yet	

(*Continued*)

Table 21.1 (*Continued*)

Herbs	Uses	Administration	Adverse drug reaction	Clinical issues
Licorice	Antiviral for cold and herpes simplex virus-1, premenstrual syndrome, Addison's disease, inflammation, allergy, peptic ulcer disease (PUD), eczema, canker sore, herpes simplex virus	Fluid extract: orally (1:1): 2–4 ml; solid extract: orally (4:1): 250–500 mg; PUD: orally 2–4 380 mg chewable tablet for 30 min before meals	Aldosterone-like adverse effect with >100 mg daily for >6 weeks, lethargy to quadriplegia	Avoid in hypertension, renal, liver failure, current cardiac glycosides therapy, potentiate steroids therapy
Milk thistle	Cirrhosis, chronic hepatitis, gallstone, liver protection from toxins	140 mg orally of silymarin three times a day	Loose stool; mild allergic reaction	Take Metamucil, oat bran to prevent loose stool
St. John's Wort	Mild to moderate depression	300 mg orally three times a day with 0.3% hypericin extract, with meals	Possibility of photosensitivity	Avoid selective serotonin reuptake inhibitor, food, or drugs interacting with monoamine oxidase inhibitor
Saw palmetto	Benign prostate hypertrophy	160 mg orally twice a day	Low profile, headache	Result in 4–6 weeks, no effects on prostate-specific antigen, estrogen effects, avoid in pregnancy and in breast cancer
Valerian	Sedative, anxiety, stress	150–300 mg extract orally 30–45 min before bedtime	Headache, excitability, rare morning drowsiness, cardiac disturbances	May potentiate other CNS depressants, reduce daytime naps and caffeine intake, increase exercise to increase results

- Anti-androgenic activity and estrogen activity demonstrated in rats
- Antidepressive effect

9. Valerian
   - Mild hypnotic
   - Do not combine with sedatives/barbiturates.

REPUTABLE MANUFACTURERS

Nature's Way, Solaray, Enzymatic Therapy, Phyto-Pharmica, Electric Institute, Usana are some reputable manufacturers. It is important to buy standardized products, labeled with the botanical name.

SUMMARY

Table 21.1 shows a summary of the most common herb usage and their clinical implications.

FURTHER READING

DerMarderosian, A. and J. A. Beutler. 2009. The review of natural products. St. Louis, MO: Wolters Kluwer Health.

Murray, M. 1995. The healing power of herbs. Rocklin, CA: Prima Publishing.

CHAPTER 22

Nutraceuticals: Reflections

Stephen L. DeFelice

CONTENTS

Although the three following categories of discussion are different in subject matter, they are intertwined. The intent is to demonstrate the rationale and critical need for a substantial increase in funding for both basic and clinical nutraceutical research and development (R & D), which will markedly accelerate medical-health discovery. Research on nutraceuticals, and in combination with other therapies, is urgently needed.

The "nutraceutical revolution" has stalled. This is bad news for the nutraceutical research and development communities. Conversely, the good news is that there is something that can be quickly done to establish a vigorous, well-funded research sector, and that something is the Nutraceutical Research and Education Act (NREA).

My involvement in the nutraceutical revolution began in 1965 after I conducted the first U.S. clinical study on carnitine as a "drug" or "pharmaceutical" in patients with hyperthyroidism. Carnitine is a natural substance found in practically all human cells, whose primary action is to transport fatty acids across mitochondria membranes to produce energy or ATP. It also has a number of other functions. While pursuing the thyroid lead in a clinical study, I stumbled on a serendipitous moment that focused my carnitine basic and clinical research efforts on its medical promise in cardiovascular disease.

With the enthusiasm of youth, I began my long journey with this natural substance, which continues to this day. The results of the cardiovascular studies were both broad based and extremely positive, ranging from the reversal of myocardial ischemia to toxic shock caused by Russell's viper venom.

When I tried to find pharmaceutical companies to sponsor additional clinical studies necessary to obtain an FDA-approved NDA, I ran into a stone wall. This was my first lesson in medical economics and politics. Because carnitine is a natural substance, it is extremely difficult to obtain a sufficiently strong patent to justify the enormous cost that a pharmaceutical company must expend to obtain an NDA. Lacking a strong patent, generic competitors would enter the market. Put another way, the company spends hundreds of millions of dollars on research and development, whereas the generic company spends none and charges a lower price. This is no way to run a business!

Then I encountered another huge, unexpected obstacle. Carnitine suddenly appeared on the shelves of the health food stores freely available to consumers at a low cost. Its presence virtually put the nail in the coffin of any company willing to sponsor a carnitine NDA.

The nutraceutical revolution followed shortly after. It began with a big bang after a consensus group of medical experts, under the auspices of the National Institutes of Health (NIH), recommended, primarily based on clinical data, calcium supplementation for the prevention of postmenopausal osteoporosis. This gave an image of legitimacy, heretofore virtually absent to dietary supplements in the medical community. Physicians and their patients began to discuss nutrition, which was a revolutionary event. There then followed studies on omega fatty acids for lipid reduction, *Gingko biloba* for memory loss, beta-carotene for the prevention of lung cancer, St. John's Wort for depression, vitamin E for heart disease, *Echinacea* for the common cold, and Ocean Spray Cranberry Juice for urinary tract infection, among others. The national impact of these benefits was rapid and enormous. The largest companies in the United States, regardless of the amounts of marketing dollars spent, could not achieve such product recognition.

The three basic components of this nutraceutical phenomenon were clinical data, beginning physicians' acceptance of nutraceuticals based on such data, and extensive mass media coverage. Physicians became the cornerstone of nutraceutical legitimacy, and it will remain so. It is important to note that the source of the clinical data had the largest physician impact. In the case of calcium, it was the NIH consensus group, and with vitamin E, a survey study published in the prestigious *New England Journal of Medicine* that brought them aboard.

However, there was another highly desirable impact that has gone unnoticed; the "ping-pong effect" blossomed. Significant numbers of basic and clinical studies ensued to further delineate the clinical claims. In addition, companies rushed to create new formulations to deliver the nutraceuticals to the consumer.

(Taking a step backward in time, in 1976, I established the nonprofit organization Foundation for Innovation in Medicine [FIM]. FIM is an educational foundation whose charge is to inform the key players in the health sector in ways to accelerate medical discovery or innovation. FIM has a particular interest in natural substances. The first conference was held at Columbia University entitled "The Promise and Problems of Natural Substances in Medicine".)

Getting back to nutraceuticals, because I am a physician trained in the methodology of pharmaceutical-controlled clinical trials and realizing that most of the nutraceutical clinical studies lacked statistical proofs or algorithms, I was impressed by the consistency of the findings. It seemed improbable to me that so many positive clinical findings just happened by chance.

During the late 1980s and 1990s, FIM held a series of nutraceutical conferences in Manhattan and Washington, D.C., in an attempt to create an R & D-intensive nutraceutical health sector. Although I am a firm believer in the potential medical and health benefits of nutraceuticals, I warned that, because of their wide consumption, controlled clinical trials would soon follow, which would either confirm or negate the findings of the previous ones. My enthusiasm was high, yet my pharmaceutical instincts told me to cool the optimism until more controlled studies were conducted.

I remember that, at one of the conferences, a presentation was made summarizing approximately a dozen clinical studies on the positive effect of beta-carotene in preventing pulmonary cancer. Frankly speaking, after a brief review of the data, I was surprised by the degree of consistency of the findings. One of the panel members was my physician colleague and friend who was a professor of nutrition at Harvard University. I asked him whether, based on the clinical studies, he would recommend beta-carotene as a pulmonary nutritional supplement. He would not. When I asked him why he would not, he replied that, in addition to serious flaws in the design of other studies, a lack of any response in the control group of the best study made him a doubter.

Then, as I predicted, controlled clinical trials were done on many nutraceuticals, including beta-carotene, which did not support positive findings of previous studies.

I must confess that the biggest surprise of all happened with vitamin E. Scientific and clinical studies very persuasively supported this vitamin's cardio-protective effect. Plus, there was a solid scientific rationale behind it. There was little doubt in my mind that it was an effective supplement. Then, as with beta-carotene, the controlled clinical trials that followed did not support vitamin E's clinical efficacy.

These, and other negative clinical findings, truly dampened the promise of nutraceuticals and R & D support, which continues today, but I emphatically do not believe that the results of the studies reflect the real world. The designs of these controlled clinical trials were based on a pharmaceutical philosophy of evaluating a single "magic bullet," such as insulin, a statin, and Viagra. Nutraceuticals, although they can act as single magic bullets, more often than not work in teams or combinations.

For instance, a meal, the most efficacious and king of all nutraceuticals, is a combination of nutritional ingredients that work together to sustain life by maintaining health. No matter where you are in the world, from Madagascar to Brooklyn, the nutraceutical meal, no matter what the ingredients, keeps you alive! The clinical studies with vitamin E should have included other active cardiovascular nutraceuticals, such as carnitine, magnesium, folic acid, and other ingredients, which work together to achieve a common goal. The combination approach represents the real world, in which consumers usually take multiple dietary supplements daily. This could explain why a number of survey types of studies on the cardiovascular benefits of vitamin E were positive.

What was truly disappointing was a lack of response from all segments of the nutrition community to criticize both the design and the results of negative magic bullet nutraceutical clinical studies, calling for more logically designed combination ones. Despite my efforts, I could not find a single leader or group to lead the charge.

Getting back to the pharmaceutical and nutraceutical carnitine, let's begin with the former. During the early 1980s, sparked by the television series *The Odd Couple*, the main theme being the unavailability of drugs for patients with rare diseases, Congress enacted the Orphan Drug Act, which significantly reduced the amount of preclinical and clinical data and, therefore, the cost, to obtain an NDA. More importantly, however, it granted the sponsor of the NDA the exclusive right to make a medical claim for a seven-year period, which, in effect, is similar to a methods patent. Although not the strongest type, it is oftentimes good enough. The combination of these two provisions led to an enormous success. There were very few approved orphan drugs before the act. Today, there are hundreds, either approved or in the process of development. (Note that the FDA has defined an orphan population as 200,000 patients or less).

Because of the Orphan Drug Act, I managed to convince a friend and philanthropist, Dr. Claudio Cavazza, the proprietor of the pharmaceutical company Sigma Tau S.p.A. (Rome, Italy), to sponsor the NDA effort to develop carnitine for the treatment of primary carnitine deficiency, a rare and often fatal disease in children. A subsequent carnitine orphan drug NDA was approved for renal dialysis patients.

The carnitine NDA clinical studies and numerous additional ones sparked an enormous number of basic research studies, which then sparked additional clinical research studies. I called this the ping-pong effect. Basic research begets clinical research, and clinical research begets basic research. There are now literally thousands of both basic and clinical papers published on carnitine, and the number continues to rapidly grow.

Now let's turn to the nutraceutical carnitine. It was in Greenwich Village that I first saw the dietary supplement carnitine on the shelves of a large health food store. I was very much surprised by the vast array of products displayed and began to wonder what this world was all about. I stayed for about an hour and eavesdropped on the conversations of a number of customers and employees who exchanged information on the medical-health benefits of various products ranging from fatigue relief to cancer prevention. I then began to review the dietary supplement literature primarily to determine whether there was sufficient clinical evidence of certain dietary supplements, which supported the claims. I was also surprised and disappointed by the paucity of even reasonably conducted clinical trials. In addition, most of the publications reported highly favorable response rates. After speaking to friends, colleagues, and others who were loyal dietary supplement takers, I don't recall anyone responding in a negative way. I was struck by our overwhelming blind acceptance of and faith in dietary supplements, with meager evidence regarding their actual clinical benefits. Contrast this to our national suspicion, sometimes bordering on hostility, of pharmaceuticals, which undergo multiple controlled clinical studies evaluating both effectiveness and safety before FDA approval.

After wondering about it all, I drew several conclusions. Many of the products sold are placebos, which, in my opinion, are oftentimes a good thing. These substances deliver a national placebo response that immensely benefits the people. This leads to a significant reduction in physician and hospital visits and substantially lowers healthcare costs. Also, these substances are relatively safe. Think of this: if we were to remove all the dietary supplements from the shelves of health food stores and replace them with FDA-approved pharmaceuticals where customers could purchase them without a physician's prescription, half of America would be either dead or on its way within six months.

I decided to try to conceive a new nutraceutical regulatory system that would lead to a quantum leap in both basic and clinical research to bring about the ping-pong effect.

First, I decided that an act of Congress was necessary, and second, it was essential to specifically define this entity necessary for congressional consideration.

In 1989, while in Rome, after a very good meal, good wine, and an excellent grappa, I decided to talk a stroll in the Piazza Navona to try to come up with both the magic name and its definition. The uplifting vibrations of late night Roman life were everywhere. Maestro Stefano—owner of a restaurant by the same name, whom I knew for many years—saw me, called me over, and invited me in for another grappa. This was beyond my limit, but I thought "Why not?" in one of those tough-to-come-by beautiful moments in life. While discussing his favorite subject, the meaning of life, the term "nutraceutical" jumped to mind, but I couldn't come up with the definition.

After I returned to the United States, I finally came up with a definition. "A nutraceutical is a food or part of a food that has a medical or health benefit, including the prevention and treatment of disease." If you think about it, this includes almost everything one consumes. It includes foods, functional foods, pharmafoods, and herbal remedies, among others. The term hit the media and became widely used. It is even now in the Oxford English Dictionary, which also credits me for coining the term. What was disturbing was that others then created their own nutraceutical definitions primarily for marketing purposes, which has had a negative impact on my efforts. Now armed with a new term and a specific definition, I needed to propose a specific law for Congress to enact. It occurred to me, no grappa this time, that the Orphan Drug Act sparked the boon in the R & D of the pharmaceutical carnitine and other drugs. I decided that the same principles and ground rules should be involved in what I termed the NREA, which can be found on my website, http://www.fimdefelice.org. Also, there is an additional critical provision of the act: the right for a company to make a disease claim based on the results of clinical studies.

Current regulations severely limit "disease" or medical claims of nutraceuticals, particularly with dietary supplements. "Health" claims, however, can be made. Thus, if carnitine is shown to prevent myocardial ischemia in patients with coronary artery disease—and it has!—the company cannot make the claim. It may, instead, be permitted to claim that carnitine maintains a healthy heart. In other words, the regulations force companies not to tell the truth. Speaking about our health care system! Also, these regulations profoundly discourage clinical research and rob us of new medical remedies. Why should a company fund both basic and then clinical research on a nutraceutical if it cannot make the claim?

The NREA has a provision that permits a company to make the claim, be it health, disease, medical, or whatever, based on the results of the clinical studies. In addition, it, as with an orphan drug, grants a seven-year exclusivity period to the company to make the claim.

During one of my FIM nutraceutical conferences, which were, as usual, heavily attended by influential people, I pushed hard for the need of the NREA. Congressman Frank Pallone (D-NJ), a major player in the health sector, who was in the audience, came to the podium and announced that he would introduce the NREA in the Congress which extends the claim-exclusivity period to 10 years. And so he did, in 1999 (http://www.fimdefelice.org). Despite my full-throttled efforts to gather support for the legislation from influential elements of the nutritional community, which Mr. Pallone critically needed to convince his colleagues to join him in passing the act, not one person, company, or other organization, including nutritional scientists, came forward to support the act. Unlike the pharmaceutical industry, the food and dietary supplement industries are market driven, not R & D driven. Thus, unnoticed, it faded away, and, although it is stilled buried, it is not yet dead. My old stethoscope still detects a faint heartbeat, which tells me that it is, Lazarus-like, capable of resuscitation by an effective advocacy group.

Because there is currently no advocacy group, the next logical step is to create one. This presents a rare opportunity for nutraceutical researchers of all types. This group should be very familiar with the corridors of Congress as well as media contacts. The good news is that there are currently powerful Congressional members who would be favorably disposed to the NREA concept.

In addition, strong consumer support exists, as evidenced by the following: FIM conducted a consumer survey to determine whether there are concerns regarding the safety and effectiveness of the foods and dietary supplements that they take. Also, the importance of clinical studies to assess these concerns was outlined. More than 90% of the responders were very much concerned and strongly supported increased clinical research and the creation of the NREA.

In conclusion, the two major nutraceutical players already exist. What is needed is an effective advocacy group, perhaps supported by only a single major corporation, to coordinate the effort. The ping-pong effect would follow quickly.

THE BIOLOGICAL-MEDICAL LESSONS OF MY CARNITINE EXPERIENCE

The carnitine experience continues to open new ideas and avenues, for both basic and clinical research. My primary concern is attacking disease, and I view both types of research with that goal in mind. I, however, fully appreciate pure scientific research because it often leads to ideas regarding goal-oriented medical research. The following are examples of fundamental lessons learned:

1. The nutraceutical rejection need-acceptance concept: Carnitine performs few biological functions when given to both healthy animals and humans. For example, only by the administration of enormous intravenous doses in animals can a cardiac inotropic effect be elicited. This early observation, about 30 years ago, got me thinking about this biologic principle.

   Applying this concept, I am currently planning a clinical study to attempt to partially reverse both the mental and physical deterioration of aging. Although the cause of aging is not known, there is little doubt that the loss of energy (or ATP production) is the clinical hallmark of this inexorable process. Young folks run faster than old ones and take to the computer much more easily. Tens of millions of U.S. consumers take dietary supplements daily that are, either directly or indirectly, involved in ATP production. Still, the aging process continues unabated. I have concluded that, for reasons not yet known, aging cells do not perceive a "need" for them and, therefore, "reject" their use, whereas, in scorbutic patients, the cells perceive a need for vitamin C and do not reject, but instead "accept" and use it.

   Perhaps the aging cell can be stimulated to create acceptance and use of the ATP-producing supplements. For example, testosterone and growth hormone stimulate muscle cells to grow and enlarge in humans, increasing energy requirements. If the proper ATP-producing supplements are also given, these cells might recognize a need and use them, further increasing ATP production.

   This condition most likely holds true with dietary supplements in many diverse conditions.
2. Acute part of organ deficiency: Sufficient cardiac cell levels of carnitine are critical to maintain normal cardiac function. In acute myocardial ischemia, the ischemic portion of the myocardium only loses carnitine. Although the blood levels are normal, a sufficient amount does not exist to replenish the ischemic cells. After ischemia occurs, cardiac arrhythmias appear, followed by cardiac arrest and death. In both animal and clinical studies, carnitine administration restores carnitine levels in the ischemic cells and dramatically prevents or reverses the ischemic changes. (Too often, dietary supplements are considered in terms of prevention, whereas—hold your breath—I believe their use as treatment has greater promise).
3. Whole organ carnitine deficiency: In chronic cardiomyopathy and congestive heart failure, the entire myocardium is carnitine deficient. In animal studies, carnitine increases cardiac output. The clinical studies to date have reported mixed results.
4. Subacute carnitine deficiency: In burn patients, for example, large amounts of carnitine are lost. The reported clinical benefits of carnitine administration are mixed, most likely attributable to inadequate dosing.
5. Chronic carnitine deficiency: In primary carnitine deficiency, a rare and oftentimes fatal disease in children, carnitine blood and tissue levels are very low. Patients respond dramatically to carnitine administration (an FDA-approved indication).
6. Pharmaceutically-induced carnitine deficiency: The chemotherapeutic agent doxorubicin is cardiotoxic. It causes both an absolute and relative carnitine deficiency. In numerous preclinical studies, the administration of carnitine, both as prevention and treatment, dramatically eliminates these toxic effects. Although my former colleague, James Vick, and I made the initial observation using isolated Langendorf heart models during the late 1960s, since that time, for reasons that are inexplicable, no definitive clinical trial has been done. I learned a sad lesson from this and similar experiences. Physicians are not aware of nutraceutical studies because, unlike the pharmaceutical industry, which makes available educational material, the nutritional industry does not.

If carnitine were a strongly patented pharmaceutical, every cardiologist and oncologist would have known about it soon after the initial laboratory studies and perhaps saved or prolonged cancer patients' lives.

7. Pharmaceutical-nutraceutical additive or synergistic effects: In human ovarian cancer cell cultures, carnitine itself destroyed more than 50% of the cells. Doxorubicin, itself, did the same. When given together, the cell kill capacity of the combination was greater than either alone. I was very fortunate to find an oncologist who was willing to conduct an early Phase II clinical trial in late-stage ovarian cancer. The study will begin shortly. Cross your fingers!
8. Medical device-induced carnitine deficiency: The renal dialysis procedure removes carnitine from the body. Dialysis patients have many problems, ranging from severe fatigue to cardiac failure. Carnitine administration ameliorates some of these manifestations (an FDA-approved indication).
9. Herbal remedies: Although included in the nutraceutical definition, herbal remedies are sufficiently distinct to warrant special consideration. Unlike carnitine and other nutraceuticals, the cell need condition does not necessarily hold. These natural remedies are active under most conditions. What is an herbal remedy? Are St. John's Wort, broccoli rabe, and the cocoa plant herbal remedies all? An epistemologist is urgently needed to delineate the category. Basic science research clearly tells us that these substances are active pumping up, for example, neurotransmitters and the immune system and doing other things. But not so in man. To date, the clinical data are woefully disappointing. I am as puzzled as anyone regarding the reasons. It could be a simple matter of dosing and kinetics. There are other concerns regarding negative clinical effects. St. John's Wort increases hepatic systems, which are used to detoxify drugs, one result being lowered blood levels of certain drugs, including those given to AIDS patients. Not too long ago, approximately 20% of FDA-approved drugs were of plant origin. These substances work as pharmaceuticals but, to date, not as nutraceuticals. Go figure it!.

MY PERSONAL EXPERIENCE TO FORM THE FIRST NUTRACEUTICAL COMPANY

The carnitine experience convinced me that, in many conditions, nutraceutical blood levels do not reflect intracellular ones and also that normal blood levels are insufficient to replenish them. Diabetics frequently have normal magnesium blood levels but low cell levels. The same pattern holds true with potassium in patients on certain diuretics. Supplementation is needed in such cases.

In the past, I was highly connected in both the food and pharmaceutical industries, and I tried to convince companies, particularly as joint ventures, to enter this health sector. There was little interest. Then, I decided to form my own company, Intracellular Health (ICH), and find private investors whom I thought would be more bullish. Not so. Interestingly enough, their major concern was FDA's regulatory position on prohibiting disease claims. The FDA allows claims to be made about how certain supplements affect the body, but it prohibits disease-related claims. Money was not the problem. If the NREA were in place, it would have been an easy sell.

I was about to throw in the towel, when I received a call from a high-level executive of one of the world's largest and most innovative companies in the chemical, agricultural, and raw material supplies business. The level of interest in this company was high enough that the board of directors requested that I present a rationale and an outline of a plan to enter the nutraceutical health sector.

The board was quite receptive to innovative ideas and asked the right questions, which I thought I answered satisfactorily enough.

I proposed that the company sign an agreement with one of the world's most prestigious hospitals to measure the nutraceutical intracellular levels, including biopsies, of patients with specific high-prevalence diseases. Those with specific deficiencies would then receive tailored design nutraceutical supplements to correct such deficiencies. Clinical outcomes would then be evaluated in studies. I pointed out that unique formulation technology would be one key to success, ranging from once-a-day dosing to taste masking. Because, in many cases, large doses of nutraceuticals would often be required, food and not pill formulations would be critical for patient compliance. Also, because of daily dosing limitations in some cases, even with food technology, it would not be necessary to administer the same high daily dose but a varied one, taking advantage of the kinetics and distribution of fat- and water-soluble nutraceuticals. With the proper know-how, it would be possible to obtain formulation patents that offer lead time in the marketplace.

The give-and-take exchange lasted for precisely 3 h. I detected a tone of high-level curiosity during the last hour. I thought that I had finally accomplished my goal of establishing the first bona fide nutraceutical company and departed a happy trooper. I was wrong. There were just too many unknowns and risks for a public company to take. Once more, as with ICH, if the NREA existed, it would have been a different story. The ping-pong effect would have been in full swing, and patients—we are all patients in one form or another—would have benefited enormously.

Index

C

D

E

F

G

H

N

O

P

Q

R